Lecture Notes in Mechanical Engineering

Series Editors

Fakher Chaari, *National School of Engineers, University of Sfax, Sfax, Tunisia*
Francesco Gherardini, *Dipartimento di Ingegneria "Enzo Ferrari", Università di Modena e Reggio Emilia, Modena, Italy*
Vitalii Ivanov, *Department of Manufacturing Engineering, Machines and Tools, Sumy State University, Sumy, Ukraine*
Mohamed Haddar, *National School of Engineers of Sfax (ENIS), Sfax, Tunisia*
Loon Ching Tang, *Department of Industrial Systems Engineering and Management, National University of Singapore, Singapore, Singapore*

Editorial Board Members

Francisco Cavas-Martínez, *Departamento de Estructuras, Construcción y Expresión Gráfica Universidad Politécnica de Cartagena, Cartagena, Spain*
Francesca di Mare, *Institute of Energy Technology, Ruhr-Universität Bochum, Bochum, Germany*
Young W. Kwon, *Department of Manufacturing Engineering and Aerospace Engineering, Graduate School of Engineering and Applied Science, Monterey, USA*
Tullio A. M. Tolio, *Department of Mechanical Engineering, Politecnico di Milano, Milano, Italy*
Justyna Trojanowska, *Poznan University of Technology, Poznan, Poland*
Robert Schmitt, *RWTH Aachen University, Aachen, Germany*
Jinyang Xu, *Faculty of Intelligence Technology, Shanghai Institute of Technology, Shanghhai, China*

AF888998

Lecture Notes in Mechanical Engineering (LNME) publishes the latest developments in Mechanical Engineering—quickly, informally and with high quality. Original research or contributions reported in proceedings and post-proceedings represents the core of LNME. Volumes published in LNME embrace all aspects, subfields and new challenges of mechanical engineering.

To submit a proposal or request further information, please contact the Springer Editor of your location:

Europe, USA, Africa: Leontina Di Cecco at Leontina.dicecco@springer.com
China: Ella Zhang at ella.zhang@cn.springernature.com
India, Rest of Asia, Australia, New Zealand: Swati Meherishi at swati.meherishi@springer.com

Topics in the series include:

- Engineering Design
- Machinery and Machine Elements
- Mechanical Structures and Stress Analysis
- Automotive Engineering
- Engine Technology
- Aerospace Technology and Astronautics
- Nanotechnology and Microengineering
- Control, Robotics, Mechatronics
- MEMS
- Theoretical and Applied Mechanics
- Dynamical Systems, Control
- Fluid Mechanics
- Engineering Thermodynamics, Heat and Mass Transfer
- Manufacturing Engineering and Smart Manufacturing
- Precision Engineering, Instrumentation, Measurement
- Materials Engineering
- Tribology and Surface Technology

Indexed by SCOPUS, EI Compendex, and INSPEC.

All books published in the series are evaluated by Web of Science for the Conference Proceedings Citation Index (CPCI).

To submit a proposal for a monograph, please check our Springer Tracts in Mechanical Engineering at https://link.springer.com/bookseries/11693.

Marcelo Calvete Gaspar ·
Maria Leopoldina Alves · Luis Serrano ·
Daniel Raposo · Luis Farinha
Editors

Sustainable Innovations in Mechanical Engineering

Selected Contributions to the 8th International Conference on Sustainable Innovation, Entrepreneurship and Technology (RegionalHelix 2024), November 14–16, 2024, Leiria, Portugal

Springer

Editors
Marcelo Calvete Gaspar
Instituto Politécnico de Leiria
Leiria, Portugal

Maria Leopoldina Alves
Instituto Politécnico de Leiria
Leiria, Portugal

Luis Serrano
Instituto Politécnico de Leiria
Leiria, Portugal

Daniel Raposo
Instituto Politécnico de Castelo Branco
Castelo Branco, Portugal

Luis Farinha
Instituto Politecnico de Castelo Branco
Castelo Branco, Portugal

ISSN 2195-4356 ISSN 2195-4364 (electronic)
Lecture Notes in Mechanical Engineering
ISBN 978-3-032-36364-0 ISBN 978-3-032-36365-7 (eBook)
https://doi.org/10.1007/978-3-032-36365-7

© The Editor(s) (if applicable) and The Author(s), under exclusive license to Springer Nature Switzerland AG 2027

This work is subject to copyright. All rights are solely and exclusively licensed by the Publisher, whether the whole or part of the material is concerned, specifically the rights of translation, reprinting, reuse of illustrations, recitation, broadcasting, reproduction on microfilms or in any other physical way, and transmission or information storage and retrieval, electronic adaptation, computer software, or by similar or dissimilar methodology now known or hereafter developed. Unless otherwise expressly agreed by the Publisher in writing, the Publisher does not authorize use or reproduction of all or any part of this work in any manner for the purpose of training, fine-tuning, evaluating, or developing artificial intelligence models, technologies, or systems. In accordance with Article 4(3) of the EU Directive 2019/790 on copyright and related rights in the Digital Single Market, the Publisher expressly reserves all rights with respect to this work under the text and data mining exception of the directive.
The use of general descriptive names, registered names, trademarks, service marks, etc. in this publication does not imply, even in the absence of a specific statement, that such names are exempt from the relevant protective laws and regulations and therefore free for general use.
The publisher, the authors and the editors are safe to assume that the advice and information in this book are believed to be true and accurate at the date of publication. Neither the publisher nor the authors or the editors give a warranty, expressed or implied, with respect to the material contained herein or for any errors or omissions that may have been made. The publisher remains neutral with regard to jurisdictional claims in published maps and institutional affiliations.

This Springer imprint is published by the registered company Springer Nature Switzerland AG
The registered company address is: Gewerbestrasse 11, 6330 Cham, Switzerland

If disposing of this product, please recycle the paper.

Preface

This volume of the Lecture Notes in Mechanical Engineering (LNME) presents peer-reviewed contributions within the Regional Helix framework under the theme of Sustainable Innovation in Mechanical Engineering. The chapters collectively articulate contemporary mechanical engineering applied research that reshape design, production and mobility systems to innovate through and within sustainability guidelines. Beyond presenting incremental advances through sustainability-oriented innovation, the chapters emphasize maintenance and repairability case studies, life-extension of assets, resource and energy efficiency in processes, interoperability across cyber-physical production systems, and data-informed decision-making at component, process, and system levels.

The Regional Helix framework is intentionally multidisciplinary and application-oriented while integrating academic, industrial, and public-sector perspectives. This integration is reflected in the contributions, which balance methodological development with application in applied operational contexts. The resultant body of work illustrates a consistent concern for measurability, replicability, and the translatability of findings into engineering practice, thus helping researchers, students and practitioners to implement more sustainability-oriented strategies and solutions.

The organizers received 29 contributions for this volume. After a thorough peer-review process (double-blind review), the committee accepted 20 papers, with an acceptance rate 69%. At least three independent reviewers evaluated each paper. To evaluate each paper, reviewers were asked to answer, among others, questions about the originality of the paper, the significance of presented results, comprehensibility, scientific merit and relevance for the Regional Helix framework. All submitted papers were checked for similarity using Turnitin.

This volume is organized into four technical subsections that range from the implementation of sustainable and innovative technologies and solutions to system-level sustainability outcomes: (I) Advanced Manufacturing and Digital Systems; (II) Maintenance and Asset Management; (III) Energy and Environment; and (IV) Sustainable Mobility and Transportation.

The first part entitled Advanced Manufacturing and Digital Systems assembles contributions on robotics-enabled additive manufacturing (AM), digital readiness, and interoperability as foundational enablers of sustainable production. The section encompasses a state-of-the-art review of robot-assisted AM platforms and control architectures; a bibliometric analysis of digital-maturity assessment models relevant to Industry 4.0 transitions; and an interoperability case using OPC UA within injection-moulding environments to enable traceability, efficient data exchange, and energy-aware operation. A case study on polymer injection moulds highlights and discusses welding-based reconditioning strategies. Together these studies evidence how advanced manufacturing and data infrastructures shorten the path from concept to circular products.

The second part dedicated to Maintenance and Asset Management focuses on sustaining performance through predictive, standardized, and decision-intelligent approaches.

Methodological contributions include a proportional hazards (Cox) modelling framework for predictive maintenance in Industry 5.0 settings and a decision-support architecture that formalizes multi-criteria evaluation for complex industrial systems. Alongside these, dedicated case studies show how disciplined improvement translates directly into uptime and quality. This section shows that sustainability-oriented strategies allow for extend asset life, improved maintenance policies, and reduced environmental impacts related with waste and premature replacements.

The third part focusing on Energy and Environment addresses process-scale resource performance and environmental impact mitigation. Contributions include energy-aware control of suspended-load movement, thermal optimization of pellet boilers, and two materials-processing studies that are related to circular value chains: the valorization of limestone-quarry waste into precipitated calcium carbonate and a model for separating plastics by dissolution/precipitation. These studies pair modelling with empirical validation, evidencing quantifiable gains in energy efficiency, material recovery, and substitution potential, while preserving or enhancing functional performance.

The fourth part dedicated to Sustainable Mobility and Transportation considers the design and operation of mobility systems with attention to performance metrics and environmental accounting. The section includes ride-comfort analytics for electric scooters aligned with ISO 2631, a genetic-algorithm framework for the optimization of vehicle suspension parameters, fleet-level CO_2 assessment tools for heavy-duty vehicles, and analyses of maintenance-related environmental impacts under biofuel operation. A complementary case study on aircraft landing-gear trends rounds out the set. The section demonstrates that sustainable mobility outcomes emerge from the interaction of component design, control strategies, operational practices, and robust measurement infrastructures.

Three transversal trends run through the volume. First, circularity is being operationalized: from welding-based repair and reconditioning of tooling to the upcycling of Nylon-6 fishing nets via large-scale additive manufacturing and micro-factories, the papers illustrate how product life-extension and local production can be engineered as systematic capabilities rather than one-off projects. Second, digitalization is maturing beyond pilots: interoperability standards, digital-maturity assessment and asset-intelligence methods are being applied to create trustworthy data flows that support optimization across the life cycle. Third, measurement is becoming integral to design and management decisions, thus closing the loop between targets, interventions and outcomes.

The contributions further indicate that environmental and economic objectives can be jointly satisfied when interventions are formulated as engineering optimizations subject to operational constraints. Interoperable control reduces information latency and improves utilization; predictive maintenance extends asset life and mitigates unplanned losses; process-level energy optimization lowers operating costs while reducing emissions; and transport analytics provide a basis for accountable decarbonization trajectories. In aggregate, the studies offer a transferable set of methods that organizations may adopt incrementally with cumulative effect.

The editors extend their appreciation to all authors for entrusting us with their work and for engaging constructively with the review process. We are equally indebted to the anonymous reviewers and track organizers for their time, expertise and careful assessments and to the Regional Helix organizing committee for the platform that enabled these exchanges. We acknowledge our institutions and partners for their support throughout the editorial process and extend our appreciation to the Springer LNME team for guidance during production.

We hope this collection serves researchers, students and practitioners as a practical reference for designing mechanical systems that are efficient, repairable and resilient. The editors trust that the ideas and tools presented here will inform future research projects, industrial deployments and curricula devoted to sustainable innovation in mechanical engineering.

Leiria, Portugal

Marcelo Calvete Gaspar
Maria Leopoldina Alves
Luis Serrano
Daniel Raposo
Luis Farinha

Organization

Chairs

Marcelo Gaspar	Polytechnic Institute of Leiria
Maria Leopoldina Alves	Polytechnic Institute of Leiria
Luís Serrano	Polytechnic Institute of Leiria
Daniel Raposo	Polytechnic Institute of Castelo Branco
Luís Farinha	Polytechnic Institute of Castelo Branco

Scientific Committee

Aidin Salamzadeh	University of Tehran
Alexandre Guedes	University of Trás-os-Montes and Alto Douro
Ana Rodrigues	University of Trás-os-Montes and Alto Douro
Ana Melo	CIPES/ESTGA—University of Aveiro
Ana Claudia Machado Padilha	University of Passo Fundo
Ana Patricia Vilha	UFABC—Federal University of ABC
Anderson Galvão	CETRAD and University of Trás-os-Montes and Alto Douro
Arminda Paço	NECE and University of Beira Interior
Artur Pires	University of Aveiro
Beatriz Casais	University of Minho
Benjamim Fonseca	University of Trás-os-Montes and Alto Douro
Carina Silva	CIICESI, School of Technology and Management, Polytechnic Institute of Porto
Carla Mascarenhas	University of Beira Interior
Carla S. Marques	CETRAD and University of Trás-os-Montes and Alto Douro
Carlos Marques	CETRAD and University of Trás-os-Montes and Alto Douro
Carlos Rodrigues	University of Aveiro
Carlos Sampaio	Polytechnic Institute of Castelo Branco
Carmem Leal	CETRAD and University of Trás-os-Montes and Alto Douro
Cristina Fernandes	FEP—School of Economics and Management of the University of Porto

Daniel Raposo	Polytechnic Institute of Castelo Branco
David Rodeiro Pazos	University of Santiago de Compostela
Elisabete Figueiredo	University of Aveiro
Fernando Ferreira	BRU and ISCTE/University Institute of Lisbon
Helena Alves	University of Beira Interior
Joana Lima	University of Évora
João Ferreira	NECE and University of Beira Interior
João Neves	Polytechnic Institute of Castelo Branco
João Nuno Lopes	CETRAD and Lusófona University
João Renato Sebastião	Polytechnic Institute of Castelo Branco
Jorge Remondes	ISVOUGA, Higher Institute of Entre Douro and Vougaa
Jorge Cunha	University of Minho
Jorge Julião	Catholic University of Portugal
José Machado	University of Minho
Jose Manuel Hernandez Mogollón	University of Extremadura
Liliana Alves	University of Aveiro
Liliana Neves	Polytechnic Institute of Castelo Branco
Luís Farinha	NECE and Polytechnic Institute of Castelo Branco
Luís Serrano	Polytechnic Institute of Leiria
Marcelo Gaspar	Polytechnic Institute of Leiria
Márcio Jacometti	Federal University of Technology, Paraná
Marcus Dejardin	University of Namur/Catholic University of Louvain
Maria José Rainho	UTAD—University of Trás-os-Montes and Alto Douro
Maria Leopoldina Alves	Polytechnic Institute of Leiria
Marina Ranga	University of Warsaw
Marisa Ferreira	CIICESI, School of Technology and Management, Polytechnic Institute of Porto
Mete Basar Baypinar	University of Istanbul
Octávio Sacramento	UTAD—University of Trás-os-Montes and Alto Douro
Orlando Rua	Polytechnic Institute of Porto—ISCAP
Paula Odete Fernandes	UNIAG, Polytechnic Institute of Bragança
Paulo Mourão	University of Minho
Pedro Torres	Polytechnic Institute of Castelo Branco
Ricardo Bento	UTAD—University of Trás-os-Montes and Alto Douro
Ricardo G. Rodrigues	NECE and University of Beira Interior
Rui Silva	CETRAD and University of Trás-os-Montes and Alto Douro

Rui Patrício	IADE—Faculty of Design, Technology and Communication
Sérgio Teixeira	ISAL—Higher Institute of Administration and Languages
Teresa Felgueira	Polytechnic Institute of Guarda
Teresa Paiva	Polytechnic Institute of Guarda
Tomaz Dentinho	University of the Azores
Vanda Lima	CIICESI, School of Technology and Management, Polytechnic Institute of Porto
Vitor Braga	CIICESI, School of Technology and Management, Polytechnic Institute of Porto

Section Contributors

Armando Ramalho	Polytechnic Institute of Castelo Branco
Carlos Catorze Pereira	Polytechnic Institute of Viseu
Carlos Neves	Polytechnic Institute of Leiria
Daniel Gaspar	Polytechnic Institute of Viseu
Helder Santos	Polytechnic Institute of Leiria
Henrique Almeida	Polytechnic Institute of Leiria
Jorge Siopa	Polytechnic Institute of Leiria
Jorge Julião	Católica Porto Business School
Luís Coelho	Polytechnic Institute of Leiria
Mário Correia	Polytechnic Institute of Leiria
Milena Vieira	Polytechnic Institute of Leiria
Olivier Gouveia	Polytechnic Institute of Leiria
Paulo Carvalho	Polytechnic Institute of Leiria
Pedro Custódio	Polytechnic Institute of Leiria
Rui Rúben	Polytechnic Institute of Leiria
Silvia Monteiro	Polytechnic Institute of Leiria

Contents

Advanced Manufacturing and Digital Systems

Maintenance and Asset Management

Advanced Manufacturing and Digital Systems

A Review on Robot-Assisted Additive Manufacturing Systems and Technologies

Marcella Cavalcanti[1,2,3(✉)], Hugo Costelha[1,2], and Carlos Neves[1,2]

[1] Polytechnic University of Leiria, Leiria, Portugal
marcella.cavalcanti@ipleiria.pt
[2] INESC Coimbra, Coimbra, Portugal
[3] Technological University of the Shannon: Midlands Midwest, Athlone, Ireland

Abstract. The general use of robot manipulators in the Additive Manufacturing (AM) world could cause a paradigm shift on how these technologies are used today. Adding more degrees of freedom to the AM systems decreases the limitations of current mainstream additive technologies, such as restricted build volume, high manufacturing times, and the use of support structures. However, existing traditional techniques for slicing 3D models and path planning generation do not translate smoothly into the requirements and constraints of robot manipulator systems. This paper presents a state-of-the-art review on the current systems and technologies, as well as advantages and challenges on the use of robot manipulators in AM, focusing on extrusion-based processes.

Keywords: Additive manufacturing · Robots · Robot-assisted manufacturing

1 Introduction

Additive Manufacturing (AM) can be defined as the process of producing parts through the deposition of material in a layer-by-layer manner [1]. Compared to subtractive methods, which involve removing material from a solid block to achieve the desired shape, additive manufacturing offers unique advantages such as reduced material waste, greater design freedom, and the ability to more easily create complex geometries.

Starting in the 1980s as rapid prototyping, AM evolved into a versatile manufacturing technology with applications spanning various industries, including aerospace, automotive, healthcare, and consumer goods [2]. Many different techniques were created, such as Fused Deposition Modelling (FDM), Stereolithography (SLA), Selective Laser Sintering (SLS), Electron Beam Melting (EBM), among others, each offering unique capabilities and advantages. Fig. 1 shows the classification of AM technologies based on the raw material used.

In spite of the many advances of AM technologies over the past years, there are still limitations that drive researchers to develop new manufacturing strategies, especially to deal with problems regarding product size, slow build rates, and the need of support structures for regions with overhangs [3]. Some of those problems can be solved by

© The Author(s), under exclusive license to Springer Nature Switzerland AG 2027
M. Calvete Gaspar et al. (Eds.): Regional Helix 2024, LNME, pp. 3–13, 2027.
https://doi.org/10.1007/978-3-032-36365-7_1

adding extra degrees of freedom (DOF) to these systems, and the use of robotics may play a part in this new approach.

This paper presents a state-of-the-art review on current systems and technologies, as well as the advantages, requirements and challenges of the use of robotics in the context of additive manufacturing. The methodology for identifying the works reviewed in this paper involved a search of the literature published within the last seven years, in the Springer Link, ScienceDirect, and IEEE Xplore repositories, for papers including the keywords "multi-axis additive manufacturing" (41 results), "robotic additive manufacturing" (329 results), and "multi-axis slicing" (23 results). These keywords were selected to ensure a comprehensive exploration of the most relevant literature, related to the integration of robotics in additive manufacturing processes.

After reading the abstracts of the retrieved papers, a subset was chosen based on whether they presented a novel approach to either the slicing or toolpath planning methods used. Subsequently, these selected papers were examined in greater detail, with the final selection presented in Sect. 3, prioritizing works that performed real tests producing parts and analysing their structural integrity.

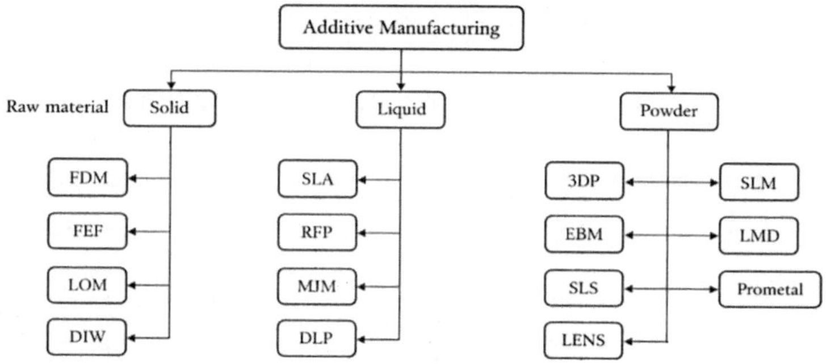

FDM Fused Deposition Modeling
FEF Freeze-form Extrusion Fabrication
LOM Laminated Object Manufacturing
DIW Direct Ink Writing or Robocasting

SLA Stereolithography
RFP Rapid Freeze Prototyping
MJM Multi-Jet Modeling
DLP Digital Light Processing

3DP Inkjet 3D Printer
EBM Electron Beam Melting
SLS Selective Laser Sintering
LENS Laser Engineered Net Shaping
SLM Selective Laser Melting
LMD Laser Metal Deposition

Fig. 1. Classification of AM technologies based on the type of raw material used (adapted from [1]).

2 Multi-Directional Additive Manufacturing

In the context of AM, multi-directionality refers to the possibility to build layers in all directions (i.e., multiple DOF) [4], in contrast with the traditional methods, where the layers are always parallel. For extrusion-based processes, this means the material deposition can be oriented dynamically to adapt to the geometry of the part, instead of having the extrusion performed along a predetermined axis. This flexibility allows for more complex geometries and designs to be built, as well as greater efficiency in the use of materials, because it minimizes the need for support structures, and reduces waste during the manufacturing process [5].

Additionally, the possibility of changing the print orientation along the part, impacts on the structural integrity, surface finish, and mechanical properties of the final product. Fry et al. [6] performed some tests by printing a structure with a 90° overhang using both traditional and multi-directional additive manufacturing (MDAM) methods using a robotic manipulator. The results show that changing the print orientation along the part makes it possible to print all overhangs with no rough edges, no drooping, and no support material. A quantitative testing of the roughness of each overhang section was also performed, supporting the qualitative visual results.

Yao et al. [7] performed tensile and flexural tests in four identical parts: one built with curved layers in a multi-DOF printing platform, and the other three built in a traditional FDM printer, each in a different (XYZ) direction. All parts were printed with the same material, nozzle diameter, temperature, speed, layer thickness, and infill density. The results showed that the curved printed part could withstand the maximum load when compared to the others. Moreover, the surface quality of the method was proved satisfactory, with all layers aligned and no staircase effect.

One can conclude that, as shown in the works cited previously, when compared to the traditional 3-DOF AM process, MDAM stands out for its distinctive features across three primary categories: hardware, slicing, and path planning.

2.1 Hardware

Most of the works reviewed in this paper were able to achieve MDAM by the use of robot manipulators, which will be discussed in Sect. 3. Here we describe other approaches for MDAM that do not use robotic manipulators.

Bhatt et al. [8] believe that robots will play a key role in the next generation of AM systems, as they are able to overcome some limitations of traditional AM, and can be used to reduce build time, improve part quality and process performance, and allow manufacturing of larger parts.

A disadvantage of the use of robots in AM, according to Jiang et al. [5], is the high cost of these manipulators, which prevent their use in a low-cost MDAM system. In their work, Hong et al. [9] create a solution for upgrading an off-the-shelf 3-axis 3D printer to a 5-axis system, by adding a 3D printed 2-axis rotating gantry, as shown in Fig. 1 of their work. Their main objective was to lower the entry barrier for MDAM, through reduced costs and increased availability.

The work by Tamir et al. [10] proposed a Stewart platform-assisted AM system, where the build plate is replaced by a 6-DOF platform capable of moving the print object to a desired position and orientation, related to the extrusion tool. They were able to demonstrate through simulation the feasibility of this configuration in allowing multi-directional printing, but no real experiments were conducted.

2.2 Slicing

In AM, slicing refers to the process of converting digital 3D models into stackable layers, that will, one by one, create a real three-dimensional object. In a multi-directional context, this process undergoes significant adaptation when compared to the traditional approach of stacking layers along a single axis, also known as 2.5D slicing.

The work by Xiao and Joshi [11] proposes a model decomposition strategy to allow the elimination of support structures. Their approach assumes a given initial build direction, and automatically decomposes the part according to the geometry of its overhangs. However, their slicing algorithm does not allow for non-planar layers or non-uniform layer thickness.

Lettori et al. [12] developed an algorithm for non-uniform planar slicing, allowing the variation of the deposition direction and thickness within each single layer. The inputs for this algorithm are: initial geometry, layer height, and maximum and minimum layer thickness that can be achieved by the AM technology being used. The results have shown the success of the slicing method, but no real experiments were conducted.

The works by Xu et al. [13] and Zhang et al. [14] both implement adaptive slicing, also called curved layer decomposition, to achieve a general multi-axis process for freeform part manufacturing. These slicing strategies require more computational power, and a more careful and optimized path planning for FDM systems when building each layer, to avoid collisions between the part being built and the printing nozzle, as well as to mitigate the increased overall manufacturing time, due to the constant change in nozzle orientation.

2.3 Path Planning

In traditional 3-DOF AM, path planning is required in some technologies, like FDM, to describe how each layer will be constructed in a two dimensional plane (usually XY, parallel to the base/floor). The focus is mainly related to generating a component with the highest possible accuracy, surface quality, and good mechanical properties [4]. Users input several parameters, such as, wall thickness, layer height, infill strategy, printing speed, among others, into slicer software like Cura[1] or Slic3r,[2] and paths for each layer will be generated and converted to G-Code.[3]

In multi-directional FDM systems, the paths are built in a non-planar three dimensional space, whereas an open-source software that allow users to prepare files for MDAM is not readily available [9]. Ramos et al. [15] proposes a strategy to decompose complex objects into simpler parts, and manufacture them in a 5-axis printer. They have proposed an heuristic to address optimal printing sequences and path planning strategies to minimize the use of support structures.

When using robot manipulators for MDAM, the path planning algorithm must take additional factors into consideration, such as the reachability of consecutive poses without configuration changes, acceleration and velocity limits, singularities [4]. Some path planning strategies for robot manipulators will be discussed in the next section.

[1] https://ultimaker.com/software/ultimaker-cura/

[2] https://slic3r.org/

[3] https://reprap.org/wiki/G-code

3 Robot-Assisted Additive Manufacturing

As stated in the previous section, the majority of the works reviewed were able to achieve MDAM by using robot manipulators. Apart from one work [16], where the results were obtained only through simulations, all approaches used material extrusion related technologies. Table 1 summarizes the results of this state-of-the-art review, regarding the use of robotic manipulators for addictive manufacturing.

Regarding the setup of the analysed solutions, most had the robot manipulator holding the extrusion tool. Only two works [13, 26] were found where the manipulator was holding the build plate. Holding the extrusion tool is a simpler approach, as it is expected that manipulators use tools attached to their flanges. It is also a less limiting solution, given the robot's reachability, because holding the build plate with the part being built may add more constraints to the robot's motion. Keeping the part stationary leads to less perturbations while printing. Furthermore, holding the build plate means that, with time, the extrusion occurs further away from the robot's flange, as the parts build up, leading to decreased accuracy in the deposition location. The work by Fry et al. [6] used a two manipulators cell, where one held the extrusion tool and the other held the build plate, but no considerations were made regarding the implications of that setup.

Regarding the materials used to produce the parts, almost all worked with polylactic acid (PLA), which is easier to work with, as it does require lower nozzle temperatures when compared to other thermoplastic materials, and it does not always require a heated print bed. The other used materials found in this review were clay [23] and concrete [24], which were used for specific applications.

Different slicing and path planning methods were found in reviewed papers. Some works used common 2.5D slicing software such as Slic3r, Cura and Simplify3D,[4] and created approaches for G-code post processing in order to adapt the generated toolpaths in order to be used with the robot manipulators. Castelli et al. [19] redefined the toolpath from the G-code by using geometric primitives (lines and arcs), already implemented in the industrial robot controller as linear and circular motion, respectively. The works by Luu et al. [18] and Safeea et al. [21] described a similar technique, dividing the G-commands into line and arc segments, and applying path smoothing algorithms to mitigate jerky motion.

Watson et al. [24] also used a 2.5D common slicing software (Cura), but developed a G-code post-processing algorithm to adapt the generated toolpath and minimize travel moves (ie. moves where no material is deposited) because, when working with concrete deposition, it is more time consuming to turn the concrete pump on and off, and because common FDM techniques, like retraction, are not possible. The work by Fry et al. [6] used a manual approach to divide a 3D model, slice each part separately, and apply G-code post-processing (offsets and rotation) for dynamically changing build orientation. Wu et al. [26] developed an automated model decomposition for non-planar slicing, but each section was still sliced by a common planar technique. Other works implemented some type of conformal slicing, or toolpath generation techniques very specific for each use case.

[4] https://www.simplify3d.com/

Two works by Bhatt et al. [20, 22] focused on multi-resolution printing. In the first one, two robot manipulators were used in a cell, each one equipped with a different size nozzle extrusion tool. In the second work, the methods for model decomposition and toolpath planning were improved, and a new double nozzle extrusion tool was developed, so a single manipulator could be used for multi-resolution printing. Manufacturing a part with multi-resolution can significantly improve the build time by using a larger nozzle in inner layers, while maintaining surface quality through the use of the smaller diameter nozzle on the external surfaces.

Yigit et al. [28] developed a novel approach for building cylindrical parts using an external turning axis, implementing cylindrical slicing and path planning. The parts produced using this method performed 30% better in tensile tests when compared to the same parts built in a conventional AM machine. That is possible due to the additively turned part having linear infills that reinforce it in the axial loading direction.

Table 1. Main results of literature review.

[7]	**Setup**: UR3 robot manipulator holds the extrusion tool. **Software**: Slicer: CurviSlicer [17]. **Slicing**: The model is flattened and sliced. As each vertex's deformation is saved, the sliced model is then restored to its original shape. **Path planning**: The path is generated to be continuous, avoiding collisions and singularities, according to how the 2D regions (layers) are interconnected. The robot and extruder speeds are set to be constant throughout the printing motions
[18]	**Setup**: A Denso VP6242 robot manipulator holds extrusion tool. **Software**: Slicer: Slic3r. Robot trajectory: Custom script for G-code post-processing. **Slicing**: 2.5D slicing. **Path planning**: Smooth trajectory generation through the refinement of a set of point-to-point (joint motion) commands, exported from the slicer software as G-Code. The G-commands are divided into two kinds of segments: long straight segments and curves. A method is applied to smooth the jerky motion around the curved segments
[19]	**Setup**: An Epson T3 robot manipulator holds the extrusion tool. **Software**: Slicer: Slic3r. Robot trajectory: Matlab for G-code post processing. **Slicing**: 2.5D slicing. **Path planning**: The G-code from the slicing software is post processed, and the toolpath is redefined by using geometric primitives (lines, circles) already implemented in the robot controller (as linear and circular motions, respectively)
[20]	**Setup**: A cell with two robot manipulators (IRB2600 and IRB120), each holding a different nozzle size extrusion tool. **Software**: Slicer: custom. Robot trajectory: custom. **Slicing**: The model is decomposed into outer and inner layers. Outer layers are non-planar and printed with a smaller nozzle, while inner layers could be planar or non-planar and printed with a larger nozzle. **Path planning**: A sample of waypoints is created for a discrete parameter optimization problem to generate a smooth trajectory based on constraints such as the robot's kinematics, collision avoidance, desired tool path and velocity

(continued)

Table 1. (*continued*)

[16]	**Setup**: Work assumes the depositing of a fast-curing material that enables mid-air printing. **Software**: Python script running as a macro in FreeCAD for the path generation. **Slicing**: None. **Path planning**: A point cloud sampling method based on a CAD surface, for path generation using search-based algorithms
[21]	**Setup**: A KUKA iiwa 7 R800/14 R820 (both were tested) robot manipulator holds the extrusion tool. **Software**: MATLAB for integrated G-code post-processing, path generation and motion control. **Slicing**: 2.5D slicing. **Path planning**: G-code post-processing with the parametrization of linear and circular motions for smooth path planning. The generated path is then translated to the robot's joint space, while avoiding singularities
[22]	**Setup**: An Yaskawa GP 12 robot manipulator holds a custom double nozzle extrusion tool. **Software**: Custom-made. **Slicing**: Automated multi-resolution hybrid slicing: The part's exterior is built using conformal layers with a smaller filament (nozzle) width. The part's interior is built using planar layers with bigger filament width. **Path planning**: Graph search based on the Dijkstra algorithm to generate the shortest collision free path where each node of the path is the desired TCP position. The robot trajectory is defined by formulating a non-linear optimization problem that computes the inverse kinematics for each waypoint based on a series of constraints that include continuity, singularities, collisions, and the robot's joint limits and velocity
[13]	**Setup**: A UR5 robot manipulator holds the build plate. **Software**: Slicer and Robot trajectory: custom-made in MATLAB. **Slicing**: Adaptive slicing based of curved layer decomposition. **Path planning**: Once the model is decomposed into sequential curved layers, the path is generated based on the pattern of contour-parallel curves. The orientation of the extruder in each point of the path is also optimized for correct material deposition
[23]	**Setup**: An IRB 6640 robot manipulator holds the clay extrusion tool. **Software**: Rhino[5], Grasshopper[6], and HAL[7] for slicing and toolpath planning. **Slicing**: Consecutive polylines generated from the 3D model's contours. **Path planning**: Toolpath planning based on the successive points of contour polylines generated from the model. An algorithm compares the distance between consecutive polylines start and end points to determine the extruder activation/deactivation between each layer

(*continued*)

[5] https://www.rhino3d.com/.

[6] https://www.grasshopper3d.com/.

[7] https://hal-robotics.com/technology/.

Table 1. (*continued*)

[24]	**Setup**: An IRB 6640 robot manipulator holds the concrete extrusion tool. **Software**: Slicer: Cura. Toolpath planning: Rhino + Grasshopper + HAL. **Slicing**: 2.5D slicing. **Path planning**: Using Rhino + Grasshopper + HAL, the toolpath generated from the 2.5 slicing is modified to minimize travel moves
[25]	**Setup**: A Sawyer robot manipulator holds the extrusion tool. **Software**: Rhino + Grasshopper for slicing and toolpath planning. **Slicing**: Non-planar and volumetric custom-made slicing software implemented in the Grasshopper plug-in for Rhino. **Path planning**: G-code post processing with path smoothing and motion optimization tested in different software (MoveIt[8], PILZ Planner[9], and Descartes Planner[10])
[26]	**Setup**: A UR3 robot manipulator holds the build plate. **Software**: Custom-made using C++. **Slicing**: Shape decomposition eliminates support structures and optimizes print sequence to avoid collisions. Each region is sliced in a conventional software. **Path planning**: Toolpaths from the planar sliced regions in G-code format are post-processed to determine each TCP position, according to the given position and orientation of the material deposition in the Euclidian space. Then these poses are transformed to the robot's joint space via inverse kinematics
[27]	**Setup**: A Motoman SV3X robot manipulator holds the extrusion tool. **Software**: MotoCom SDK[11] libraries. Repetier Firmware[12] on Arduino. **Slicing**: None. **Path planning**: The toolpath planning is based on the geometry features of the lattices structure
[6]	**Setup**: A cell with two Denso VS-068 robot manipulators: one holding the build plate and the other holding the extrusion tool. **Software**: Simplify3D. Custom LabVIEW program for G-code post-processing. **Slicing**: 2.5D slicing of manually decomposed sections of a model. **Path planning**: G-code post-processing (offsetting and rotation) for different printing orientations. The orientation of the build plate is set to change before the construction of each new section
[28]	**Setup**: A Melfa RV-2F robotic manipulator holds the extrusion tool. **Software**: Custom-made application for cylindrical slicing and toolpath planning, post-processing and translation to Melfa Basic V programming language. **Slicing**: Cylindrical slicing with circular and linear infill strategies. **Path planning**: Generated based on the island boundaries and infill of every layer. Travel moves follow a helical motion to avoid colliding with previously deposited material. Filament retraction is also implemented

(*continued*)

8 https://moveit.ros.org/

9 https://moveit.ros.org/documentation/planners/#pilz-industrial-motion-planner

10 https://wiki.ros.org/descartes

11 https://www.motoman.com/en-us/products/software/development/motocom-sdk

12 https://www.repetier.com/

Table 1. (*continued*)

[29]	**Setup**: A UR10e robot manipulator holds the coextrusion (fiber and polymer) tool. **Software**: Custom-made. **Slicing**: Conformal slicing of a shell shaped structure. The surface is mapped, and the offset layers are generated. **Path planning**: Surface mapping of the shell structure and toolpath generation from different strategies, such as zigzag, arc offset, and contouring. The conformal paths are converted to TCP poses and the inverse kinematics is computed
[14]	**Setup**: A UR5e robot manipulator holds the extrusion tool. **Software**: Custom-made C++ application for slicing and toolpath planning. **Slicing**: Adaptive slicing based of curved layer decomposition. **Path planning**: Once the model is decomposed into sequential curved layers, the path is generated based on a hybrid strategy of contour-parallel and direct-parallel patterns

4 Conclusion

Robot-assisted additive manufacturing has emerged as a promising approach to address the limitations of traditional layer-by-layer AM processes, enabling the use of multi-directional AM techniques. This paper reviewed works about robot manipulators use in the material extrusion AM context.

Multi-directional AM, characterized by its ability to build layers in all directions, presents a paradigm shift from conventional AM methods. By allowing for the printing of complex geometries without the constraints of fixed layer orientation, multi-directional AM is a key enabling technology for the fabrication of parts with superior structural integrity, surface finish, and mechanical properties, while eliminating the use of support structures, and lowering material consumption.

Industrial robot manipulators are versatile platforms already used in most manufacturing industries. They are flexible, have multiple DOF and can be used in a variety of applications, such as welding, pick and place, painting and assembly. While the higher cost of these manipulators can create a barrier, the innovation in slicing and path planning methods presented in the works reviewed demonstrates the feasibility and advantages in using this technology for multi-directional additive manufacturing.

Further state of the art review work could delve into the utilization of robots across a broader spectrum of AM technologies beyond the scope of material extrusion-based methods. Additionally, the use of mobile robots in the context of AM could be investigated. A robot with these capabilities could autonomously navigate manufacturing environments, enabling on-site fabrication of complex structures or repairs in remote or hazardous locations.

Acknowledgements. This work is carried out within the framework of the "Agendas para a Inovação Empresarial" (Project no 49, acronym "INOV.AM", with reference PRR/49/INOV.AM/EE, operation code 02/C05-i01.01/2022.PC644865234-00000004), supported by the RRP - Recovery and Resilience Plan and by the European Funds NextGeneration EU.

References

1. Abdulhameed, O., Al-Ahmari, A., Ameen, W., Mian, S.H.: Additive manufacturing: challenges, trends, and applications. Adv. Mech. Eng. **11**(2), 1687814018822880 (2019). https://doi.org/10.1177/1687814018822880
2. Wong, K.V., Hernandez, A.: A review of additive manufacturing. Int. Sch. Res. Not. **2012**, e208760 (2012). https://doi.org/10.5402/2012/208760
3. Urhal, P., Weightman, A., Diver, C., Bartolo, P.: Robot assisted additive manufacturing: a review. Robot. Comput.-Integr. Manuf. **59**, 335–345 (2019). https://doi.org/10.1016/j.rcim.2019.05.005
4. Schmitz, M., Wiartalla, J., Gelfgren, M., Mann, S., Corves, B., Hüsing, M.: A robot-centered path-planning algorithm for multidirectional additive manufacturing for WAAM processes and pure object manipulation. Appl. Sci. **11**(13), 13 (2021). https://doi.org/10.3390/app11135759
5. Jiang, J., Zhong, R., Newman, S.: A review of multiple degrees of freedom for additive manufacturing machines. Int. J. Comput. Integr. Manuf. **34** (2020). https://doi.org/10.1080/0951192X.2020.1858510
6. Fry, N.R., Richardson, R.C., Boyle, J.H.: Robotic additive manufacturing system for dynamic build orientations. Rapid Prototyp. J. **26**(4), 659–667 (2020). https://doi.org/10.1108/RPJ-09-2019-0243
7. Yao, Y., Zhang, Y., Aburaia, M., Lackner, M.: 3D printing of objects with continuous spatial paths by a multi-axis robotic FFF platform. Appl. Sci. **11**, 4825 (2021). https://doi.org/10.3390/app11114825
8. Bhatt, P.M., Malhan, R.K., Shembekar, A.V., Yoon, Y.J., Gupta, S.K.: Expanding capabilities of additive manufacturing through use of robotics technologies: a survey. Addit. Manuf. **31**, 100933 (2020). https://doi.org/10.1016/j.addma.2019.100933
9. Hong, F., Hodges, S., Myant, C., Boyle, D.E.: Open5x: accessible 5-axis 3D printing and conformal slicing. In: Extended Abstracts of the 2022 CHI Conference on Human Factors in Computing Systems, in CHI EA '22, pp. 1–6. Association for Computing Machinery, Apr, New York, NY, USA (2022). https://doi.org/10.1145/3491101.3519782
10. Tamir, T., et al.: Design and optimization of a control framework for robot assisted additive manufacturing based on the Stewart platform. Int. J. Control. Autom. Syst. **20**, 968–982 (2022). https://doi.org/10.1007/s12555-021-0058-4
11. Xiao, X., Joshi, S.: Process planning for five-axis support free additive manufacturing. Addit. Manuf. **36**, 101569 (2020). https://doi.org/10.1016/j.addma.2020.101569
12. Lettori, J., Raffaeli, R., Borsato, M., Pellicciari, M., Peruzzini, M.: Non-uniform planar slicing for robot-based additive manufacturing. Comput.-Aided Des. Appl., 104–118 (2023). https://doi.org/10.14733/cadaps.2024.104-118
13. Xu, K., Li, Y., Chen, L., Tang, K.: Curved layer based process planning for multi-axis volume printing of freeform parts. Comput. Aided Des. **114**, 51–63 (2019). https://doi.org/10.1016/j.cad.2019.05.007
14. Zhang, T., et al.: S3-slicer: a general slicing framework for multi-axis 3D printing. ACM Trans. Graph. **41**(6), 1–15 (2022). https://doi.org/10.1145/3550454.3555516
15. Ramos, B., Pinho, D., Martins, D., Vaz, A.I.F., Vicente, L.N.: Optimal 3D printing of complex objects in a 5–axis printer. Optim. Eng. **23**(2), 1085–1116 (2022). https://doi.org/10.1007/s11081-021-09624-0
16. Onstein, I.F., Evjemo, L.D., Gravdahl, J.T.: Additive manufacturing path generation for robot manipulators based on CAD models. IFAC-Pap. **53**(2), 10037–10043 (2020). https://doi.org/10.1016/j.ifacol.2020.12.2724

17. Etienne, J., et al.: CurviSlicer: slightly curved slicing for 3-axis printers. ACM Trans. Graph. **38**(4), 81:1–81:11 (2019). https://doi.org/10.1145/3306346.3323022
18. Luu, Q.K., La, H.M., Ho, V.A.: A 3-dimensional printing system using an industrial robotic arm. In: 2021 IEEE/SICE International Symposium on System Integration (SII), pp. 443–448 (2021). https://doi.org/10.1109/IEEECONF49454.2021.9382645
19. Castelli, K., Zaki, A.M.A., Balakrishnappa, A.Y., Carnevale, M., Giberti, H.: A path planning method for robotic additive manufacturing. In: 2021 3rd International Congress on Human-Computer Interaction, pp. 1–5. Optimization and Robotic Applications (HORA) (2021). https://doi.org/10.1109/HORA52670.2021.9461380
20. Bhatt, P.M., et al.: A robotic cell for multi-resolution additive manufacturing. In: 2019 International Conference on Robotics and Automation (ICRA), pp. 2800–2807 (2019). https://doi.org/10.1109/ICRA.2019.8793730
21. Safeea, M., Bearee, R., Neto, P.: An integrated framework for collaborative robot-assisted additive manufacturing. J. Manuf. Process. **81**, 406–413 (2022). https://doi.org/10.1016/j.jmapro.2022.06.067
22. Bhatt, P.M., Kulkarni, A., Malhan, R.K., Shah, B.C., Yoon, Y.J., Gupta, S.K.: Automated planning for robotic multi-resolution additive manufacturing. J. Comput. Inf. Sci. Eng. **22**(021006) (2021). https://doi.org/10.1115/1.4052083
23. Kontovourkis, O., Tryfonos, G.: Integrating parametric design with robotic additive manufacturing for 3D clay printing: an experimental study. ISARC Proc., 918–925 (2018)
24. Watson, N., Meisel, N., Bilén, S., Duarte, J., Nazarian, S.: Large-Scale Additive Manufacturing of Concrete Using a 6-Axis Robotic Arm for Autonomous Habitat Construction (2019)
25. P. Fucile, V. David, M. Kalogeropoulou, A. Gloria, and L. Moroni, RAVEN: development of a novel volumetric extrusion-based system for small-scale Addit. Manuf. 2023. doi:https://doi.org/10.1101/2023.03.30.534759.
26. Wu, C., Dai, C., Fang, G., Liu, Y.-J., Wang, C.C.L.: RoboFDM: a robotic system for support-free fabrication using FDM. In: 2017 IEEE International Conference on Robotics and Automation (ICRA), pp. 1175–1180 (2017). https://doi.org/10.1109/ICRA.2017.7989140
27. Ishak, I., Larochelle, P.: Robot Arm Platform for Additive Manufacturing: 3 D Lattice Structures (2017). Last accessed 29 Jan 2024. https://www.semanticscholar.org/paper/Robot-Arm-Platform-for-Additive-Manufacturing-%3A-3-D-Ishak-Larochelle/1d99b64c16cbd0a1ece2c4571de8a60c2b53b490
28. Yigit, I.E., Khan, S.A., Lazoglu, I.: Robotic additive turning with a novel cylindrical slicing method. Int. J. Adv. Manuf. Technol. **119**(11), 7641–7651 (2022). https://doi.org/10.1007/s00170-021-08567-1
29. Zhang, G., Wang, Y., Chen, Z., Xu, X., Dong, K., Xiong, Y.: Robot-assisted conformal additive manufacturing for continuous fibre-reinforced grid-stiffened shell structures. Virtual Phys. Prototyp. **18**(1), e2203695 (2023). https://doi.org/10.1080/17452759.2023.2203695

Twin Transitioning to Local Circular Production Systems: Upcycling Fishing Nets through Large-Scale Additive Manufacturing and Microfactories

Robin Teigland[1](✉), Mohammad Rouhi[2], Woodrow Wiest[2], Markus Eriksson[2], Jon Erik Borgen[3], and Karoline Teigland[4]

[1] Chalmers University of Technology, Department of Technology Management and Economics, Gothenburg, Sweden
robin.teigland@chalmers.se
[2] RISE—Research Institutes of Sweden, Mölndal, Sweden
[3] Ocean Tech Hub Lda, Peniche, Portugal
[4] Brandworks AB, Stockholm, Sweden

Abstract. The production of large-scale products is currently undergoing a considerable shift in the manufacturing sector in favor of local additive manufacturing (AM). At the same time, using recycled materials in the production process has also risen to the top of the industry's priority list due to a growing sustainability focus. In this context, this paper presents the initial outcomes of the SuRFLSAM project—a twin transition project aimed at transforming global linear manufacturing value chains into circular value networks of distributed local microfactories using recycled materials. More specifically, the SuRF-LSAM project is developing two pilot microfactories in Sweden and Portugal and has conducted several activities required in a local microfactory: recycling used Nylon6 (PA6) fishing nets into secondary material composites, manufacturing products through large-scale additive manufacturing (LSAM), and conducting LSAM simulation. The SuRF-LSAM project continues to investigate how to most efficiently transform used fishing nets into the secondary material composite for LSAM and the most appropriate microfactory layout as well as conducting feasibility studies using supply chain models and parameters related to circular scenarios to determine which products are optimal for local LSAM production.

Keywords: Circular economy · Microfactory · Large scale additive manufacturing · Recycling · Ocean plastics · Polymer compounding · Digital twins · Process simulation

1 Introduction

Today's dominant manufacturing model is a linear one driven by economies of scale and characterized by centralized manufacturing factories using primarily virgin materials within global supply chains. Not only is this model showing an increasing lack of

© The Author(s), under exclusive license to Springer Nature Switzerland AG 2027
M. Calvete Gaspar et al. (Eds.): Regional Helix 2024, LNME, pp. 14–29, 2027.
https://doi.org/10.1007/978-3-032-36365-7_2

resiliency to pandemics, geopolitics, and climate change, but it has also led to an unsustainable "take-make-dispose" pattern of production and consumption. In response, the EU's twin transition of manufacturing offers great promise, in which digital technologies facilitate sustainability and the circular economy through designing new products from waste and used materials and keeping products and materials in use longer. The challenge is, however, that the linear model is so deeply ingrained in business ecosystems that our progress toward a truly circular economy model remains hindered. To tackle this challenge, we take a novel holistic approach with the overarching purpose of transforming a global linear value chain into a circular value network.

To achieve this, we initiated two subsequent projects—(1) the feasibility/stage one six-month project OCEAN-LSAM (Organizing the Circular Economy Through A Network of LSAM Microfactories Recycling Ocean Plastics) and (2) the stage two three-year project SuRF-LSAM (Developing Sustainable and Resilient circular economy microFactories) in Sweden and Portugal. The overarching objective of these projects is to develop a network of distributed local microfactories, in which local waste or used materials are recycled into composites for the additive manufacturing (AM) of final products for local markets within the same manufacturing facility. Today such efforts are limited, thus this project is a global frontrunner in the twin transition to a truly circular economy model as it focuses on piloting two microfactories in Sweden and Portugal. In this paper, we present the projects along with some initial results.

2 Background

2.1 Circular Economy and the Twin Transition

The drawbacks from the linear consumption of materials and products are now instigating more sustainable solutions based on circular economy principles [1, 2]. The circular economy (CE) can be defined as a "regenerative system in which resource input and waste, emission, and energy leakage are minimized by slowing, closing, and narrowing material and energy loops" [3]. This can be achieved through activities such as long-lasting design, maintenance, repair, reuse, remanufacturing, refurbishing, and recycling. As such, industry has moved beyond its focus of recent decades on merely reducing emissions from factories and transportation to also exploring how to reduce the use of resources from a product lifecycle perspective, such as through new forms of production and the use of recycled materials [4]. One means to achieve the circular economy and reduced resource usage is to leverage digital technologies, e.g., additive manufacturing, digital twins, industry 4.0, in what has been coined the twin transition—or the interconnected focus on digital transformation and sustainability.

2.2 Additive Manufacturing (AM) Microfactories

In recent years, the concept of distributed manufacturing (also known as distributed production) has emerged as an alternative to centralized mass production due to production and infrastructure technology advances enabling smaller (and micro scale) manufacturing units (i.e., microfactories) located near the point of consumption [5, 6]. Microfactories can respond more quickly to external changes than traditional factories, while microfactories using AM are able to not only print on demand but can also stimulate innovation through customizing products to the local market. As a result, microfactories are argued to potentially be more sustainable forms of production than more traditional production such as injection molding for some use cases [6].

As the concept of microfactories using AM is still emerging, the primary focus has been on the distributed production of spare parts. Some industrial sectors are decentralizing their spare parts production. Renault uses distributed AM to produce spare parts, and Fieldmade in the defense sector is digitalizing for on-demand spare part production using metal AM. Part information is delivered in digital format through a secured digital platform and 3D printed in a mobile microfactory, which can easily be disassembled and re-located. However, successful examples of microfactories producing spare or even final parts from recycled materials are limited. US-based Local Motors created low-volume microfactories and developed a driverless shuttle made with recycled materials through LSAM. The process was optimized to be profitable with as few as 200 units a year at each microfactory [7].

However, Local Motors closed in 2022 seemingly due to a lack of funding [8]. Similarly, UK-based Arrival filed for bankruptcy in early 2024. Although Arrival did not use AM, it wanted to revolutionize electric vehicle production through microfactories in city centers. Local Motors' and Arrival's failures may be more related to driving a radical transition to electric vehicles in a highly regulated industry than to microfactories. More recent microfactory efforts include Haddy in St Petersburg, USA and the Arizona microfactory facility in Phoenix, which is a collaboration among Arizona State University, the City of Phoenix, Goodwill, and Hustle PHX. This microfactory collects and processes discarded plastics from Goodwill of Central and Northern Arizona and turns them into pellets for manufacturing circular economy products such as pet bowls, skateboards, and plastic lumber. However, to date just how to integrate the various processes of recycling and compounding waste and used materials with AM within one facility remains empirically unexplored.

2.3 Recycled Plastics for AM

Plastics recycling can be performed both mechanically and chemically [9]. As chemical recycling is still emerging and considered to be more energy intensive and requiring large-scale efforts, mechanical recycling is more common. The first step in recycling is usually sorting by polymer and cleaning to remove dirt, residues, and other impurities that could affect the recycled polymer quality. The plastic is then run through a mechanical shredder to produce smaller, uniform pieces before converting them into pellets for manufacturing. Today's current practice is to conduct the various activities on separate sites, i.e., recycling plants, plastics factories, part/final product factories. As such, materials are transported from one site to another, often across the globe, before being reintroduced into a mass-produced product, and therefore merely sustaining the linear manufacturing model.

While recycled materials are used in traditional manufacturing such as injection molding, the use of recycled materials in AM is lagging. Doing so, however, would align with sustainability goals, support material reuse and recycling innovation and could offer cost savings [10]. Today several recycled thermoplastic polymers, e.g., PLA, PETG,

ABS, in filament and pellet forms for AM and large-scale AM (LSAM) are sourced from pre-consumer waste streams (e.g., production scrap materials) and from post-consumer waste streams (e.g., discarded water bottles). Further, other waste materials are being investigated as potential recycled materials for printing, such as ocean plastics, office paper, tires, and electronic and consumer goods [11].

For AM and LSAM, the plastic compounding process consists of melting and mixing the shredded plastics with ingredients, such as fillers/reinforcements and compatibilizers, to form a homogeneous blend with tailored properties. The compound is then extruded into a filament or pelletized for production. While successful results have been achieved with other polymers such as ABS, several technical challenges exist when using nylon6/PA6 in high-quality recyclate for AM and LSAM [9]. First, PA6 shrinks during the cooling process, which may compromise the printed product's structural performance. Adding fillers or other reinforcements to the compound can increase its mechanical properties, and these materials can be sourced from various waste streams. For example, in the project Repair3D [12], lightweight and strong carbon fibers were reclaimed from end-of-life aircraft structures and used as reinforcements to produce high-added-value 3D-printed products.

Second, recycled materials often have varying compositions and impurities, which can affect the printed parts' quality and mechanical properties. To ensure consistent quality, efficient techniques for sorting, sampling, tracing, cleaning, material analysis and quality control must be implemented throughout the recycling process, which was addressed in the EU Project Precycling [13]. Additives and compatibilizers may improve material consistency and printability; however, every thermoplastic waste material has different characteristics, and optimizing the compounding for any given material heavily relies on trial-and-error. The EU project TRANSFORM-CE [14] demonstrated the potential value of single use plastic waste for AM; however, the project did not investigate how to combine recycling and manufacturing of new product activities into one facility.

2.4 Digital Twins for more Sustainable AM

A second component of AM microfactories is digital twins. Digital tools can be used to ensure the best possible AM outcome for a given material, process, equipment, and final product, and these tools can be integrated into a digital twin, i.e., a virtual representation of the manufacturing process from the design phase to the final part production. The digital twin incorporates data, such as computer-aided design (CAD) models, process parameters, material properties, and sensor data, to mimic the AM system's real-world behavior and to predict outcomes. Digital twins help optimize AM processes by improving productivity, reducing costs and waste, and ensuring the quality and consistency of printed parts, which is even more critical for recycled materials as they tend to be more sensitive to printability issues.

AM simulations are often combined with in-situ monitoring for real-time data. Sensors and high-resolution cameras are used to detect defects in real-time, e.g., warpage. Data from sensor signals and processed images can also be combined with machine learning (ML) techniques to predict defects in the early stages of printing process [15].

In-situ monitoring can be used to stop a print that is failing, while live data can be interpreted in real time and used to adjust the print parameters, therefore potentially "saving" a print, which would otherwise have failed, thereby reducing wasted resources such as material, energy, and human resource time. Finally, data collected during a print can be used to develop AM models for simulation and provide experimental data to correlate the models. Due to so many parameters, optimizing for achieving desired part properties without extensive material characterization testing, in-situ monitoring with the help of sensors, and simulation tools can be extremely expensive and time consuming if they are needed for every recycled material recipe and part geometry. Therefore, there is growing interest in developing AI/ML tools to handle this data complexity [16].

2.5 Connecting AM Microfactories through a Digital Platform

Several more visionary forms of digital manufacturing systems (DMS) have recently emerged, such as connected mobile factories, production franchising, and cloud production [17] with cloud production the most relevant as it involves transferring CAD data to local AM production hubs. Digital platforms already play a crucial role in connecting factories, and the industry 4.0 (I4.0) concept is being tested across the globe. For example, the automotive sector has achieved significant productivity and product quality improvements from I4.0 digital platforms that enable data and information sharing both within factories and across business units and supply chains [18], such as in the Renault example noted above. Further, other digital technologies facilitate distributed data and information sharing in material traceability (e.g., digital product passports [19]), secure data exchange (e.g., MainChain blockchain protocol [20]), IP protection (e.g., CodeMeter Technology [21]), and AM data management (e.g., Materialisé Streamics [22]). However, digital twins for digital manufacturing systems to streamline processes and lower maintenance costs across units has only been implemented in a handful of use cases [23]. For AM, tools such as CAD, process simulation, part slicing, FE analysis, and in-situ monitoring with sensors are important parts of the platform. How these technologies can be integrated into a digital platform that connects AM microfactories across different countries using recycled materials for customized products for local markets remains unexplored.

3 SuRF-LSAM Project Description

3.1 Project Background

In order to transform a global linear value chain into a circular value network, as noted above we initiated two projects—OCEAN-LSAM (Organizing the Circular Economy Through A Network of LSAM Microfactories Recycling Ocean Plastics) and SuRFLSAM (Developing sustainable and resilient circular economy microfactories) in Sweden and Portugal. These projects were grounded in the Peniche Ocean Watch Initiative [24] that was founded in 2019 with the purpose of creating a blue circular economy model pilot through digitalization in Peniche, a commercial fishing village one hour northwest of Lisbon.

The second stage SuRF-LSAM project is a three-year project involving partners in Sweden and Portugal that resulted from the OCEAN-LSAM project - a six-month feasibility project in Sweden. Both projects were funded under Production2030—a strategic innovation program supported by Vinnova—Sweden's Innovation Agency, the Swedish Energy Agency, and Formas—Sweden's government research council for sustainable development. The vision of the two projects is to help fulfill the EU's goal of a twin transition in manufacturing by taking a holistic approach with the overarching purpose of transforming a global linear value chain into a circular value network of local microfactories. Thus, the overarching objective is to develop a pilot network of distributed microfactories, in which local waste and used materials are recycled into composites for additive manufacturing (AM) of final products for local markets within the same manufacturing facility, thus going beyond the state-of-the-art distributed manufacturing systems (DMS) concept [23].

SuRF-LSAM merged the OCEAN-LSAM project with SPARSAM—another previous Vinnova/Production2030 Step 1 feasibility project, to *implement the circular economy microfactory concept*. Chalmers University of Technology in Gothenburg, Sweden coordinates and manages the SuRF-LSAM under Professor Robin Teigland. Swedish partners originally included RISE, Adaxis, add:north AB, Artex AB, Brandworks AB, Ekbacken AB, PaperShell AB, and The Industry AB, while Portuguese partners include AGIX Lda, Instituto Politécnico de Leiria/CDRSP – Center for Rapid and Sustainable Product Development, Ocean Tech Hub Lda, Penimar Lda, ScalesOceans Lda, and University of Coimbra. Unfortunately, due to financial issues unrelated to this project, both Ekbacken Studios and The Industry filed for bankruptcy in 2024 and are no longer part of the project. The SuRF-LSAM project is to run from May 2023 to May 2026 and originally had a total budget of SEK 22 Mkr (approximately EUR 2 mln) with SEK 8 Mkr (approximately EUR 800,000) support from Production2030.

3.2 Project Work Packages and Activities

Figure 1 shows the project's five work packages. SuRF-LSAM leverages its partners' expertise and networks as well as Swedish industrial advances to establish two LSAM microfactories: (1) in Peniche, Portugal for SuRF-LSAM partner, Ocean Tech Hub Lda, that participated in Step 1 Ocean-LSAM, and (2) in Stockholm for SuRF-LSAM partner Brandworks AB.

SuRF-LSAM encourages interaction and knowledge sharing between the microfactories, thereby extending the microfactory concept by establishing an initial network of the first two microfactories. Driven by the identified use cases and supported by its partners, SuRF-LSAM also facilitates development of a network of microfactories by developing a model to calculate the actual values of LSAM for circular material production. As such, SuRF-LSAM is conducting feasibility studies using supply chain models and parameters related to circular scenarios in order to determine which products are optimal for local LSAM production. This enables SuRF-LSAM to go beyond the state-of-the-art to facilitate an LSAM microfactory network where global and local (glocal) partners will contribute with their competencies and establish further microfactories.

Further, SuRF-LSAM is creating *a computational model to replace real-world LSAM experiments*. Concentrating on material and process level properties, SuRFLSAM also

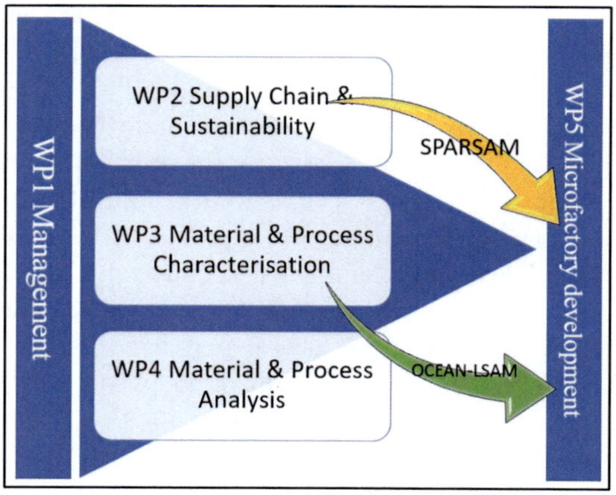

Fig. 1. SuRF-LSAM project work packages (Vinnova application)

seeks to advance earlier studies on the AM process and digitalization. Results from digital twins enhance existing designs and facilitate an effective evaluation of new ideas and products. These activities are complemented by extensive research at the material level to optimize the compounding process of recycled polymers followed by necessary multi-physical characterization in different levels, enabling microfactory efficiency.

3.3 Project Goals and Outcomes

Figure 2 shows the project goals for 2026, the effect goals for 2026, and outcomes for Production2030. As such, SuRF-LSAM contributes to improving today's low plastic recycling rate of 32.5% in the EU [25] and to decreasing the future use of virgin plastic, which is of dire importance as plastic production is expected to double within the next 20 years [26]. Thus, SuRF-LSAM also contributes to four UN SDGs: SDG 9 Industry, Innovation, and Infrastructure; SDG 11 Sustainable Cities and Communities; SDG 12 Responsible Consumption and Production; SDG 14 Life Below Water.

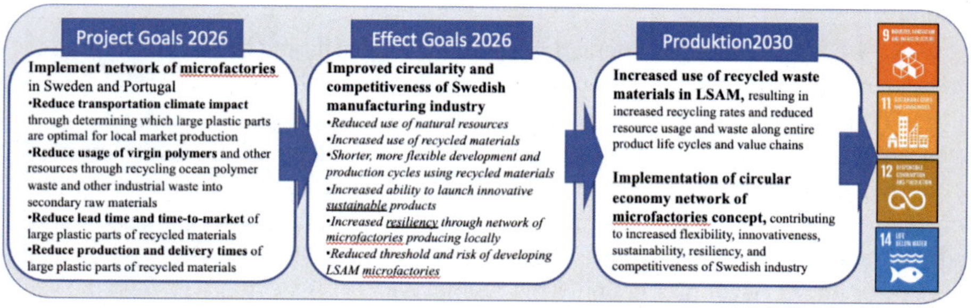

Fig. 2. Project goals, effect goals, and production2030 impact (Vinnova application)

4 Initial Project Outcomes

As the project is just over halfway, only our initial project outcomes are reported here.

4.1 Polyamide Material Development

The SuRF-LSAM project is developing a qualified recycled ocean polyamide composite for LSAM from end-of-life fishing nets provided by project partner Ocean Tech Hub LDA in Portugal and compounded with waste fillers from Swedish industry. Challenges have included ensuring the appropriate type of filler and the correct filler-torecycled polyamide ratio to enable the successful printing of the recycled ocean polyamide material composite through LSAM in different printing environments and for different products. A variety of different minerals have been tried as fillers and various material test methods were used, e.g., ISO 527-2, ISO 179, TGA, DSC, ISO 1183, Method A.

More specifically, the PENYLON™ material from Ocean Tech Hub LDA was characterized regarding its mechanical and other properties for both the pure recycled net material and the compounds created within the project. As a result, a qualified recycled polyamide compound was developed through adding mineral powder (material recipe) to enhance stiffness and LSAM printability, c.f. Fig. 3 (left).

Fig. 3. Developing and testing recycled polyamide (left), LSAM demonstrator (right) at RISE

4.2 LSAM Process

The initial methodology that we developed for LSAM of the above recycled ocean polyamide composite faces many challenges, such as (1) uncertain material properties for large format AM extrusion, (2) unknown process parameters, (3) unknown quality of final printed part, (4) poor adhesion of the part to the printing surface, (5) deformation of the part due to temperature gradients vs. printing time, (6) the degree to which the real printing differs from a process simulation and what could be done to improve it, (7) how the part design affects the resulting printed part, and finally (8) how minimally processed recycled materials could be directly integrated into the printing process.

To tackle these challenges, several steps were taken. First, we selected a somewhat complex digital part and prepared the printing through actions such as inspecting the 3D model to assure printability; preparing the tool path in Robot Studio, 3DP; defining the robot printing process parameters, e.g., extrusion rate, printing speed, part location in physical space; and testing different adhesion solutions. Second, the robot cell printing process parameters were defined, e.g., temperature, process speed, print surface adhesion, based first on known industry standards and expert opinions and then refined through trial and error. Third, the robotic printing cell was improved through building an enclosure around the heated printing surface and applying water soluble adhesive to the printing bed surface. Fourth, the actual printing was improved through actions such as further drying the material pellets before printing, fine tuning adjustments based on observations, improving part adhesion to the printing surface by heating PEI coated aluminum to various temperatures and test printing, printing extra layers. SuRF LSAM led to the successful printing of various pieces of furniture, c.f. Fig. 3 (right), which received recognition in various international media, e.g., *Vogue Magazine*. Furthermore, Research Engineer Woodrow Wiest at partner RISE successfully printed for the first time ever a boat from the Penylon material of recycled fishing nets, (Fig. 4). Of additional interest is that the boat was a design for a traditional Swedish wooden rowboat from the year 1896.

Fig. 4. Rowboat from 1896 printed by Research Engineer Woodrow Wiest at RISE on an ABB printer (left) and Wiest rowing the boat in the Gothenburg archipelago (right)

4.3 LSAM Process Simulation

The AM process simulation helps to virtually examine the printing process and ideally to improve the process parameters and material characteristics before printing. The goal is to reduce the trial-and-error process when printing by learning from the process model and material behavior during simulation. In this project, the LSAM computational process simulation was performed using the commercial software ABAQUS, and the thermal simulation model was validated against experimental data.

For this task, first, we designed a means to collect real-time data. Through several iterations, members of the team placed infrared thermometers and simple oven thermometers in key locations in the part to measure the temperature change over time during the printing process as well as documenting the process through timestamping, filming, and photographing. Second, we determined which parameters to include in the simulation model, as there were numerous (e.g., temperature, speed, air temperature, base temperature, layer thickness).

For the modeling analysis, RISE Senior Scientist Mohammad Rouhi used the thermo-mechanical approach. Thermo-mechanical simulations are a standard approach involving sequential thermal-stress analyses consisting of a transient heat transfer step followed by a static step that uses the temperatures from the first stage as it loads. Figure 5 shows the part that was selected for demonstration. The geometry (left) is converted into a finite element mesh (middle) for simulation, and the gcode from the slicer software is also necessary for the printing simulation which shows the printing path and layers (right).

Fig. 5. Preparing the digital part for simulation by RISE Senior Scientist Mohammad Rouhi

Senior Scientist Mohammad Rouhi at partner RISE performed a simulation using the virgin material nylon PA6 and nylon PA6 with reinforcements to compare the material performance. The process parameters such as printing speed, layer thicknesses, layer width, etc., were collected from experiments to set up the model. The initial simulation results for virgin material are shown in Fig. 6.

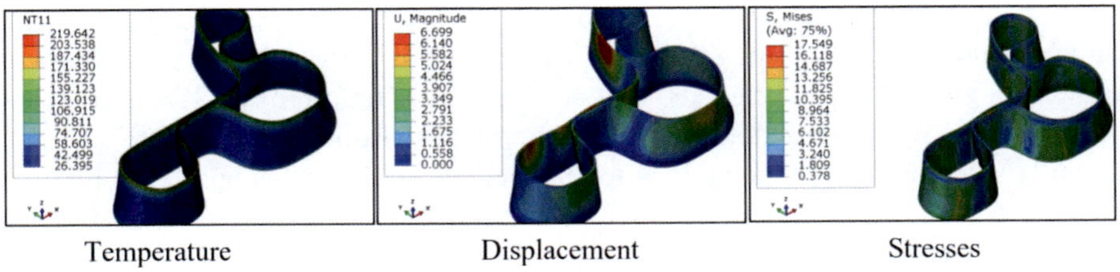

Fig. 6. Initial simulation results for virgin material by RISE Senior Scientist Mohammad Rouhi

Temperature data were collected at two different points of the printing part during the printing to validate the simulation results in the thermal step. Figure 7 (left) shows that the temperature readings did not match the simulation predictions at the point (seating area in middle of part) where data were collected due to very complex physics involving heat transfer between the two adjacent printing sections of the part. One reason for this is that accurate thermal modeling in this region requires using models in which experimental data were difficult to collect. However, the trend on simulation predictions follows the experiment very well with a good match in the second data point (Fig. 7, right). In summary, our initial results are satisfying as printed parts generally showed temperature comparisons to the simulated model parts, and the printed parts showed similarities in deformation as the simulation. However, more work and data are needed for comparisons.

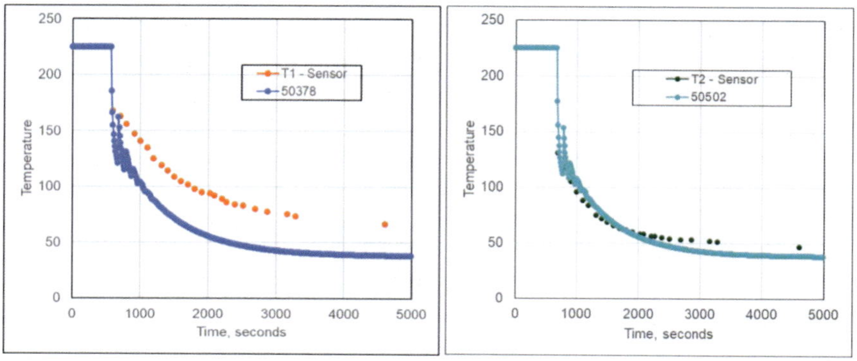

Fig. 7. Temperature validation by Rouhi at RISE

4.4 Microfactory Establishment

As noted above, to truly reach a circular economy, a systems-level transformation of manufacturing is needed [27], yet knowledge of how to integrate the various processes of recycling and compounding polymer waste and used materials with LSAM for an optimized circular economy manufacturing process within one facility remains empirically unexplored. Today SuRF-LSAM is leveraging the process that has been developed in collaboration with project partner Ocean Tech Hub Lda and the Peniche Ocean Watch Initiative under the lead of co-founder Jon Erik Borgen. Figure 8 shows the chain of transforming used PA6 fishing nets into a final product (furniture) already in progress for a microfactory under establishment by the project consortium.

SuRF-LSAM is investigating how to (1) integrate activities within one facility – a microfactory, and (2) to standardize and automate the process going from product idea to final 3D printed product such that (3) a network of microfactories could be established. To move towards these intents, the project team has developed and discussed potential layouts for a microfactory for partner Ocean Tech Hub LDA – integrating the team's competences. Several alternative solutions have been discussed as seen in Fig. 9.

Further, SuRF-LSAM is developing a use case in the furniture/exhibition industry with partner Brandworks AB in Sweden and in the marine industry with partner Penimar Lda in Portugal. As a first proof of concept, Brandworks Design Engineer Karoline

Fig. 8. Transforming used PA6 fishing nets into 3D printed products

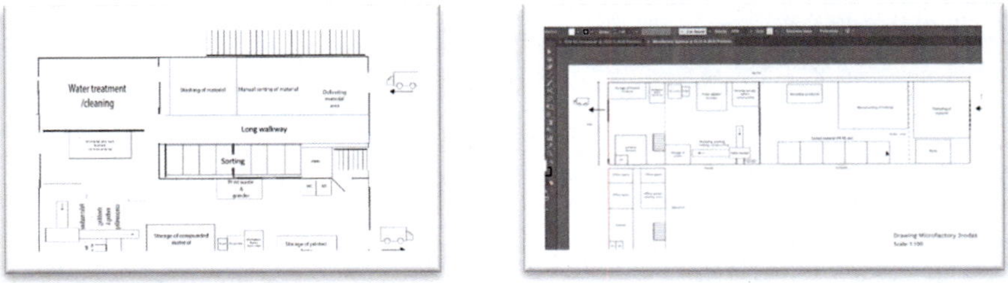

Fig. 9. Potential layouts for partner Ocean Tech Hub LDA's microfactory

Teigland designed a table base that was printed not only in both Portugal and in Sweden but also using two different LSAM machines: (1) at partner AGIX's facility in Portugal using The Industry's XYZ machine and (2) at partner Brandworks' facility in Sweden using ABB's industrial robot (Fig. 10).

4.5 Sustainability Analysis

SuRF-LSAM is making important contributions toward sustainability through the development of supply chain models that assess the environmental and economic implications

Fig. 10. Table base by Brandworks Design Engineer Karoline Teigland printed at AGIX Portugal on The Industry machine (left) and at Brandworks Sweden on an ABB robot (right).

of LSAM microfactories. Using established reference data, it is known that the production of one kg of virgin fossil-based PA6 resin (footprint to factory) emits up to 9 kgs of CO_2 [28], while incineration of one kg of plastic results in an additional 2.9 kg of CO_2 emissions [29]. Within the SuRF-LSAM project, feasibility studies are being conducted to quantify the potential cost, lead time, and sustainability impacts of local LSAM production compared to conventional alternatives. These efforts have enabled the development of measurable contributions to four UN Sustainable Development Goals, as outlined in Table 1.

Table 1. Quantitative impact measures for four UN SDGs

SDG 9 industry, innovation, and infrastructure SDG 11 sustainable cities and communities SDG 12 responsible consumption and production	Quantitative impact measure
Increased recycling of polymer and other waste/used materials into secondary raw materials for AM and improved	Kgs recycled per cycle
Circularity of recycled polymer and other waste/used material composites through repelletizing for further AM	
Reduced material, energy, transports, and other resource usage in prototyping and production through an AM process optimized for recycled polymer and other waste/used material composites, compared to traditional subtractive manufacturing techniques that result in significant spill material	Emissions reduction based on LCA of products compared to traditional manufacturing

(continued)

Table 1. (*continued*)

SDG 9 industry, innovation, and infrastructure SDG 11 sustainable cities and communities SDG 12 responsible consumption and production	Quantitative impact measure
Extending the service life of products and production systems through circular design and additive manufacturing of new components from recycled polymer and other waste/used material composites that modify the product's functions and/or its operation	Emissions reduction based on LCA of products compared to traditional manufacturing
Creation of a network of microfactories, thereby reducing resources along entire product life cycles and value chains by Swedish and Portuguese industry through awareness creation, competence development, and business model innovation	Emissions reduction based on LCA of products compared to traditional manufacturing
Improved community resilience through the creation of microfactories that both create new jobs and enable upskilling of workers in both urban and rural areas.	Number of new jobs in microfactories, number of hours in training
SDG 14 Life Below Water	*Quantitative Impact Measure*
Increased protection of marine ecosystems through reduced marine litter from fishing gear and increased recycling of polymer fishing nets into high-quality material as today only 1.5% of fishing gear is recycled	Kgs recycled
Increased ocean literacy through working with local communities to facilitate a transition to sustainable practices, circular material flows, and litter reduction through the increased valuation of waste and used materials, such as fishing nets	Number of collaborations, visits with local organizations

5 Conclusion

To date, SuRF-LSAM has conducted several activities required in a local microfactory: recycling used PA6 fishing nets into secondary material composites, producing products through large-scale additive manufacturing (LSAM) at different locations and on different machines, and conducting LSAM simulation. The project continues to investigate how to most efficiently transform shredded nets into the secondary material composite for LSAM and the most appropriate microfactory layout as well as conducting feasibility studies using supply chain models and circular economy parameters to quantify costs, lead time, and sustainability impacts of local LSAM microfactory production compared to conventional alternatives under partner RISE Researcher Markus Eriksson.

Acknowledgements. SuRF-LSAM project is co-funded under the Production2030 strategic innovation program by Vinnova—Sweden's Innovation Agency, Energimyndigheten—The Swedish Energy System, and FORMAS—Sweden's Government Research Council for Sustainable Development (https://Produktion2030.se/).

References

1. MacArthur Foundation.L Towards the Circular Economy (2013)
2. Tukker, A., Tischner, U.: New Business for Old Europe: Product-Service Development, Competitiveness and Sustainability. Greenleaf Publishing, Sheffield (2006)
3. Geissdoerfer, M., Savaget, P., Bocken, N.M, Hultink, E.J.: The circular economy: a new sustainability paradigm? J. Clean. Prod., 143, 757–768 (2017).
4. European Commission.: Circular Economy Action Plan: for a Cleaner and more Competitive Europe (2020)
5. Srai, J.S. et al.: Distributed manufacturing: scope, challenges and opportunities. Int. J. Prod. Res. **54**(23), 6917–6935 (2016)
6. Kohtala, C.: Addressing sustainability in research on distributed production: an integrated literature review. J. Clean. Prod. **106**, 654–668 (2015)
7. Manufacturing USA homepage. https://www.manufacturingusa.com/studies/3d-printedself-driving-vehicles. Last accessed 22 Jun 2025
8. The Drive homepage. https://www.thedrive.com/news/43857/local-motors-shuts-downafter-pivoting-from-rally-fighters-to-self-driving-buses. Last accessed 22 Jun 2025
9. Al Rashid, A., Koç, M.: Additive manufacturing for sustainability and circular economy: needs, challenges, and opportunities for 3D printing of recycled polymeric waste. Mater. Today Sustain. **24**, 100529 (2023)
10. Sanchez, F.A.C., Boudaoud, H., Camargo, M., Pearce, J.M.: Plastic recycling in additive manufacturing: a systematic literature review and opportunities for the circular economy. J. Clean. Prod. **264**, 121602 (2020)
11. Lodha, S. et al.: Sustainable 3D printing with recycled materials: a review. J. Mech. Sci. Technol. **37**(11), 5481–5507 (2023)
12. European Commission homepage. https://cordis.europa.eu/project/id/814588. Last accessed 22 Jun 2025
13. European Commission homepage. https://cordis.europa.eu/project/id/101058670. Last accessed 22 Jun 2025
14. Interreg homepage. https://vb.nweurope.eu/projects/project-search/transform-ce-transforming-single-use-plastic-waste-into-additive-manufacturing-and-intrusion-extrusionmoulding-feedstocks-and-creating-a-new-circular-economy-model-for-nwe/. Last accessed 22 Jun 2025
15. Nasrin, T., Pourkamali-Anaraki, F., Peterson, A.M.: Application of machine learning in polymer additive manufacturing: a review. J. Polym. Sci. **62**(12), 1–31 (2023)
16. Goh, G.D., Sing, S.L., Yeong, W.Y.: A review on machine learning in 3D printing: applications, potential, and challenges. Artif. Intell. Rev. **54**(1), 63–94 (2020)
17. Matt, D.T., Rauch, E., Dallasega, P.: Trends towards distributed manufacturing systems and modern forms for their design. Procedia Cirp. **33**, 185–190 (2015)
18. Piepoli, A., Arcidiacono, F., Basile, L.J., Pellegrino, R., Schupp, F., Zuehlke, T.: The impact of industry 4.0 on business performance: a multiple case study in the automotive sector. Procedia Comput. Sci. **232**, 2117–2126 (2024)
19. Additive marking homepage. https://additive-marking.de/digital-product-passport/. Last accessed 22 Jun 2025

20. 3D Adept Media homepage. https://3dadept.com/vistory-and-its-blockchain-protocolfor-additive-manufacturing/. Last accessed 22 Jun 2025
21. WIBU Systems homepage. https://www.wibu.com/solutions/ip-protection/additive-manufacturing.html. Last accessed 22 Jun 2025
22. Materialise homepage. https://www.materialise.com/en/inspiration/cases/3d-printer-control-industry-4-0. Last accessed 22 Jun 2025
23. Risling, M., Oberle, M., Bauernhansl, T.: Analyzing the purpose and technologies of digital twins in distributed manufacturing: a systematic literature review. Procedia Comput. Sci. **232**, 368–376 (2024)
24. Peniche Ocean Watch homepage. www.penicheoceanwatch.com. Last accessed 22 Jun 2025
25. Chawla, S., Varghese, B.S., Hussain, C.G., Keçili, R., Hussain, C.M.: Environmental impacts of post-consumer plastic wastes: treatment technologies towards eco-sustainability and circular economy. Chemosphere. **308**, 135867 (2022)
26. Lebreton, L., Andrady, A.: Future scenarios of global plastic waste generation and disposal. Palgrave Commun. **5**(1), 1–11 (2019)
27. Unruh, G.: Circular economy, 3D printing, and the biosphere rules. Calif. Manag. Rev. **60**(3), 95–111 (2018)
28. CarbonCloud ClimateHub homepage. https://apps.carboncloud.com/climatehub/productreports/id/480206512146. Last accessed 01 Aug 2025
29. GAIA—No Burn homepage. www.no-burn.org/wp-content/uploads/Plastic-is-CarbonOct2021.pdf. Last accessed 01 Aug 2025

Bibliometric Analysis of Digital Maturity Assessment Models for Industry 4.0

G. G. M. P. da Silva[1](✉), A. Guimarães[2,3], Marisa G. Pereira[4], F. A. Ferreira[5], and M. T. Pereira[4]

[1] Federal University of Santa Catarina, Florianópolis SC 88040-900, Brazil
glauco.silva@ufsc.br
[2] CISeD—Research Centre for Digital Services—Polytechnic Institute of Viseu, 3504-510 Viseu, Portugal
[3] CISE—Electomechatronic Systems Research Centre—University of Beira Interior, 6201-001 Covilhã, Portugal
[4] (LAETA-INEGI), ISEP, Polytechinc of Porto, 4249-015 Porto, Portugal
[5] UNIAG, ESHT, Polytechnic of Porto, 4480-876 Vila do Conde, Portugal

Abstract. This article presents a bibliometric review of the literature on maturity models applied to Industry 4.0 and digital transformation. The study aims to map and analyze research trends, key authors, and influential works in this emerging field. It focuses on publications indexed in the Web of Science and Scopus databases over the past decade. It uses bibliometric analysis tools to explore publication patterns, co-authorship networks, and thematic evolution. The findings reveal that global scientific production in Industry 4.0 research experienced slow growth between 2016 and 2019, followed by a significant increase from 2020 to 2021, peaking in 2022. Countries such as Italy, Portugal, and the UK lead citations, while Brazil plays a central role in international research collaborations. Thematic maps show developing trends, including growing interest in "green" technologies and emerging topics like "virtual reality". This bibliometric analysis highlights key research dynamics and identifies potential future directions for studies in Industry 4.0 and digital transformation.

Keywords: Industry 4.0 · Maturity models · Digital transformation

1 Introduction

Digital transformation and Industry 4.0 represent a new era of innovation and competitiveness in the industrial landscape, driven by the integration of emerging technologies such as the Internet of Things (IoT), artificial intelligence (AI), big data, and cyber-physical systems [1, 2]. These advancements enable unprecedented connectivity, intelligent automation, and greater organizational operational efficiency [3]. However, transitioning to this new reality is not instantaneous, as it requires structural, technological, and cultural changes across organizations [4, 5]. In this context, challenges arise in understanding the current level of readiness and establishing concrete pathways for successfully implementing these innovations [6].

© The Author(s), under exclusive license to Springer Nature Switzerland AG 2027
M. Calvete Gaspar et al. (Eds.): Regional Helix 2024, LNME, pp. 30–47, 2027.
https://doi.org/10.1007/978-3-032-36365-7_3

In recent years, numerous maturity models have emerged, offering different perspectives and frameworks to assess digital readiness and alignment with Industry 4.0 principles. Understanding the specificities and structures of these models is crucial for their proper and most suitable application, allowing for the selection of models that best fit organizational needs. Maturity models serve as vital strategic tools, providing a detailed diagnosis of companies' current states, offering insights into their capabilities, identifying gaps, and assisting in defining action plans oriented toward continuous improvement [7]. Moreover, these models provide a structured view of progress over time, supporting informed decision-making and realistic goal-setting for technological development and innovation. The application and development of maturity models in the literature are explored to assess their importance in guiding companies toward digital transformation and Industry 4.0. This bibliometric study on maturity models for Industry 4.0 systematically map the evolution of research in this field, identify key trends, influential contributions, and thematic clusters, and provide insights into the conceptual development and future directions of Industry 4.0 adoption and digital transformation.

2 Literature Review

2.1 Digital Transformation and Industry 4.0

Digital Transformation (DT) and Industry 4.0 are closely related concepts, yet they have distinct focuses and scopes [1]. Digital transformation refers to the comprehensive process of integrating digital technologies across all areas of an organization, leading to fundamental changes in how businesses operate and deliver value to customers [8]. It goes beyond the simple adoption of new technologies, instead involving a complete shift in business models, operational processes, and customer engagement strategies. The goal is to create more agile, responsive, and data-driven organizations that can adapt to rapid changes in the market [9].

Industry 4.0, on the other hand, refers to the Fourth Industrial Revolution, which focuses explicitly on transforming manufacturing by integrating technologies to create smart factories. These factories are characterized by highly automated processes, where machines and systems communicate autonomously, monitor physical processes, and make decentralized decisions. The critical focus of Industry 4.0 is increasing efficiency, flexibility, and innovation in production, emphasizing real-time insights that optimize industrial performance [2].

The relationship between digital transformation and Industry 4.0 can be understood in terms of complementarity. Industry 4.0 can be seen as a subset of digital transformation, providing the technological backbone for achieving it in industrial settings [10]. While digital transformation is a broader concept applicable to various sectors such as finance, healthcare, and services [11], Industry 4.0 is specifically concerned with manufacturing and industrial processes. In this sense, Industry 4.0 technologies like IoT, robotics, and big data analytics enable the goals of digital transformation by making production processes more efficient, flexible, and data-driven.

Moreover, both concepts reinforce each other. Digital transformation's focus on data as a core organizational asset aligns with Industry 4.0's reliance on real-time data for decision-making. For instance, the data-driven insights used in smart manufacturing optimize production and can be applied across the broader business ecosystem, enhancing customer experiences and supply chain management [12].

Industry 4.0 and digital transformation also emphasize the need for a customer-centric approach. In Industry 4.0, this is reflected in the growing demand for personalized products and just-in-time production enabled by digital technologies [13]. Digital transformation, in turn, seeks to transform businesses to become more responsive to customer needs by leveraging these technologies.

Despite their potential, both digital transformation and Industry 4.0 face challenges in adoption. Implementing these changes requires significant investment in technology, training, and cultural shifts. For many organizations, the interdisciplinary nature of these transformations, which involve IT and operational technology (OT) teams working closely together, can be a barrier [14]. However, the benefits of aligning digital transformation with Industry 4.0 are substantial. These efforts drive operational efficiencies and enable new business models, such as servitization, where manufacturers provide services based on real-time data from their products [15].

2.2 Digital Transformation and Industry 4.0 Maturity Models

Maturity models have emerged as essential tools that enable organizations to assess their readiness for digital transformation and Industry 4.0, set strategic goals, and prioritize investments in technology [16]. The significance of these models lies in their ability to provide organizations with a structured framework for evaluating their current capabilities and planning their digital transformation journey. These models allow businesses to benchmark their progress against industry standards and identify improvement areas. By doing so, organizations can set realistic and measurable goals that align with their digital transformation strategies [6]. Maturity models also assist organizations in prioritizing investments, ensuring that resources are allocated to critical areas such as technology integration, workforce development, or process automation. Furthermore, these models help businesses identify potential risks associated with digital transformation, such as cybersecurity threats or integration challenges, allowing them to mitigate these risks before they become problematic [17].

Maturity models typically consist of multiple stages or levels representing varying degrees of organizational competence in digital transformation and Industry 4.0 adoption. These stages range from basic or informal practices to fully integrated and optimized systems. Initially, organizations may only use digital technologies in a limited or fragmented way, without a strong focus on digital transformation. As they move through the stages, they increasingly adopt digital technologies in their processes, rely more on data, and use advanced analytics for decision-making. In the final stage, digital technologies are fully embedded across the organization, driving greater efficiency, innovation, and real-time decision-making [12].

In addition to structured stages, maturity models often incorporate multidimensional evaluations that assess various aspects of an organization's digital transformation readiness [18]. These dimensions can include technological integration, data management, organizational culture, leadership, and innovation. Maturity models typically use qualitative and quantitative metrics to comprehensively assess an organization's maturity.

For instance, they may evaluate the percentage of automated processes, the extent of real-time data utilization, or leadership's commitment to digital transformation efforts. This holistic approach ensures that organizations assess both the technical and organizational factors that contribute to successful digital transformation and Industry 4.0 adoption [19].

By assessing organizations across different areas, like technology integration, data management, organizational culture, and process automation, maturity models help businesses create detailed strategies for their digital transformation. As digital technologies evolve, these models will remain crucial in helping organizations stay competitive and succeed in the growing digital economy [6].

Various frameworks have been developed to assess digital transformation and Industry 4.0 maturity, created by both academic and non-academic contributors [1], each offering different methodologies and focus areas. Despite the growing body of research on digital maturity in the context of Industry 4.0, there remains a lack of comprehensive analyses that consolidate existing knowledge and identify critical gaps in theoretical development and practical implementation. Additionally, there is limited bibliometric research that examines the evolution of digital maturity models, their methodological approaches, and their impact across different industrial sectors. This paper aims to address these gaps to provide a roadmap for future research and support more effective digital transformation strategies in industry.

3 Methodology

3.1 Data

This study uses a bibliometric approach with quantitative methods applied to bibliometric data, following several steps. First, Scopus and Web of Science (WoS) were selected as the databases for this study due to their extensive coverage of high-quality, peer-reviewed academic literature, their robust citation indexing, and their widespread use in bibliometric analyses, ensuring a comprehensive and reliable mapping of research on digital maturity for Industry 4.0. The data collected covers the period from 2016 to 2024, focusing on research about digital maturity assessment models for Industry 4.0. The reliability of the bibliometric analysis depends on the credibility of these research platforms. The next step was a review of scientific literature, linking Industry 4.0 principles with digital maturity assessment models. Finally, data was collected from the databases, followed by detailed processing, analysis, and discussion.

3.2 Methods

Articles were collected from the Scopus and Web of Science (WoS) databases, which are widely used for scientific research. An initial exploratory analysis was performed to define the search strategy and identify standard terms in studies related to Industry 4.0 and digital maturity. Based on this, keywords such as "Industry 4.0", "Digital Transformation", and "Maturity Model" were selected. These terms were combined using Boolean operators (AND, OR) to create two search strings, one for each database, resulting in

Searches 1 and 2, as shown in Table 1. The search was limited to peer-reviewed articles published in the last ten years to ensure the relevance and quality of the data. Data analysis was carried out using VOSviewer (version 1.6.18) [20] and the Bibliometrix R package [21], which enabled the visualization of relationships between studies, keyword trends, and collaboration networks. VOSviewer, in particular, helped create and interpret bibliometric maps, even with large datasets [22, 23].

As highlighted by Hood and Wilson [24], bibliometric analysis has become a standard method in various scientific areas. This study used it to identify key authors, map research development, and assess the current relevance of the topic. This method gave a clear view of research trends, productivity, and collaborations, helping to track the progress of digital maturity models in Industry 4.0 (I4.0). The study addressed these research questions:

RQ1: How has scientific productivity evolved about digital maturity models in I4.0?

RQ2: What is the structure of the collaborative research network in this field?

RQ3: What are the main research trends in this area?

The study followed a methodology based on Donthu et al. [25] and the PRISMA-ScR guidelines Tricco et al. [26] to ensure transparency and avoid duplication in the review process, were also used, as outlined in Fig. 1.

This bibliometric study collected 98 articles from Web of Science and 89 from Scopus, focusing on models for assessing digital maturity in Industry 4.0. The initial results were screened for relevance based on titles, abstracts, and keywords. The analysis identified key publications, influential authors, trends, and emerging interests in certain countries. Data was organized by author, title, publication year, citations, abstracts, keywords, document type, and references.

Out of the 139 screened papers, 14 were excluded after full-text review because they focused on sector-specific applications, technical frameworks, or implementation strategies rather than comprehensive maturity models. Many papers addressed digital transformation and risk assessment or present case studies. Still, they lacked a structured approach to maturity progression and did not integrate core Industry 4.0 dimensions, essential requirements for inclusion. As a result, 125 relevant publications were identified and included in the final dataset, forming the basis for the bibliometric analysis. The study identified key publications, influential authors, research trends, and emerging thematic areas, providing a consolidated foundation for future research.

4 Results

4.1 Bibliometric Analysis

The analysis examines scientific production in a specific field and identifies key questions for future research. Its goal is to enhance the understanding of Industry 4.0 maturity assessment models, highlighting authors through co-citation maps, co-occurrence analysis, and identifying the most cited articles and authors. The next section presents the results of the bibliomtric analysis of the 125 selected articles, following the PRISMA-ScR methodology for selection and filtering.

Table 1. Gathering data for the bibliometric analysis.

Source	Query	# of publications
Scopus	(TITLE-ABS-KEY ("industry 4.0" OR "smart manufacturing" OR "manufacturing 4.0" OR "smart factory" OR "factory 4.0" OR "fourth industrial revolution" OR "fourth industrial" OR "4th industrial") AND TITLE-ABS-KEY ("maturity model" OR "readiness model" OR "readiness assessment" OR "assessment model" OR "maturity model development" OR "maturity assessment" OR "evaluation model" OR "foresight maturity" OR "maturity index" OR "maturity framework") AND TITLE-ABS-KEY ("digital transformation" OR "digitalization" OR "digital shift" OR "digital innovation" OR "digital evolution" OR "digital transition" OR "technological transformation" OR "tech-driven change" OR "digital disruption" OR "digital reengineering")) AND (LIMIT-TO (DOCTYPE,"ar"))	89
WoS	((TS = ("industry 4.0") OR TS = ("smart manufacturing") OR TS = ("manufacturing 4.0") OR TS = ("smart factory") OR TS = ("factory 4.0") OR TS = ("fourth industrial revolution") OR TS = ("fourth industrial") OR TS = ("4th industrial")) AND (TS = ("maturity model") OR TS = ("readiness model") OR TS = ("readiness assessment") OR TS = ("assessment model") OR TS = ("maturity model development") OR TS = ("maturity assessment") OR TS = ("evaluation model") OR TS = ("foresight maturity") OR TS = ("maturity index") OR TS = ("maturity framework")) AND (TS = ("digital transformation") OR TS = ("digitalization") OR TS = ("digital shift") OR TS = ("digital innovation") OR TS = ("digital evolution") OR TS = ("digital transition") OR TS = ("technological transformation") OR TS = ("tech-driven change") OR TS = ("digital disruption") OR TS = ("digital reengineering"))) and Article (Document Types)	98

Global Publications by Year

Figure 2 presents the annual global scientific production in a specific research area from 2016 to 2024. The significant increase in academic publications since 2020 indicates a growing research interest in the subject from that time to the present. On the other hand, the number of citations per article per year peaked between 2018 and 2019, suggesting that the limited number of publications until 2019 served as a foundation for the numerous studies published from 2020 onward. The decline in citations per article since 2020 can be attributed to a greater dispersion of references due to the expanding body of literature on the topic. Additionally, the relatively short time since publication for more recent articles has contributed to a lower citation count per article, as these studies are still in the process of being cited in ongoing research.

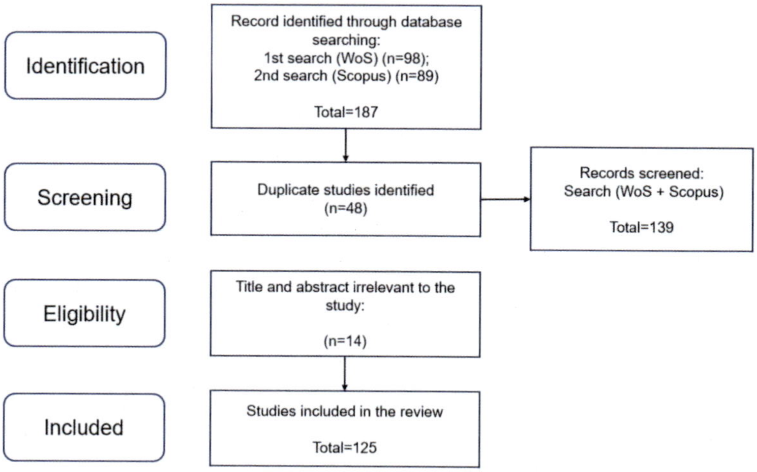

Fig. 1. Bibliometric analysis procedure of the PRISMA-ScR flowchart.

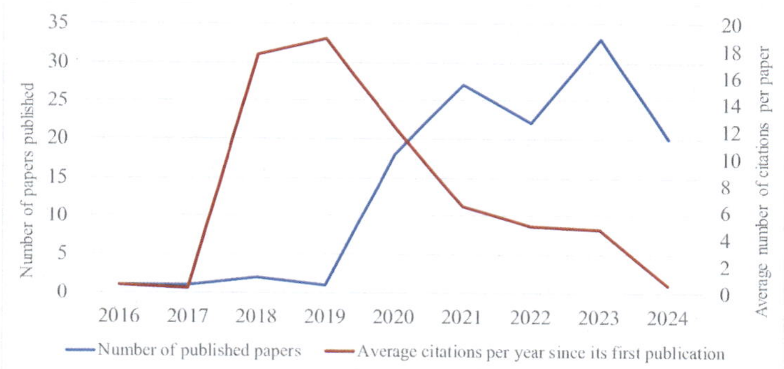

Fig. 2. Annual global scientific production obtained using bibliometrix R software.

Distribution of the Most Cited Publications by Countries

Figure 3 shows a network of research collaborations in Industry 4.0, with Brazil at the center, showing strong connections to Germany, Portugal, and Italy. The color gradient represents the timeline of collaborations, with darker shades for earlier partnerships and lighter ones for more recent collaborations. Germany was involved in earlier collaborations, while Turkey and Italy are more recent partners. This highlights Brazil's growing role in global research, forming new partnerships and expanding its influence in Industry 4.0.

Distribution of the Most Cited Publications by Country

Figure 4 shows the global distribution of scientific production by country, with Brazil leading as the most active contributor. Moderate contributions come from countries in Europe (such as Italy, Portugal, and the United Kingdom), Asia (China, India), North America, and Oceania (Australia). In contrast, regions in Africa, Southeast Asia, and

Fig. 3. Visualization of research collaborations in industry 4.0.

parts of Eastern Europe display lower research activity. This map highlights the concentration of research efforts in specific areas, particularly Brazil, and suggests a need for broader global collaboration.

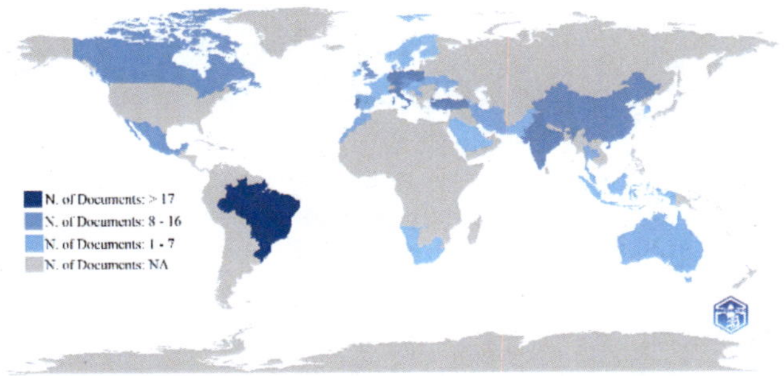

Fig. 4. Scientific production by country obtained using bibliometrix R software.

Global most Relevant Publications of Recent Years at the Global Level

Table 2 highlights the significant contributions that have shaped the discourse on digital transformation and maturity models in recent years. At the top, Culot et al. [1], with 337 citations, stands out for offering a foundational analysis of Industry 4.0 concepts, addressing critical questions that have made it a key reference in the field. Sjodin et al. [2], with 243 citations, follows closely by focusing on smart factory implementation and process innovation, providing one of the early and influential explorations of digitalization in manufacturing. Caiado et al. [12], with 146 citations, contributes to supply chain management through a fuzzy rule-based maturity model, offering adaptable frameworks that address the complex needs of Industry 4.0.

Further down the list, Santos and Martinho [27] and Colli et al. [28], with 124 and 113 citations respectively, offer structured approaches to assessing digital readiness, helping organizations measure their Industry 4.0 integration. Other publications, mostly from 2020 and 2021, show growing interest in areas like logistics, SMEs, and machine tool companies. Works like Facchini et al. [30] and Dutta et al. [29] provide practical insights for specific sectors, emphasizing the need for tailored frameworks for Industry 4.0 adoption.

In summary, this analysis highlights the diverse and multidimensional nature of Industry 4.0 research, from conceptual studies to industry-specific applications. The

Table 2. Most relevant publications of recent years at the global level.

Ranking	Author	Title	Year	Citations
One	Culot et al. [1]	Behind the definition of industry 4.0: analysis and open questions	2020	337
Two	Jodin et al. [2]	Smart factory implementation and process innovation is a preliminary maturity model for leveraging digitalization in manufacturing	2018	243
Three	Caiado et al. [12]	A fuzzy rule-based industry 4.0 maturity model for operations and supply chain management	2021	146
Four	Santos and Martinho [27]	An industry 4.0 maturity model proposal	2020	124
Five	Colli et al. [28]	A maturity assessment approach for conceiving context-specific roadmaps in the industry 4.0 era	2019	113
Six	Dutta et al. [29]	Digital transformation priorities of india's discrete manufacturing SMES - a conceptual study in perspective of industry 4.0	2020	98
Seven	Facchini et al. [30]	A maturity model for logistics 4.0: an empirical analysis and a roadmap for future research	2020	93
Eight	Rafael et al. [31]	An industry 4.0 maturity model for machine tool companies	2020	85
Nine	Amaral and Peças [32]	SMES and industry 4.0: two case studies of digitalization for a smoother integration	2021	85
Ten	Vuksanović Herceg et al. [33]	Challenges and driving forces for industry 4.0 implementation	2020	75

high citation counts, especially for works from 2020 and 2021, show a recent surge in interest, reflecting the growing importance of Industry 4.0 in both research and industry.

Global Keyword Co-occurrence Analysis

Figure 5 presents a keyword map illustrating the co-occurrence relationships among key terms in the research field. The nodes represent keywords extracted from the analyzed literature, while the edges indicate the strength of their co-occurrence. The map is divided into multiple clusters, each represented by a different color, highlighting thematic groupings within the research domain.

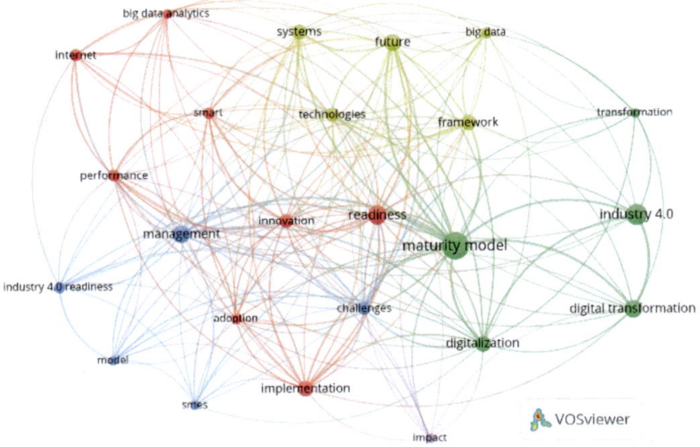

Fig. 5. Keyword map obtained using VOSviewer software.

The green cluster centered around "maturity model", "industry 4.0", and "digital transformation", suggests a strong focus on frameworks for assessing technological advancement and digitalization processes. The red cluster, which includes "readiness", "innovation", and "implementation", reflects discussions on preparedness and the challenges associated with adoption of technology. The blue cluster, featuring "management" and "industry 4.0 readiness", highlights the strategic and managerial aspects of industrial transformation. Finally, the yellow cluster, which includes terms like "future", "systems", and "technologies", indicates research trends focused on emerging developments and the future direction of digitalization.

The interconnections between these clusters suggest a multidisciplinary research approach, where discussions on readiness and implementation are closely linked to broader topics such as maturity models and digital transformation. The density of connections also indicates the growing academic interest in Industry 4.0 and its associated challenges, demonstrating the increasing complexity of research in this field.

Figure 6 presents a temporal keyword map illustrating the evolution of key research themes over time. The colors of the nodes correspond to the average publication year of the respective keywords, as indicated by the color gradient at the bottom of the figure. Darker shades (blue) represent earlier occurrences in the dataset, whereas lighter shades (yellow) indicate more recent developments.

The central theme of the map revolves around the term "maturity model", which serves as a focal point connecting various research domains. The evolution of the field suggests an early emphasis on "framework", "systems", and "management", as seen in the blue-toned nodes. These early studies provided the foundational theories and methodologies for assessing technological progress and organizational readiness.

Over time, research has shifted towards "readiness", "innovation", and "digitalization" (green to yellow tones), indicating a growing focus on implementation strategies

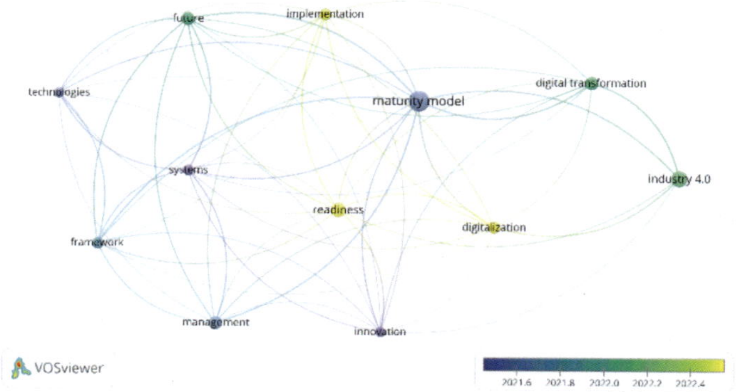

Fig. 6. Overlay of the keyword co-occurrence network obtained using VOSviewer software.

and the practical challenges of digital transformation. More recent research trends, highlighted in yellow, emphasize "future", "implementation", and "digital transformation", suggesting an increasing interest in the forward-looking aspects of Industry 4.0 adoption and its implications.

The connections between these keywords illustrate how different research streams have evolved and influenced one another. The transition from theoretical frameworks to more applied discussions reflect the maturing nature of the field, with current research focusing on how digital transformation can be effectively implemented across industries.

Analysis of the Keyword Cloud Map
The keyword map, illustrated in Fig. 7, highlights the primary Industry 4.0 research topics. "Maturity Model", "Industry 4.0", and "Digital Transformation" are the most prominent terms, indicating a strong focus on assessing digital maturity and the implementation of Industry 4.0 practices. Other significant keywords like "Management", "Implementation", "Readiness", and "Digitalization" suggest that research explores both the technological and organizational aspects of digital transformation. The presence of terms such as "Innovation", "Framework", and "Challenges" shows a comprehensive approach to addressing advancements and hurdles within the digitalization process. This map captures the key topics shaping current research in the field.

Thematic Map
The thematic map, illustrated in Fig. 8, offers an overview of Industry 4.0 research, categorizing themes by their relevance and development. "Readiness", "management", and "future" are identified as motor themes, driving current research and actively evolving. Basic themes like "maturity model", "Industry 4.0", and "digital transformation" are central to the field but less developed, indicating they form the foundation of research with potential for further exploration. Niche themes such as "assessment models" and "green" are well-developed but less central, representing specialized areas of study. "Virtual reality" is shown as an emerging or declining theme, suggesting it is either gaining attention or losing relevance. This map highlights key focus areas and potential opportunities for future research in Industry 4.0.

Fig. 7. Keyword map obtained using bibliometrix R software.

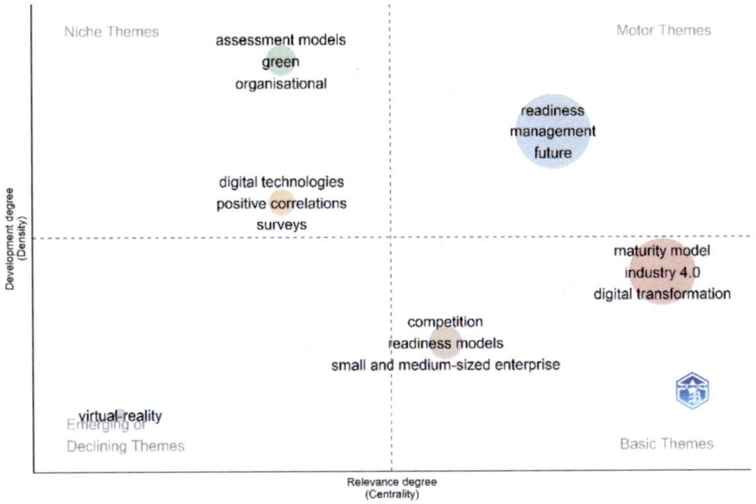

Fig. 8. Thematic map.

Analysis of Maturity Models

The most cited Maturity Models found in the research are presented in Table 3 and offer structured frameworks for assessing the progress and implementation of Industry 4.0 technologies across various domains. These models range from general frameworks applicable to manufacturing and logistics to specialized approaches for specific organizational structures or processes. Each model outlines progressive levels, with the lowest stages representing minimal or no engagement with Industry 4.0 concepts and the highest stages indicating full integration and optimization, requiring both technological upgrades and significant changes in organizational culture, strategy, and workforce skills.

A common trend among the models is their focus on multidimensional assessment criteria. For example, the Smart Factory Maturity Model [2] evaluates maturity based on People, Processes, and Technology, taking a holistic approach that includes both human and technological factors. Similarly, the OSCM4.0 Maturity Model [12] and the Industry 4.0 Maturity Model [27] cover many dimensions such as Customer, Logistics,

Table 3. Maturity models.

Name	Paper	Levels	Dimensions
Smart facture maturity model	Jodin et al. [2]	Level 1—Connected technologies; Level 2—Structured data gathering and sharing; Level 3—Real-time process analytics and optimization; Level 4—Smart, predictable manufacturing;	People; Process; Technology;
OSCM4.0 maturity level	Caiado et al. [12]	Level 0—Nonexistence; Level 1—Conceptual; Level 2—Managed; Level 3—Advanced; Level 4—Self-optimized;	Customer; Logistics; Supplier; Integration; Production, Planning and Control (PPC); Quality; Maintenance

(*continued*)

Table 3. *(continued)*

Name	Paper	Levels	Dimensions
Industry 4.0 maturity model	Santos and Martinho [27]	Level 0—low or none degree of implementation; Level 1—pilot actions being planned or being developed; Level 2—implementation of actions initiated, with some benefits being observed; Level 3—partial implementation of actions, that enhance the competitiveness of the company; Level 4—advanced implementation of actions, with clear economic returns; Level 5—reference in applying the concepts and implementing the technologies of Industry 4.0.	Organizational strategy; Structure and culture Workforce; Smart factories; Smart processes; Smart products and services.
Model based on acatech maturity model	Colli et al. [28]	None; Basic; Transparent; Aware; Autonomous; Integrated.	Governance; Technology; Connectivity; Value creation; Competences.

(continued)

Table 3. *(continued)*

Name	Paper	Levels	Dimensions
Small and medium discrete manufacturing establishments (SMME) maturity assessment	Dutta et al. [29]	Level of adoption from 1 to 5, where 5 represents the highest level of sophistication	Product design; Production planning; Production engineering; Production execution; Services.
Maturity model for logistics 4.0	Facchini et al. [30]	Ignoring; Defining; Adopting; Managing; Integrated;	Knowledge; Strategy and Leadership (S&L); Employees; IT Systems; Smart Products; Smart Warehouses; Technologies

Organizational Strategy, and Workforce, emphasizing the need for integration across various operational areas.

Some models, like the Acatech-based Maturity Model [28] and the SMME Maturity Assessment [29], focus on specific contexts. The Acatech model prioritizes governance and connectivity, crucial for aligning digital transformation with organizational strategy. The SMME model is tailored for small and medium-sized enterprises (SMEs), simplifying the framework to key areas like Product Design and Production Planning, making it more accessible to smaller companies with limited resources.

Additionally, models like the Maturity Model for Logistics 4.0 [30] focus on specific sectors, such as logistics. This model emphasizes Smart Products, IT Systems, and Leadership, showing the importance of integrating digital tools in logistics to achieve Industry 4.0 readiness.

In conclusion, these maturity models provide valuable tools for organizations to assess their progress in adopting Industry 4.0 technologies. By outlining clear levels of development and addressing multiple dimensions, they help businesses identify gaps and guide their digital transformation strategies. The varied focus areas across the models reflect the diverse applications of Industry 4.0 technologies, highlighting the need for flexible and tailored approaches in different industrial contexts.

5 Conclusions

This bibliometric review highlights the rapid growth of research on Industry 4.0 maturity models and digital transformation, with a notable surge in publications since 2020. The study reveals key trends, such as the emergence of "green" technologies and "virtual reality," which point to new research directions aligned with sustainability and technological innovation. The central role of countries like Italy, Portugal, and the UK in citations, alongside Brazil's strong international collaborations, underscores the global nature of this field.

Despite its contributions, this study has limitations, including potential biases from relying on Web of Science and Scopus databases, which may exclude publications and regional perspectives. Future research should expand to include diverse sources and focus on sector-specific models, longitudinal studies, and cross-cultural comparisons to enhance the applicability of maturity frameworks. By addressing these gaps, researchers can develop more robust and inclusive tools to guide organizations through the complexities of digital transformation, ensuring they remain competitive in the Industry 4.0 era.

Acknowledgments. This study was financed in part by the Coordenação de Aperfeiçoamento de Pessoal de Nível Superior—Brasil (CAPES)—Finance Code 001. We thank the Foundation for Science and Technology (FCT) for grant UI/BD/154507/2022.

References

1. Culot, G., Nassimbeni, G., Orzes, G., Sartor, M.: Behind the definition of industry 4.0: analysis and open questions. Int. J. Prod. Econ. **226**, 107617 (2020)

2. Sjödin, D., Parida, V., Leksell, M., Petrovic, A.: Smart factory implementation and process innovation a preliminary maturity model for leveraging digitalization in manufacturing. Res.-Technol. Manag. **61**(5), 22–31 (2018)
3. El Hamdi, S., Abouabdellah, A.: Logistics: impact of industry 4.0. Appl. Sci. **12**, 4209 (2022)
4. Ronaghi, M.H.: Toward a model for assessing smart hospital readiness within the industry 4.0 paradigm. J. Sci. Technol. Policy Manag. **15**(2), 353–373 (2024)
5. Brodny, J., Tutak, M.: Assessing the level of digital maturity in the three seas initiative countries. Technol. Forecast. Soc. Change. **190**, 122462 (2023)
6. Ansari, I., Barati, M., Sadeghi Moghadam, M.R., Ghobakhloo, M.: An industry 4.0 readiness model for new technology exploitation. Int. J. Qual. Reliab. Manag. **40**(10), 2519–2538 (2023)
7. Senna, P.P., Barros, A.C., Roca, J.B., Azevedo, A.: development of a digital maturity model for industry 4.0 based on the technology-organization-environment framework. Comput. Ind. Eng. **185**, 109645 (2023)
8. Kırmızı, M., Kocaoglu, B.: Digital transformation maturity model development framework based on design science: case studies in manufacturing industry. J. Manuf. Technol. Manag. **33**(7), 1319–1346 (2022)
9. Gökalp, E., Martinez, V.: Digital transformation capability maturity model enabling the assessment of industrial manufacturers. Comput. Ind. **132**, 103522 (2021)
10. Cognet, B., Pernot, J.P., Rivest, L., Danjou, C.: Systematic comparison of digital maturity assessment models. J. Ind. Prod. Eng. **40**(7), 519–537 (2023)
11. Khuntia, J., Saldanha, T., Kathuria, A., Tanniru, M.R.: Digital service flexibility: a conceptual framework and roadmap for digital business transformation. Eur. J. Inf. Syst. **33**(1), 61–79 (2022)
12. Caiado, R.G.G., Scavarda, L.F., Gavião, L.O., Ivson, P., de Mattos Nascimento, D.L., Garza-Reyes, J.A.: A fuzzy rule-based industry 4.0 maturity model for operations and supply chain management. Int. J. Prod. Econ. **231**, 107883 (2021)
13. Lee, C.H., Li, L., Li, F., Chen, C.: H.: requirement-driven evolution and strategy-enabled service design for new customized quick-response product order fulfillment process. Technol. Forecast. Soc. Change. **176**, 121464 (2022)
14. Treviño-Elizondo, B.L.: García-Reyes H.: an employee competency development maturity model for industry 4.0 adoption. Sustainability. **15**(14), 11371 (2023)
15. Frank, A.G., Mendes, G.H.S., Ayala, N.F., Ghezzi, A.: Servitization and industry 4.0 convergence in the digital transformation of product firms: a business model innovation perspective. Technol. Forecast. Soc. Chang. **141**, 341–351 (2019)
16. Hortovanyi, L. et al.: Assessment of digital maturity: the role of resources and capabilities in digital transformation in B2B firms. Int. J. Prod. Res. **61**(23), 8043–8061 (2023)
17. Ayyildiz, E., Taskin Gumus, A.: Interval-valued Pythagorean fuzzy AHP method-based supply chain performance evaluation by a new extension of SCOR model: SCOR 4.0. Complex Intell. Syst. **7**, 559–576 (2021)
18. Mohanta, P.R., Mahanty, B.: Assessing industry 4.0 implementation maturity in manufacturing MSMEs—a graph theory and matrix-based approach. TQM J. **ahead-of-print**(ahead-of-print) (2024)
19. Bhamriya, A.K., Gupta, M.A.: Holistic approach for assessment of global industry 4.0 maturity level. Eng. Manag. J., 1–27 (2024)
20. Van Eck, N.J., Waltman, L.: Manual for VOSviewer Version 1.6. 8. CWTS Meaningful Metrics. Universiteit Leiden (2018)
21. Aria, M., Cuccurullo, C.: Bibliometrix: an R-tool for comprehensive science mapping analysis. J. Inf. **11**(4), 959–975 (2017)
22. Castillo-Vergara, M., Alvarez-Marin, A., Placencio-Hidalgo, D.: A bibliometric analysis of creativity in the field of business economics. J. Bus. Res. **85**, 1–9 (2018)

23. Mas-Tur, A., Roig-Tierno, N., Sarin, S., Haon, C., et al.: Co-citation, bibliographic coupling and leading authors, institutions and countries in the 50 years of technological forecasting and social change. Technol. Forecast. Soc. Chang. **165**, 120487 (2021)
24. Hood, W.W., Wilson, C.S.: The literature of bibliometrics, scientometrics, and in-formetrics. Scientometrics. **52**, 291–314 (2001)
25. Donthu, N., Kumar, S., Mukherjee, D., Pandey, N., Lim, W.M.: How to conduct a bibliometric analysis: an overview and guidelines. J. Bus. Res. **133**, 285–296 (2021)
26. Tricco, A.C., Lillie, E., Zarin, W., O'Brien, K.K., Colquhoun, H., Levac, D., et al.: PRISMA extension for scoping reviews (PRISMA-ScR): checklist and explanation. Ann. Intern. Med. **169**(7), 467–473 (2018)
27. Santos, R.C., Martinho, J.L.: An industry 4.0 maturity model proposal. J. Manuf. Technol. Manag. **31**(5), 1023–1043 (2020)
28. Colli, M., Berger, U., Bockholt, M., Madsen, O., Møller, C., Wæhrens, B.V.: A maturity assessment approach for conceiving context-specific roadmaps in the industry 4.0 era. Annu. Rev. Control. **48**, 165–177 (2019)
29. Dutta, G., Kumar, R., Sindhwani, R., Singh, R.K.: Digital transformation priorities of India's discrete manufacturing SMEs: a conceptual study in perspective of industry 4.0. Compet. Rev. **30**(3), 289–314 (2020)
30. Facchini, F., Oleśków-Szłapka, J., Ranieri, L., Urbinati, A.: A maturity model for logistics 4.0: an empirical analysis and a roadmap for future research. Sustainability. **12**(1), 86 (2020)
31. Rafael, L.D., Jaione, G.E., Cristina, L., Ibon, S.L.: An industry 4.0 maturity model for machine tool companies. Technol. Forecast. Soc. Chang. **159**, 120203 (2020)
32. Amaral, A., Peças, P.: SMEs and industry 4.0: two case studies of digitalization for a smoother integration. Comput. Ind. **125**, 103333 (2021)
33. Vuksanović Herceg, I., Kuč, V., Mijušković, V.M., Herceg, T.: Challenges and driving forces for industry 4.0 implementation. Sustainability. **12**, 4208 (2020)

Industry 4.0 & Sustainability: A Case Study of OPC UA Injection Moulding Machine

A. Kaduk[1], L. Ferreira[1,2](✉), and C. Neves[1,3]

[1] ESTG, Polytechnic University of Leiria, Leiria, Portugal
2200318@my.ipleiria.pt, {lino.ferreira,carlos.neves}@ipleiria.pt
[2] Instituto de Telecomunicações, Coimbra, Portugal
[3] INESCC, Coimbra, Portugal

Abstract. Incorporating sustainability principles into Industry 4.0 initiatives not only promotes eco-friendly innovation and operational effectiveness in industrial processes but also addresses pressing sustainability challenges. Industry 4.0 technologies enable companies to improve flexibility, efficiency, and cost savings while also tackling sustainability issues. This paper focuses on the development of a monitoring system for an injection machine utilizing the Open Platform Communications Unified Architecture (OPC UA) protocol. This modernization effort aims to align the IMI1300S machine, built in 1973, with the principles of Industry 4.0. This transformation enables real-time monitoring and analysis of the injection machine's operational and production data. Continuous data analysis enables the evaluation of production efficiency and the generation of new recommendations for process optimization, including, for example, the reduction of the injection cycle time, lowering production costs. The proposed system consists of two complementary modules: an electronic module for acquiring sensor data from the injection process and a software module for processing and storing the data. The electronic module continuously transmits data to the software module, which utilizes the OPC UA protocol to structure and store it in a server-hosted database. This data is then accessible for analysis by any device with an OPC UA client, or directly from the database. The tests conducted in an industrial environment enabled the validation of the system in terms of its operability and applicability.

Keywords: Injection moulding machine · Industry 4.0 · OPC UA · Sustainability

1 Introduction

Plastic injection moulding is a crucial element in the development of the plastics industry, driving technological progress, the creation of innovative products and the generation of opportunities. The plastics industry is widely used in various sectors of industry, and its products are ever-present in our daily lives. In 2022, the importance of the plastics industry in the European Union alone was represented by 53,150 companies, more than 1.5 million jobs, with a turnover of more than 400 billion and profits of 9.2 billion euros [1]. In this context, research into sustainable solutions to minimize environmental impact and achieve increased efficiency in production processes is essential. Adopting advanced

© The Author(s), under exclusive license to Springer Nature Switzerland AG 2027
M. Calvete Gaspar et al. (Eds.): Regional Helix 2024, LNME, pp. 48–59, 2027.
https://doi.org/10.1007/978-3-032-36365-7_4

technologies, particularly those aligned with Industry 4.0 principles, unlocks a wealth of benefits for the plastics industry. This positions the sector for a more sustainable and prosperous future [2].

Plastic injection moulding is a complex process, influenced by various production parameters and variables that determine the quality and quantity of the produced plastic parts. Throughout the production process, the melt is subject to significant and dynamic changes in pressure and temperature. These variables are tightly coupled and difficult to analyse precisely, requiring specific sensors and elaborate data processing. Furthermore, the production of plastic products via injection moulding still predominantly relies on manual configuration and trial-and-error methods to obtain the best parts [3]. This traditional approach has disadvantages of low production efficiency, low reliability and repeatability, and is heavily dependent on operator experience. Therefore, it is imperative and crucial to develop methods that are science-based and utilize Industry 4.0 technologies like IoT to modernize and optimise plastic production.

One of the most notable trends is the increasing adoption of Industry 4.0 technologies. In this context, technologies such as OPC UA stand out, playing a crucial role in the integration of efficient communication between different systems and devices in industry. OPC UA has become an important industrial communication standard due to its platform independence, security model, information modelling capabilities, and suitability for heterogeneous manufacturing systems [4–6]. In the literature, there are several studies on the monitoring and optimization of the injection moulding process using OPC UA to ensure compliance with Industry 4.0 standards [7–11].

In this work, a case study on monitoring an Injection Moulding Machine (IMM) using OPC UA is presented. The primary objective is to instrument and digitize the machine (shown in Fig. 1) to make its process data available in a standardized, interoperable, scalable, and secure manner through OPC UA. This process data can be accessed by various software platforms, such as Enterprise Resource Planning (ERP) and (Manufacturing Execution System) MES systems. Specific process data monitored include mould cooling circuit water temperature, hydraulic oil temperature, injection pressure, energy consumption per cycle, cycle count, and cycle time.

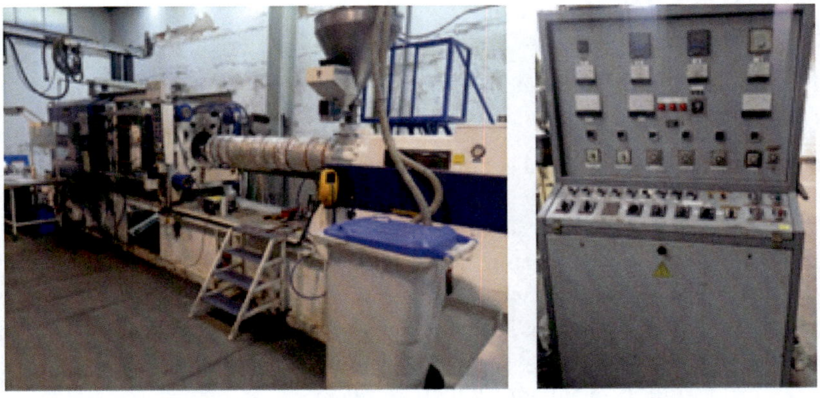

Fig. 1. Injection moulding machine (IMI1300S) and control board.

2 Developed System

This section describes details about the developed system. As the IMM (IMI1300S) did not have a digital interface for OPC UA communication, several sensors were installed to collect injection process data. This data is then structured into OPC UA variables according to EUROMAP specifications [12], making it accessible through an OPC UA server. The connection to this server by an OPC UA client allows accessing the OPC UA variables that contain the mapped IMM data.

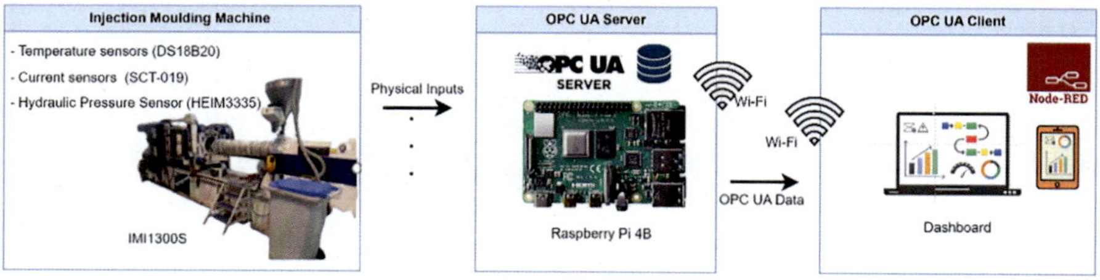

Fig. 2. System architecture.

The proposed system architecture is illustrated in Fig. 2. This system utilizes external sensors attached to the injection machine to collect data about the injection process, including temperature, hydraulic pressure, cycle count, and energy consumption per cycle. The sensors are also connected to a data acquisition module comprising two Arduino Pro Mini boards and one Raspberry Pi 4 Model B. The open62541 project library, an open-source C library freely available, was chosen to implement the OPC UA server. This server is installed on a Raspberry Pi 4 Model B, with a Wi-Fi connection. The sensor measurements are mapped to specific variables within the OPC UA server's address space, enabling data sharing with connected OPC UA clients. In this work, a customized OPC UA client was developed in Node-RED to visualize the temporal evolution of data processing.

Further details concerning the hardware and software implementations will be provided in the following subsections.

2.1 Electronic Module Implementation

The Fig. 3 illustrates the hardware block diagram of our system. It includes a Raspberry Pi 4 Model B, two Arduino Pro Mini boards, temperature sensors, pressure sensors, current sensors, and voltage sensors. One of the Arduino Pro Mini boards was responsible for acquiring temperatures from the cooling system and hydraulic oil, as well as hydraulic pressure. The DS18B20 temperature sensors used are connected to the Arduino Pro Mini via the OneWire interface, which enables multiple sensors to share a single communication cable. For measuring hydraulic pressure, the HEIM3335 sensor was used. Voltage adaptation was necessary as the HEIM3335 sensor outputs 0 to 10 V, whereas the Arduino Pro Mini input supports only 0 to 5 V.

The second Arduino Pro Mini is used to implement a module that monitors the energy consumed by the machine. Three SCT-019 current sensors and three ZMPT101B voltage sensors, one for each phase (R, S and T), were employed to calculate the energy consumed by the injection moulding machine. Cycle production information was obtained by monitoring the two relays of the machine's main board.

A logic level converter was used to connect the Arduino Pro Mini to the Raspberry Pi, as the Arduino's 0 to 5 V outputs and the Raspberry Pi's 0 to 3.3 V inputs are incompatible. The detailed sensor installation in the injection moulding machine is presented below.

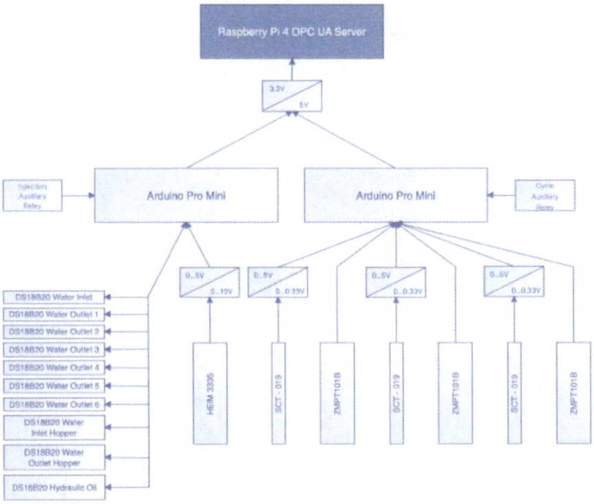

Fig. 3. Block diagram of the system hardware.

The ten DS18B20 temperature sensors were installed in the IMM as the follow:

- Seven to monitor the mould cooling circuit (input and output), see Fig. 4.
- Two to monitor the hopper cooling circuit, see Fig. 5.
- One to monitor the hydraulic oil temperature, see Fig. 6.

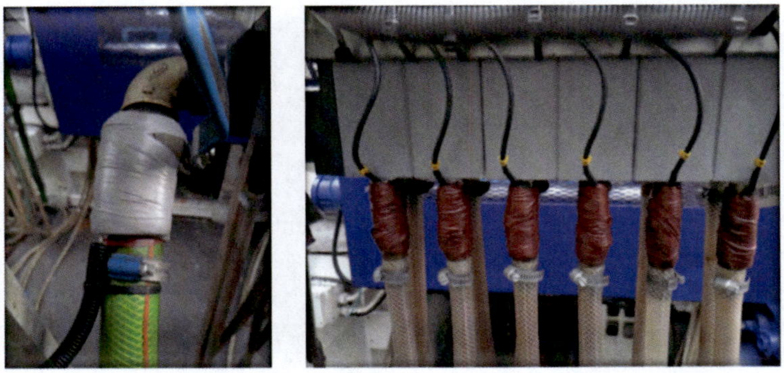

Fig. 4. Installation of the DS18B20 temperature sensors in the mould cooling circuit. **a** Input and **b** output.

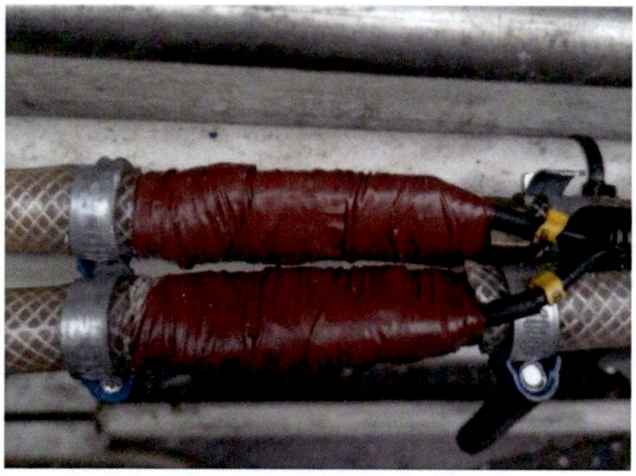

Fig. 5. Installation of the two DS18B20 sensors in the hopper cooling circuit (input and output).

Fig. 6. Installation of DS18B20 sensor to monitor the hydraulic oil temperature.

In Fig. 7, the installation of the HEIM3335 pressure sensor is shown, which measures hydraulic pressure. This sensor has a range from 0 to 160Bar, operates on 12 to 30 V power supply, and outputs from 0 to 10 V, where 0 V corresponds to 0Bar and 10 V corresponds to 160Bar. To convert the pressure sensor's output voltage to a range compatible with the Arduino Pro Mini, a voltage divider was employed. This voltage divider consists of two 10 kΩ resistors connected in series.

One of the objectives of the proposed monitoring system is to calculate the electrical energy consumed by the injection machine in real time. For this purpose, it is necessary to continuously measure the current, voltage, and power factor. Considering that the power supply circuit of the injection machine is three-phase alternating current, three current sensors and three voltage sensors are required.

After analysing the characteristics of the components of the injection machine (Motor 1, Motor 2, chamber heating resistors, nozzle heating, and mould heating), it was concluded that the maximum total machine consumption, considering a simultaneity factor of 1, is approximately 81.4 kW, which corresponds to approximately 151.35A per phase. Therefore, the non-invasive current sensor was chosen (SCT-019), with a nominal measuring current of 200A. According to the SCT-019 sensor datasheet, a current of 200A registered by the sensor corresponds to a voltage of 0.33 V at the sensor output. Since the

Fig. 7. The HEIM3335 pressure sensor mounted on the machine's hydraulic injection circuit.

Arduino Pro Mini module used to develop the electricity monitoring module operates at a voltage of 5 V, it was necessary to adjust the voltage at the sensor output. For this purpose, an amplifier module based on the LM358 integrated circuit was used, which adapts the sensor's measurement range to the input range of the microcontroller.

Figure 8 shows the electrical diagram of the energy consumption monitoring module. The module consists of three SCT-019 current sensors, three LM358 operational amplifiers, and three ZMPT101B voltage sensors (one for each phase: R, S, and T). Figure 9 illustrates the installation of the three SCT-019 current sensors in the machine's main power supply line.

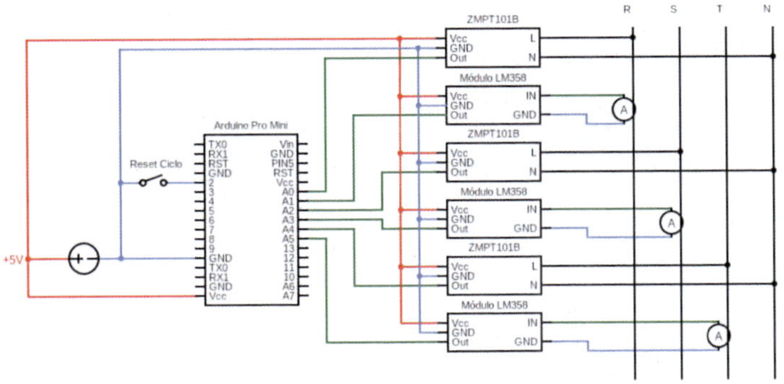

Fig. 8. Electrical diagram of the energy consumption monitoring module.

The electrical panel, installed near the plastic injection machine, can be seen in Fig. 10, interior and exterior views. To ensure the proper functioning of the hardware, it was necessary to install two fans in the electrical panel: one on the door and another at the bottom, ensuring air circulation inside the panel. This electrical panel includes a DC-DC converter to supply all hardware, i.e., sensors, amplifiers, microcontrollers. The input voltage of the converter comes from injection machine supply.

To ensure uninterrupted access to the factory's Wi-Fi network, a Mercusys MWE300RE Wi-Fi repeater was strategically placed near the machine's electrical panel

Fig. 9. Installation of the three SCT-019 current sensors

to mitigate the effects of electromagnetic interference and enhance the Raspberry Pi's wireless connection.

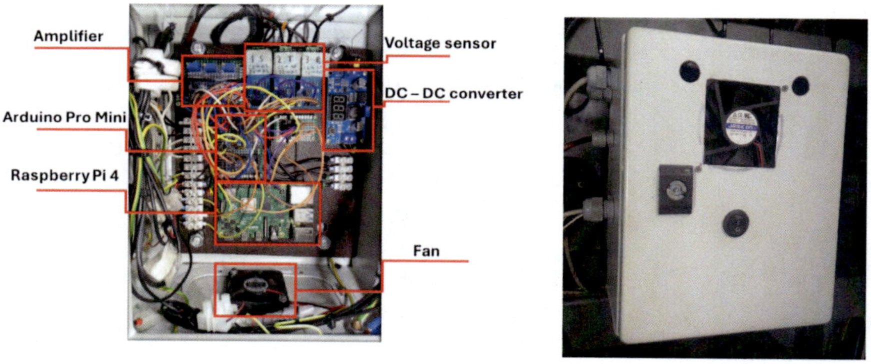

Fig. 10. Electrical panel of the developed system: **a** interior and **b** exterior views.

2.2 Software Module

The software module includes the development of firmware and software for the electronics module. First, a Raspberry Pi is configured to host an OPC UA server, enabling standardized communication with other OPC UA client devices. Then, firmware is created for the two Arduino Pro Mini to collect the overmentioned sensor data values and send them to the OPC UA server. An SQLite database is implemented to store the collected data as well as provide data history. Finally, an intuitive graphical user interface is developed using a Node-RED bases implementation of an OPC UA client.

To install and config the OPC UA server on the Raspberry Pi, the following steps were necessary:

- Raspberry Pi OS installation.

- Operating system update.
- OPC UA server installation, using the open62541 project library [13].
- Server configuration, including setting parameters such as, Wi-Fi network, IP addresses, communication ports, address space, etc.
- Testing and verification involved using OPC UA client tools to verify connectivity and communication with the server. Specifically, the Unified Automation UaExpert OPC UA client was initially used for this purpose.

The firmware developed for the Arduino Pro Mini is responsible for acquiring data from various sensors, including temperature, pressure, voltage and current during the injection molding machine's production cycle. The Arduino Pro Mini is connected directly to the Raspberry Pi 4 via the serial communication channel, enabling efficient transmission of the collected injection data to the OPC-UA server in a structured format.

A communication protocol has been implemented to ensure that all messages between the acquisition modules and the server are synchronized with the machine's production cycle. As an example, consider the message "#95#23.15," where #95 represents the cycle number and #23.15 the cycle time in seconds.

Based on the number of cycles and the temperature, pressure, voltage, and current information gathered during production, the server program calculates the average electricity consumption during each cycle, as well as the total energy consumed. The calculation results, along with the original data, are stored as variables on the OPC UA server, allowing real-time monitoring and analysis of the production process.

To further analyse the production data, a database was implemented to store the data obtained from the acquisition system. SQLite is chosen because its code is in the public domain, making it free for use for any purpose, whether commercial or private [14]. This database efficiently handles large data volumes, enabling subsequent analysis of machine operation using external big data analytics services (e.g., generating graphs and visualizing trends). The sqlite3.h library is used to implement the SQLite database on the Raspberry Pi 4. Fig. 11 presents a table detailing the variables and their corresponding data types within the SQLite database employed in the proposed system. The data storage in the SQLite database is performed at every production cycle.

```
// Create tables if they don't exist
const char *createTableSQL = "CREATE TABLE IF NOT EXISTS operating_data ("
                             "id INTEGER PRIMARY KEY AUTOINCREMENT,"
                             "temperature1 REAL,"
                             "temperature2 REAL,"
                             "temperature3 REAL,"
                             "temperature4 REAL,"
                             "temperature5 REAL,"
                             "temperature6 REAL,"
                             "temperature7 REAL,"
                             "temperature8 REAL,"
                             "temperature9 REAL,"
                             "temperature10 REAL,"
                             "cycle_counter INTEGER,"
                             "cycle_time REAL,"
                             "cycle_energy REAL,"
                             "total_energy REAL,"
                             "pression TEXT,"
                             "timestamp DATETIME DEFAULT CURRENT_TIMESTAMP"
                             ");";
```

Fig. 11. Creation of SQLITE database table

To access the values of the variables stored in the OPC UA server's address space, a generic OPC UA client, such as UaExpert, can be used. However, the interface of UaExpert is not very intuitive and does not offer graphs and diagrams for more effective visual analysis of the data. Node-RED [15] is chosen as the development tool for the customized OPC UA client due to its visual nature, which facilitates the creation of complex workflows. Additionally, its widespread use in IoT projects, systems integration, and backend application development demonstrates its versatility and reliability. In addition, it has pre-existing functionalities, such as an OPC-UA client.

Node-RED base comes with the Input, Function, and Dashboard palettes pre-installed. However, it is necessary to install the OPC UA and ChartJs palettes to develop the customised OPC UA Client, for the proposed system. The development of the OPC UA client involves creating an interface with the Dashboard Layout Editor and connecting the nodes using a specific tool. These two tools run on a web page. The dashboard shown in Fig. 13 corresponds to the output of Node-RED Dashboard Layout Editor presented in the Fig. 12.

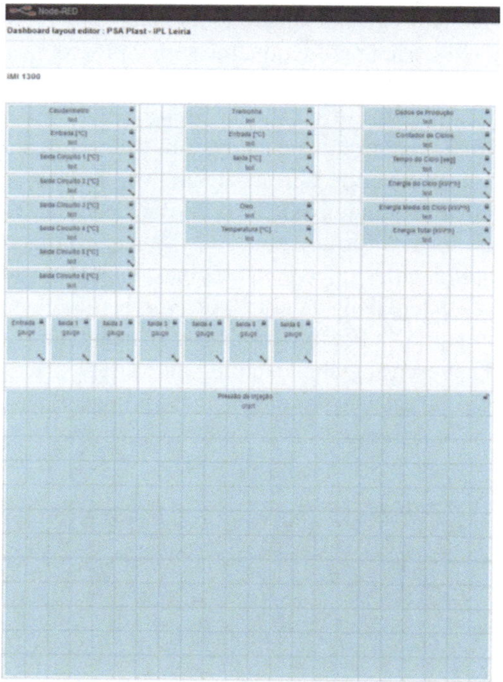

Fig. 12. OPC UA client interface implemented using the dashboard layout editor

3 Tests and Results

The proposed system was tested in an industrial environment, where plastic buckets for the paint industry were manufactured using the mould installed in the IMI300S machine. To determine energy efficiency, production costs, and overall performance of the injection molding machine, the following process parameters were collected and processed:

- Cooling system input temperature
- Output temperatures of the 6 circuits of the mold cooling system
- Input and output temperature of the hopper cooling circuit
- Hydraulic oil temperature
- Hydraulic pressure evolution during the mold injection process
- Number of production cycles
- Cycle times
- Energy consumed in each cycle
- Average energy consumed per cycle over a production period
- Total energy.

The dashboard presented in Fig. 13 provides real-time visualization of the manufacturing process variables previously enumerated. Simultaneously, these variables are stored in an SQLite database. This database can be downloaded from the server, enabling offline analysis and processing using various software tools like Excel, MATLAB, Microsoft Power BI, and others.

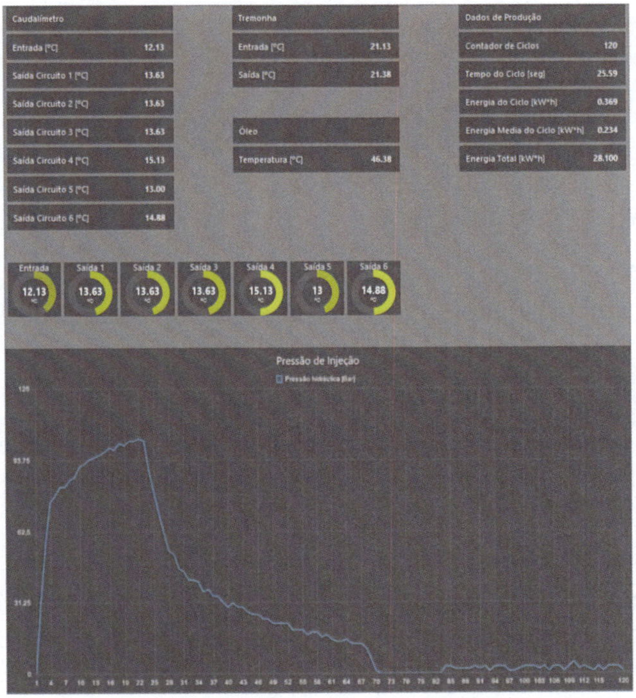

Fig. 13. The OPC UA client dashboard, created using Node-RED.

To evaluate the system's performance, it is necessary to install an OPC UA client on a device and connect it to the same network as the OPC UA server. In this case, two devices were used: an Android smartphone and a laptop. Communication between the OPC UA server and the clients was successfully tested using the following client platforms: UAExpert OPC UA client, OPC UA PROSYS OPC installed on an Android smartphone, and the solution developed in Node-RED during this development.

The proposed system demonstrated its robustness by operating continuously during the production cycles of plastic paint buckets. Using the developed dashboard, the production manager has real-time visibility into the evolution of the production process throughout the production periods. However, for more comprehensive historical analyses of process variables, conducting offline studies is essential. This necessity arises from the limitations of the implemented OPC UA server, which lacks historical data storage functionality. However, this need was superseded by the twin database implemented.

4 Conclusions and Future Work

The proposed system enables real-time data acquisition and storage for subsequent analysis. Real-time data observation, via OPC UA protocol, facilitates the immediate detection and response to machine anomalies or malfunctions during production. Furthermore, post-production data analysis enables the exploration of solutions to enhance the quality and efficiency of plastic part manufacturing in subsequent production runs. This focus on continuous improvement not only minimizes material waste but also paves the way for more sustainable manufacturing practices. Additionally, analysing machine operation data can reveal opportunities to optimize electricity consumption, leading to significant long-term cost savings and reduced environmental impact. Process optimization not only reduces energy consumption but can also contribute to the company's carbon footprint reduction.

As future work, it would be important to implement the following functionalities in the proposed system:

- Historical data: The addition of the OPC UA HDA functionality would make it possible to store historical data on the server, facilitating access to this information by other analysis and decision-making tools, eliminating the need for an intermediate database (SQLite).
- Alarm management: In addition, the implementation of OPC UA Alarms and Conditions would make it possible to create mechanisms to manage alarms and conditions in machine operation, as well as allowing alerts and reports to be sent to OPC UA clients.
- Node-RED dashboard: It would be important to incorporate each of the new functionalities (historical data and alarm management) into the interface for better equipment management and operation.

In what regards the future use of the developed system, one must address the issues of sustainability that were the main motivation for this work. The way the machine can now be operated is much more rigorous due to the ability to gather all the operation data continuously enabling both immediate action as well as further analysis.

In fact, future analysis of the collected data, which the company involved in this project will certainly undertake as soon as a substantial dataset is accumulated, will enhance process efficiency. This will be achieved by reducing the peak power demand and overall energy consumption, thanks to detailed data now available for the main energy-consuming components of such machines: hydraulics, heating, and refrigeration.

Through the continued use of this monitoring system and the detailed analysis of the operation that it allows, a finer configuration of machine parameters can be used by the company to achieve a more efficient operation of the injection process, a knowledge that may even be used across the plant in other similar machine, improving overall efficiency in the company. The success of this operation can also entice future developments and increase sustainability issues awareness within the company.

References

1. Plastics Europe.: Plastics—the fast Facts (2023). Last accessed 23 Mar 2023. https://plasticseurope.org/knowledge-hub/plastics-the-fast-facts-2023/
2. Javaid, M., Haleem, A., Singh, R.P., Khan, S., Suman, R.: Sustainability 4.0 and its applications in the field of manufacturing. Internet Things Cyber-Phys. Syst. **2**, 82–90 (2022)
3. Li, D., Zhou, H., Zhao, P., Li, Y.: A real-time process optimization system for injection molding. Polym. Eng. Sci. **49**(10), 2031–2040 (2009)
4. Mahnke, W., Leitner, S.H., Damm, M.: OPC Unified Architecture, 1st edn. Springer, Berlin Heidelberg (2009)
5. Da Silva, J.T., Dias, A.L., Da Silva, I.N.: A survey on OPC UA protocol: overview, challenges and opportunities. In: 2023 15th IEEE International Conference on Industry Applications (INDUSCON), pp. 1523–1530 (2023)
6. Busboom, A.: Automated generation of OPC UA information models — a review and outlook. J. Ind. Inf. Integr. **39** (2024). https://doi.org/10.1016/j.jii.2024.100602
7. Martins, A., Silva, B., Costelha, H., Neves, C., Lyons, J., Cosgrove, J.: An approach to integrating manufacturing data from legacy injection Moulding machines using OPC UA. In: 37th International Manufacturing Conference (IMC37) (2021)
8. Fuhrmann, F., Weber, A., Ladstätter, S., Dietrich, S., Rella, J.: Multimodal interaction in the production line—an OPC UA-based framework for injection molding machinery. In: Proceedings of the 2021 International Conference on Multimodal Interaction (ICMI '21), pp. 837–838. Association for Computing Machinery, New York, NY, USA. https://doi.org/10.1145/3462244.3481300
9. Wang, Z., Feng, W., Ye, J., Yang, J., Liu, C.: A study on intelligent manufacturing industrial internet for injection molding industry based on digital twin. Complexity. **8838914**, 16 (2021). https://doi.org/10.1155/2021/8838914
10. Aminabadi, S., Tabatabai, P., Steiner, A., Gruber, DP., Friesenbichler, W., Habersohn, C., Berger-Weber G., Industry 4.0 IN-LINE AI quality control of plastic injection molded parts, Polymers vol.14, no. 17:3551, 2022.
11. Santos, G., Lopes, M., Rodrigues, J., Ribeiro, E., Vasco, J.C.: ICM 4.0—injection moulding machine control and monitoring. In: Correia Vasco, J.O., et al. (eds.) Progress in Digital and Physical Manufacturing. ProDPM 2021. Springer Tracts in Additive Manufacturing. Springer, Cham (2023). https://doi.org/10.1007/978-3-031-33890-8_29
12. EUROMAP Homepage. https://www.euromap.org/. Last accessed 23 Mar 2024
13. Open62541 Homepage. https://github.com/open62541/open62541. Last accessed 10 Oct 2024
14. SQLite Homepage. https://www.sqlite.org/. Last accessed 10 Mar 2024
15. Node-RED Homepage. https://nodered.org/. Last accessed 10 Mar 2024

Welding Processes on AISI H11, P20 and 420 Tool Steels for Polymer Injection Molds

J. P. Rasteiro[1], A. A. Nogueira[1], M. M. Vieira[2,3,4(✉)], and M. A. Correia[2,3]

[1] Thyssenkrupp - Tratamentos Térmicos e Superficiais, Pero Neto, Marinha Grande, Portugal
{joao.rasteiro,albano.nogueira}@thyssenkrupp-materials.com
[2] School of Technology and Management of the University of Leiria and Oeste, Leiria, Portugal
{milena.vieira,mario.correia}@ipleiria.pt
[3] CDRsp-ESTG, Politécnico de Leiria, Leiria, Portugal
[4] Department of Mechanical Engineering, CEMMPRE, ARISE, University of Coimbra, Coimbra, Portugal

Abstract. This work investigates the effects of carrying out heat treatments of quenching and tempering before and after repairing polymer injection molds using TIG and Nd:YAG Laser welding technologies. For the mold base material, three types of steels commonly used in the mold industry were considered: a pre-hardened steel (AISI P20), a hot-working steel (AISI H11) and a martensitic stainless steel (AISI 420). The aim is to identify the best sequence of operations that result in a welded area that is as imperceptible as possible and with mechanical properties suitable for the function of the injection mold. The prepared samples were submitted to hardness profile and roughness measurements, microstructural analysis and visual inspection. The results obtained show that the most critical cases correspond to those in which welding is performed after the quenching heat treatment. In these cases, performing a post-weld tempering originates a more uniform microstructure and hardness between base material, heat affected zone and weld material. For each base material type, the best solution with respect to the sequence of operations was found.

Keywords: Injection molds · Welding technologies · Heat treatment · Tool steels

1 Introduction

During the manufacture of molds for polymers injection or during their useful life in operation, changes in the geometry of the tool may occur, compromising its accurate functioning, which requires repairing operations using different technological processes and, in certain cases, replacement of the tool [1–6]. Replacement is a more time-consuming and expensive option. Therefore, the most common option is to repair the existing tool. Among the various available techniques, fusion welding is considered a reliable technology that can ensure the required mechanical properties to extend service life of the tools, besides they are fast and economical technologies [1–8]. The welding process uses a heat source to melt the material locally, which followed by the melting bath solidification,

© The Author(s), under exclusive license to Springer Nature Switzerland AG 2027
M. Calvete Gaspar et al. (Eds.): Regional Helix 2024, LNME, pp. 60–72, 2027.
https://doi.org/10.1007/978-3-032-36365-7_5

originates the joint between the base and weld metal. However, despite being a quick solution with relatively low associated costs, heat introduction to the tool often originates a Heat-Affected Zone (HAZ) and a deposited material with different microstructure from the base material. These differences affect mechanical properties, leading to the development of residual stresses, which make subsequent polishing operations difficult. To achieve a satisfactory surface, filler metals being used in welding repairs should be recommended by the base metal manufacturers and must only be used with suitable welding procedures [1, 5]. For repaired surfaces to be texturized, the welding must be done using a filler metal of similar composition and carbon content slightly smaller than the one of the base metal. The pre-heating around the repairing area helps to decrease the differences of temperature and hardness between base and weld metals. Also, pre-heating allows hydrogen to escape [9]. Moreover, the high cooling rates characteristic of welding repair processes give rise to the development of hard and brittle microstructures and residual stresses. Post-repair heat treatment improves the repair itself and helps to relieve residual stresses. It is a way to reduce defects and control the mechanical properties and metallurgical aspects of the weldments. Therefore, the lifespan of the mold depends whether the quality of the welding repair but also on the performed pre and post-repair heat treatments [5, 9, 10]. Thus, the repair procedure requires knowledge of the appropriate sequence of operations for each type of base material that ensures an improved quality of repaired tools.

TIG and Laser welding processes are commonly adopted to carry out repairs in the industry of molds and dies [1, 2, 4–6, 9]. TIG welding is easy to operate, highly portable, less time consuming, highly flexible to approach in complex geometries. Repairs with TIG welding require pre-heating and post-weld heat treatment to avoid solidification cracking and residual stresses. The HAZ is a critical area, since it changes the metal properties, mechanical and metallurgical aspects, as result of being exposed to high temperatures. Nevertheless, many repairs are successfully performed in the hardened and tempered condition. When this can be done safely, it avoids the need for annealing before repairing. Laser technology leads to small HAZ, negligible undercut, very small volume weld and precise positioning and control of the spot of incidence and its movement. The high-density heat input leads to localized heating effect reducing distortions during welding, being repairs carried out without pre-heating. Other benefits include reduced post-processing requirements, reduced cycle time, increased part quality.

In this study, the effect of performing heat treatments before and after welding tool repair using TIG (Tungsten Inert Gas) and Nd:YAG Laser technologies is investigated. The aim is to identify the best sequence of operations that results in a welded area that is as imperceptible as possible and with mechanical properties suitable for the function of the polymer injection mold. For mold base material, 3 different tool steels were considered: pre-hardened steel, hot-working steel and martensitic stainless steel. The analysis of the prepared samples included visual inspection, surface roughness measurements, preparation of hardness profiles and metallographic examination. The work reports on the effect of pre and post-weld quenching and/or tempering heat treatments performed on three tool steel types repaired by TIG and Nd:YAG Laser Technologies.

2 Methodology

The study focused on three types of steels of particular relevance in the polymer injection molding industry which were used as base material: AISI 420, AISI H11, and AISI P20. A standard machining procedure, commonly practiced in the mold industries for these types of steels was followed to prepare the specimens. After cutting and machining, the used specimens have dimensions of 60 × 60 × 20 mm. All specimens were subjected to a linear and circular groove to be filled by welding, considering that the depth of this machining operation was different depending on the welding process to be applied (Fig. 1). Cleaning of samples between the machining and welding operations were performed by a standard washing procedure with Solnet-velnet R2.60 degreaser from CGM/END Portugal. The chemical composition of the base metals and the welding filler wires are indicated in Table 1.

a) b)

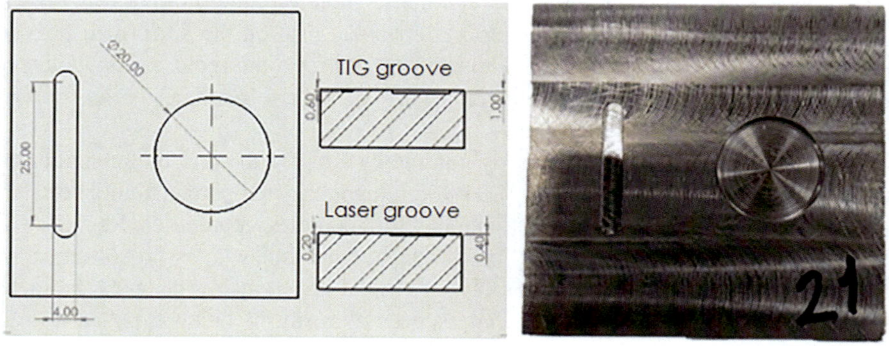

Fig. 1. Geometry of specimens for TIG and Laser Nd:YAG welding: a) Computer-aided design (CAD) Drawing; b) After machining.

TIG (Tungsten Inert Gas) welding process was carried out by a REHM welding machine, model INVERTIG.PRO digital 280 AC/DC using three current intensity values (90, 100 and 115 A) and welding speed 120 mm/min. High purity argon (99.999%) at a flow rate of 8-9 l/min was used as a shielding gas. TIG welded samples were pre-heated at temperatures of 300 °C.

Nd:YAG Laser welding was performed with an AlphaLaser welding machine, model ALM300 v4 with laser power of 300 W, pulse energy of up to 90 J, pulse power of 9 kW, variable pulse duration between 0.5 and 20 ms, pulsating electric current with 11 Hz. Filler wires had 0.4 and 0.6 mm diameter. An auxiliary laser machine from Max Photonics, model Backpack Laser Cleaning Machine was used to clean the surface to be welded.

In Fig. 2 is presented a flowchart with the sequence of operations carried out on AISI H11, P20 and 420 steel samples. Some of them were welded in the annealed state, followed by quenching and tempering heat treatment. Others were welded after quenching and tempering heat treatment and in this case some of the samples were submitted to a post-welding tempering. All quenching heat treatments were carried out

Table 1. Chemical composition of base metals and filler wires.

	C	Si	Mn	P	S	Cr	V	Mo	Ni	Ti
Base metal										
AISI 420	0.39	0.32	0.84	0.22	<0.001	13.01	–	–	–	–
AISI H11	0.39	0.94	0.44	0.013	<0.001	5.07	0.47	1.25	–	–
AISI P20[a]	0.266	0.11	1.38	–	0.002	1.37	–	0.45	0.31	–
Filler wire										
QuFe10	0.10	0.65	1.14	–	–	–	–	0.46	–	–
QuFe13	0.25	0.40	0.63	–	–	4.83	–	3.60	–	0.71
QuFe23	0.37	1.00	0.39	–	–	4.87	0.42	1.21	–	–
Stavax	0.38	0.45	0.45	–	–	13.50	–	1.00	–	–
QuFe21	0.35	0.43	0.99	–	–	6.65	–	2.10	–	0.24
QuFe15	0.09	0.55	0.98	–	–	2.30	–	0.98	–	–

[a] A modified alloy of AISI P20 was used

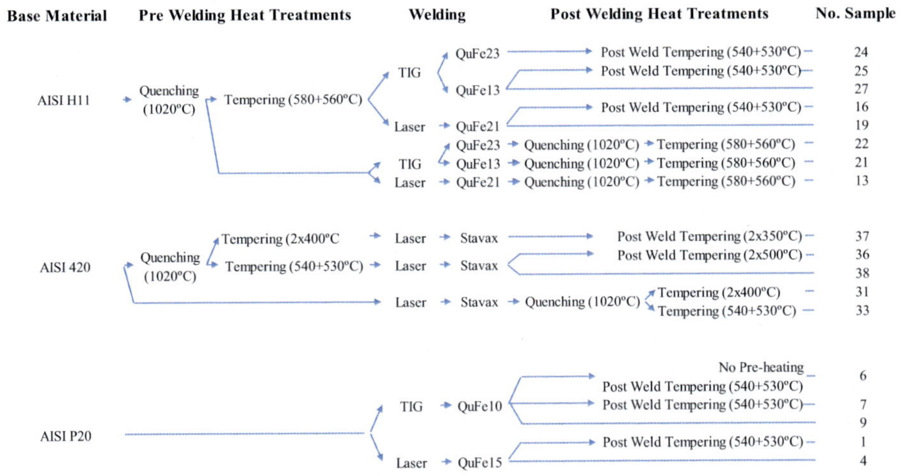

Fig. 2. Flowchart with the sequence of operations carried out on samples relating to steels AISI H11, 420 and P20.

at 1020 °C. Welded samples for quenching and tempering were previously stress-relief by annealing.

All specimens were prepared for metallographic analysis by grinding, polishing and etching procedures (Nital 5% and Kalling's No.2 were used as etching agent).

Hardness measurements were performed on the surface of samples, in the region of the welds, starting from the weld metal towards the base material over an extension of approximately 4 mm (Fig. 3) using a Hardness tester EMCO Duravision 20 Test, with 0.2 mm spacing between indentations on the Vickers hardness scale, with a load of 0.5 kgf. Microstructures of samples were analyzed with the same equipment.

The roughness measurements of the welded sample area were performed on a MarSurf PS10 Set of Mahr company. Measurements were made in compliance with the ISO 16610-21 – "Geometrical Product Specification – Filtration".

Fig. 3. Sample illustrating the region on the surface where hardness tests were carried out. Measurements started from the weld metal towards the base metal over an extension of approximately 4 mm.

3 Results

In Fig. 4 is presented the hardness profile for P20 base material specimens welded by Nd:YAG Laser and TIG processes.

For those samples Laser welded and without post-weld tempering heat treatment (sample No. 4) was observed an increase in hardness in the Heat-Affected Zone (HAZ). Such behavior was attributed to the rapid cooling after austenitization, leading to the formation of non-tempered martensite.

On the other hand, in the specimens with post-weld tempering heat treatment, the hardness values of the base material, HAZ and weld material remained quite similar suggesting no significant heat effects due to the Laser welding process (sample No.1). These samples showed a gradual transition in hardness without significant peaks, indicating a more uniform microstructure between weld and base material. Similar results were obtained in TIG welded samples (samples No. 6 and 7), where was observed uniform hardness profile in all the welded region. However, in the samples not submitted to a post-weld tempering heat treatment, a noticeable hardness peak was detected in the HAZ. This peak was also attributed to the formation of martensite (sample No. 9) responsible for high hardness and brittleness. The hardness profiles obtained for the two welding processes in which the specimens were submitted to a post-weld tempering are very similar. In the samples in which post-weld tempering was not carried out, there was an increase in hardness. For TIG welding, the HAZ was much larger, approximately

a)

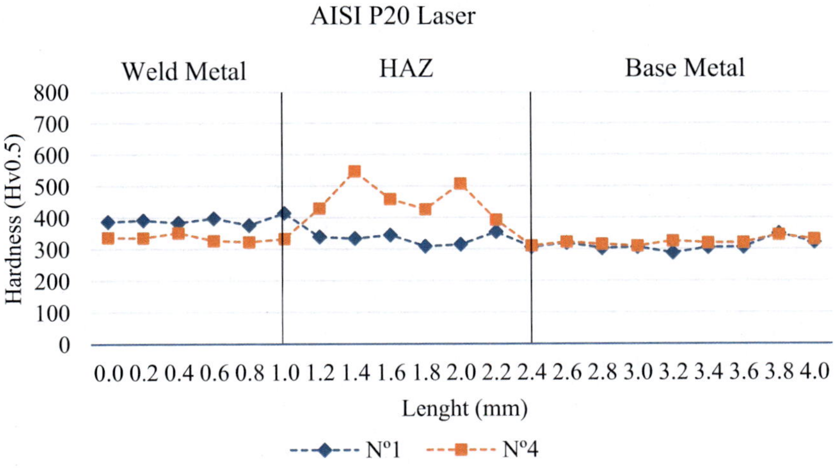

b)

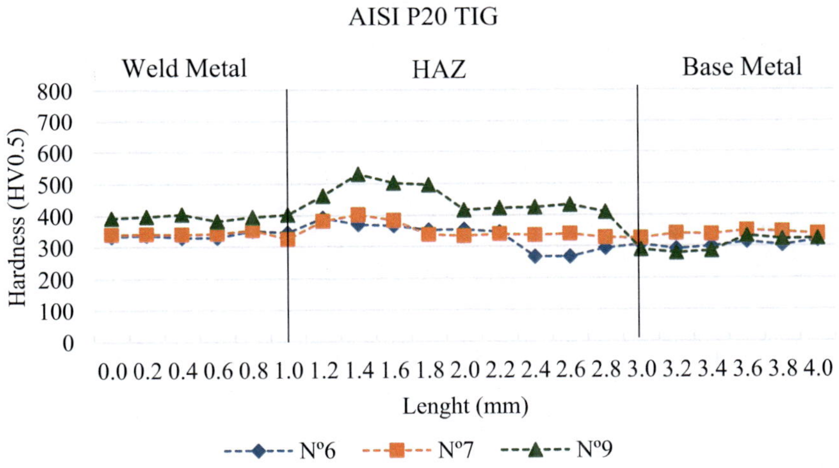

Fig. 4. Surface hardness profile of the AISI P20 steel samples: a) Laser Nd:YAG welding; b) TIG welding.

2 mm. While for Laser welding the HAZ extended for around 1.2 mm. The difference in extension in the heat-affected areas is related to the heat transferred to the parts. In the TIG process, the heat transmitted is bigger, which results in a larger area of the specimen that reaches the temperature of austenitization and as consequence an increased HAZ.

The hardness profile for H11 base material is shown in Fig. 5. The best results were obtained in the specimens quenched and tempered after welding. For these samples, a more uniform hardness profile was observed, regardless of the welding process (samples No. 13, 21, and 22). On the other hand, for samples submitted to quenching and tempering heat treatments before welding, the same results as previously described for the P20 base

a)

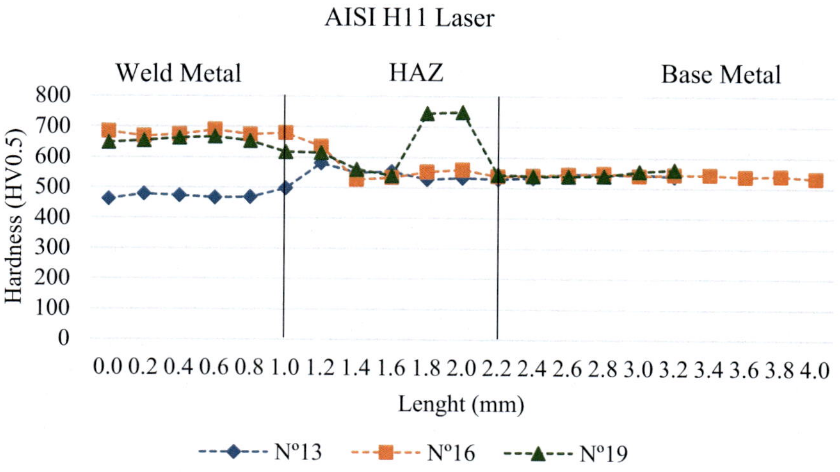

b)

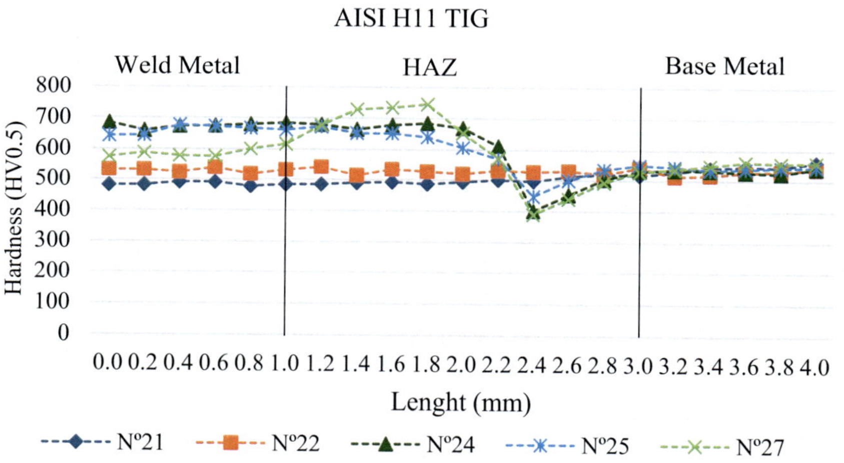

Fig. 5. Surface hardness profile of the AISI H11 steel samples: a) Laser Nd:YAG welding; b) TIG welding.

material were found. Thus, also here with the base material H11, an increase in hardness is observed in the HAZ of the specimens not subjected to post-weld tempering heat treatment after the welding process (samples No. 19 and 27). However, even when post-weld tempering heat treatment is carried out, the hardness values in the weld material are always higher than those obtained in the base metal, which may be related with the chemical composition of the filler wire. For the Laser welded samples (samples No. 16 and 19), the higher hardness of weld metal should be more related with the carbon content of the filler metal than to the occurrence of secondary hardening after post-weld tempering since the sample No. 19 was not submitted to this heat treatment after

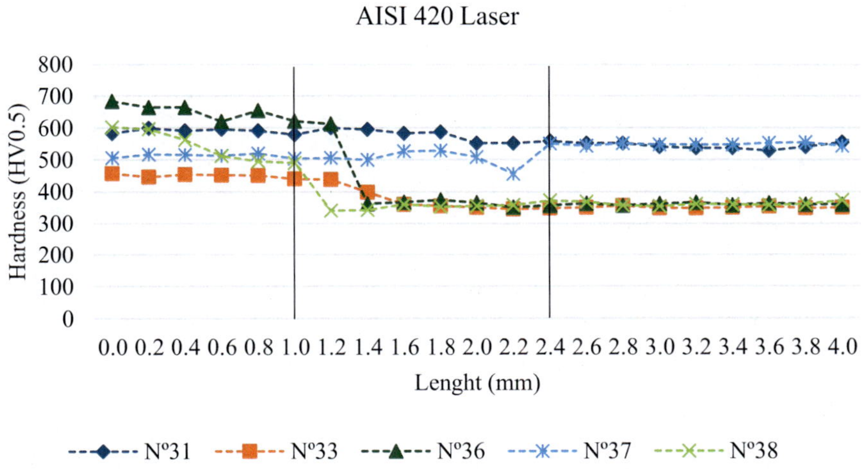

Fig. 6. Surface hardness profile of the AISI 420 steel Laser welded samples.

welding. For specimens No. 25 and 27 that were TIG welded with the same filler metal, the difference in hardness observed in the weld metal should be related to a secondary hardening phenomena after the post-weld tempering performed to the sample No. 25. While for sample No. 24, TIG welded with different filler wire, the obtained hardness can be attributed either to secondary hardening or carbon content, which combined with cooling rate led to the formation of harder martensite. Extended HAZ is still detected for TIG welded samples. The metallurgical effects in the HAZ resulting from welding processes are more pronounced when TIG welding technology is used.

Figure 6 shows the hardness profile obtained for AISI 420 specimens. A more uniform hardness profile is observed in those samples that were quenched and tempered after Laser welding (samples No. 31, 33). For the other samples, a heterogeneous hardness profile was detected between weld and base material being more irregular for the specimens not submitted to post-weld tempering heat treatment (sample No. 38). For some of the samples with post-weld tempering heat treatment, the increase in hardness detected in the weld metal (sample No. 36) may also be related with the secondary hardening effect, since tempering was carried out at around 500 °C.

The influence of the post-weld tempering treatment on the microstructural characteristics of the base metal, HAZ and weld metal is shown in Fig. 7. For P20 and H11 steel samples, the observed results were similar. Therefore, in Fig. 7 are presented optical microscopy images referring to P20 steel. Identical microstructures between base material, HAZ and weld metal were observed in those samples where a post-weld tempering was performed. On the contrary, in the specimens without post-weld tempering, the HAZ microstructure was visibly affected. The observed microstructure was characterized by the presence of narrower grains of non-tempered martensite. In TIG-welded specimens, the absence of "post-weld tempering" causes more pronounced differences between the base material and the HAZ. The AISI 420 steel samples, martensite and significant presence of carbides were detected in the base material. The white zone observed in the HAZ was attributed to the presence of residual austenite. For the specimens hardening after welding, the presence of residual austenite was not observed as result of the carried-out

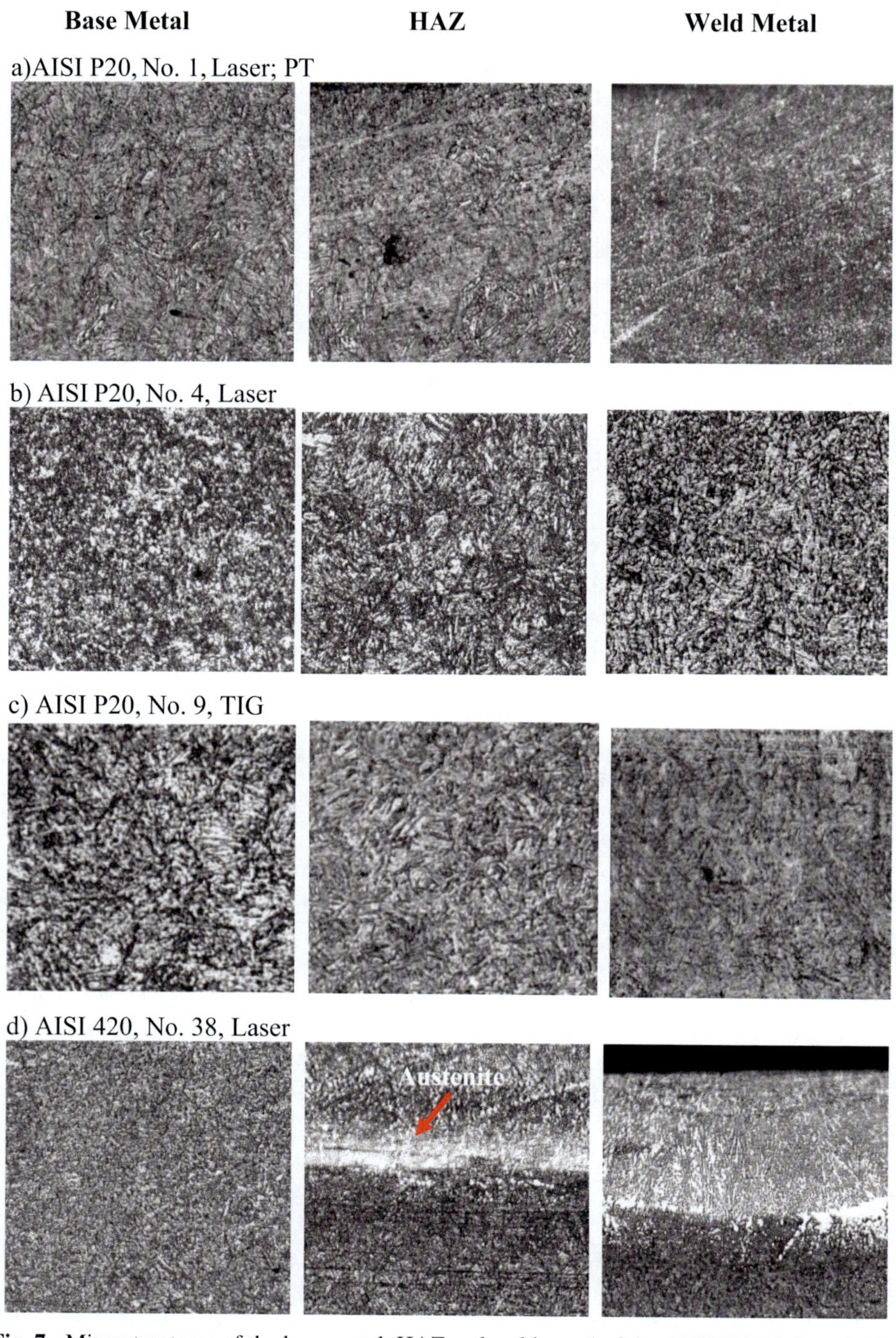

Fig. 7. Microstructures of the base metal, HAZ and weld metal of the AISI P20 (a) Sample No. 1; b) Sample No. 4; c) Sample No. 9) and AISI 420 (d) Sample No. 38) steels TIG and Laser welded with and without post-tempering (PT) (60x).

quenching heat treatment. The microstructures observed in the region of the welding of the three steels match with hardness profiles previously presented in Figs. 4, 5 and 6. Microstructural analysis has revealed that for samples quenched and tempered previously to the welding process, it is recommended to perform a post-weld tempering treatment. The aim is to achieve a more homogeneous and uniform microstructure in the whole welded zone.

Considering that injection molds must be polished after welding, it is essential that the weld metal does not have significant differences in the composition and hardness with respect to the base metal. Otherwise, in the contour of the weld could remain, after polishing, marks that would be reproduced in the injected part [1]. For this reason, the finished surface generally is evaluated by visual inspection.

Regarding the visual exam of the specimens welded area, it was found for both base materials H11 and AISI 420 that the best results were achieved in the samples that were welded in the annealed state and subsequently subjected to quenching and tempering heat treatments (Fig. 8). In the base material AISI 420, the best result in terms of visual appearance/aspect was attained in samples post-weld tempered at the lowest temperature.

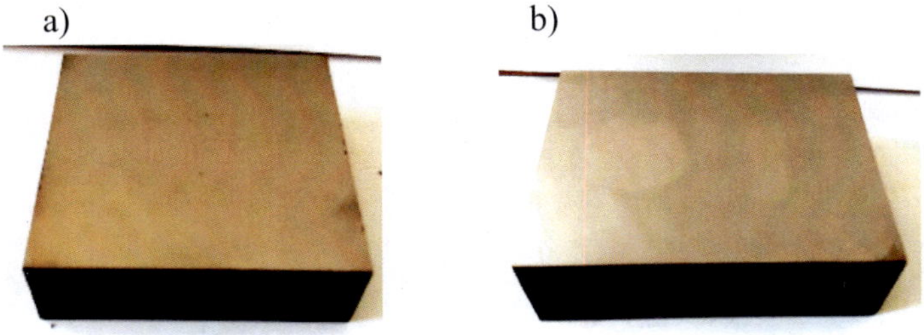

Fig. 8. Visual inspection of the weld area: a) Imperceptible; b) visible.

For the samples that were welded after quenching and tempering or for the P20 base material, the ones with the best visual appearance (without the welded area being detected) were the samples subjected to the post-weld tempering heat treatment. Those who do not undergo this treatment, the repaired area was visible. In the specimens in which TIG welding was applied, the difference in quality in terms of visual exam between the specimens with and without tempered welding was higher.

The evaluation of a molding surface is characterized by geometric precision and the quality of polishing. In turn, the quality of the polishing, in addition to defects such as scratches, pores, orange peel effect, dents and punctures, can be quantified by surface roughness. The most used roughness parameters for quantification are: Ra, Rz and Rt [11].

Figure 9 shows the results of the surface roughness tests conducted on the welded specimens with respect to the arithmetical mean roughness value (Ra). For P20 specimens no significant difference was found between the base metal, weld metal and transition zone in those samples submitted to post-weld tempering regardless of the welding process (samples No. 1, 7). Also, similar results with respect to roughness were observed in H11 and AISI 420 samples welded in the annealed state and then submitted

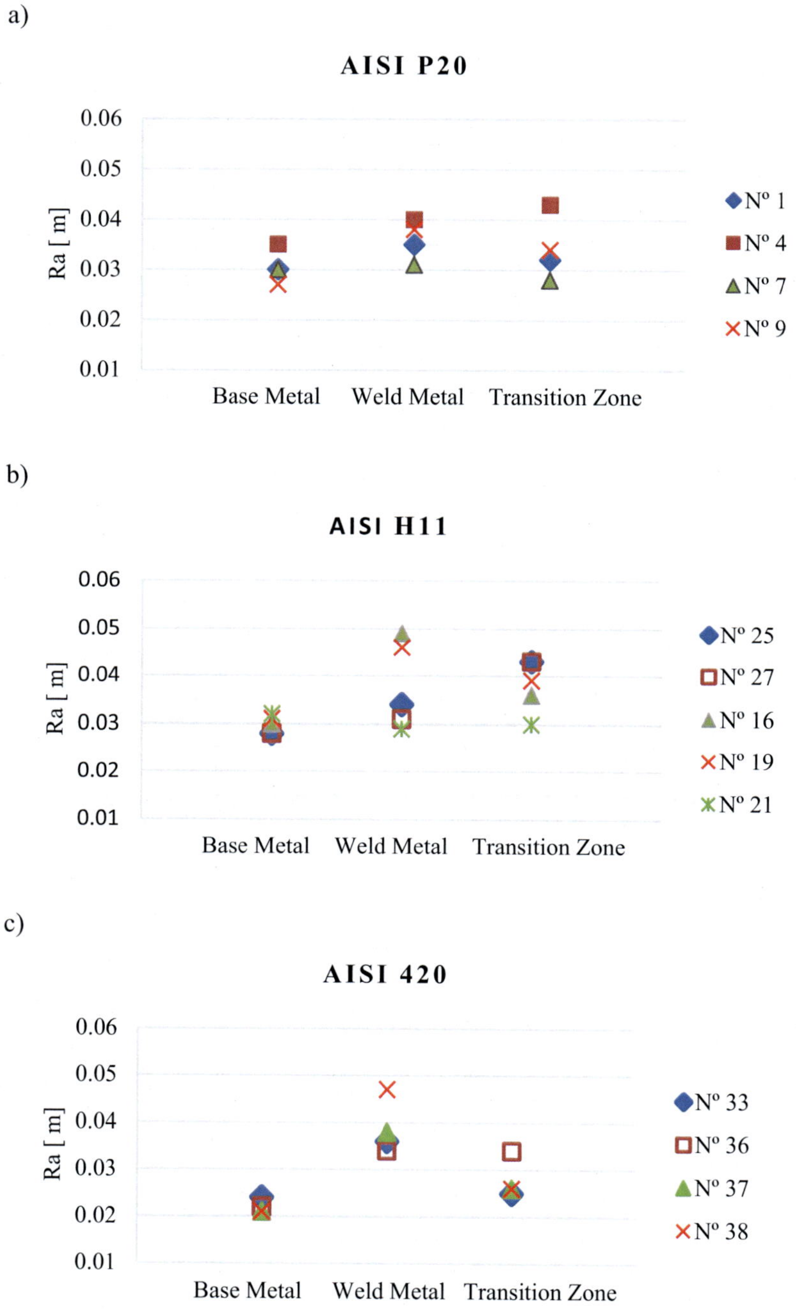

Fig. 9. Arithmetical mean roughness value (Ra) obtained from surface roughness tests: a) AISI P20; b) AISI H11; c) AISI 420.

to quenching and tempering heat treatments (sample No. 21, 33). It was noticeable a higher Ra in the weld metal for those Laser welded samples, which was attributed to porosities on the surface that probably interfered with measurements.

In addition, for H11 samples, the Ra values in the transition zone are close to those obtained in the base metal in the samples post-weld tempered (samples No. 16).

Nonetheless, for AISI 420 samples, the benefit of performing a post-weld tempering is not so evident. However, the Ra results in the transition zone were closer to the base metal when a lower tempering temperature was used (samples No. 37).

4 Conclusions

In this study, Laser and TIG welding technologies were selected to repair three types of steel (AISI 420, P20, H11) generally used for polymer injection molds. The effect on mechanical properties and microstructure of performing heat treatments before or after the welding repair were analyzed. The goal is to obtain proper sequence of operations for each type of steel achieving improved quality of repaired tools.

For H11 and AISI 420 steels, the best results were obtained for those samples repaired by welding in the annealed condition and subsequently submitted to quenching and tempering heat treatments. In cases where the welding is conducted after hardening and tempering, a post-weld tempering is required to achieve homogeneous microstructure and suitable mechanical properties in all the welding region. In the base material AISI 420, the best result in terms of visual appearance/aspect was attained in the samples with post-weld tempering at lower temperatures.

For P20 pre-hardened steel, a post-weld tempering heat treatment greatly improves the homogeneity of the welding area, promoting desirable properties and microstructure regardless of the performed welding process.

References

1. Preciado, W.T., Niño Bohorquez, C.E.: Repair welding of polymer injection molds manufactured in AISI P20 and VP50IM steels. J. Mater. Process. Technol. **179**, 244–250 (2006)
2. Vedani, B., Previtali, G.M., Vimercati, A., Sanvito, G.: Somaschini: problems in laser repair-welding a surface-treated tool steel. Surf. Coat. Technol. **201**, 4518–4525 (2007)
3. Borrego, L.P., Pires, J.T.B., Costa, J.M., Ferreira, J.M.: Mould steels repaired by laser welding. Eng. Fail. Anal. **16**(2), 596–607 (2009)
4. Pleterski, M., Tusek, J., Kosec, L., Muhic, M., Muhic, T.: Laser repair welding of molds with various pulse shapes. Metalurgija. **49**, 41–44 (2010)
5. Jhavar, S., Paul, C.P., Jain, N.K.: Causes of failure and repairing options for dies and molds: a review. Eng. Fail. Anal. **34**, 519–535 (2013)
6. Silva, B., Pires, I., Quintino, L.: Welding technologies for repairing plastic injection moulds. Mater. Sci. Forum. **587–588**, 936–940 (2008)
7. Roshanghias, A., Barzegari, M., Kokabi, A.H., Mirazizi, M.: The effects of functionally graded material structure on wear resistance. Mater. Des. **32**, 892–899 (2011)
8. Kah, P., Martikainen, J.: Current trends in welding processes and materials: improve in effectiveness. Rev. Adv. Mater. Sci. **30**, 189–200 (2012)

9. Khoshnaw, F.: Welding of Metallic Materials—Methods, Metallurgy and Performance, 1st edn. Elsevier, Amsterdam—Netherlands (2023)
10. Ul Islam, S.S., Khan, N.Z., Siddiquee, A.N.: Review of heat treatment of welded sheet metals during past 15 years. In: Comprehensive Materials Processing, vol. 2, pp. 41–56 (2024)
11. Stoeckhert, Klaus e Mennig, Günter: Polishing technology in mold making. In: Mennig, G., Stoeckhert, K. (eds.) Mold-Making Handbook, 3rd edn, chap. 4.4, pp. 251–533. Hanser Publishers, Munich (2013)

Maintenance and Asset Management

Asset Management and the Digital Transformation of Companies in Portugal: A Thematic Literature Review

André Guimarães[1,2(✉)], Samuel Messias[3], Hugo Raposo[4], and Daniel Gaspar[1]

[1] CISeD - Research Centre for Digital Services - Polytechnic Institute of Viseu, Viseu, Portugal
aguimaraes@estgv.ipv.pt

[2] CISE - Electomechatronic Systems Research Centre - University of Beira Interior, Covilhã, Portugal

[3] ISEC - Higher Institute of Engineering of Coimbra- Polytechnic Institute of Coimbra, Coimbra, Portugal

[4] RCM2+ - Research Centre for Asset Management and Systems Engineering, Polytechnic Institute of Coimbra, Coimbra, Institute of Engineering, Coimbra, Portugal

Abstract. This study is based on a bibliographic review employing bibliometric techniques, representing a valuable contribution to the organization and analysis of research trends in the scientific literature on Asset Management and Digital Transformation in companies. This work also specifically focused on visualizing the growth of these concepts in Portugal. A comprehensive literature review was conducted, encompassing articles published in journals and conferences over the past 10 years, on the Web of Science (WoS) and Scopus platforms, which were peer-reviewed and analyzed using bibliometric techniques, resulting in the formulation of conclusions. This study employed Bibliometrix R and VOSviewer software to perform a bibliometric analysis. Thus, the relevance of this study lies in the detailed analysis of existing literature, identifying the main research trends, highlighting notable articles based on citations and publications, identifying the most cited authors, and identifying research gaps. In this regard, possible directions and priorities for future research on this emerging theme are suggested. Existing literature still requires recent studies addressing Asset Management and Digital Transformation in companies.

Keywords: Asset management · Digital transformation · Bibliometric analysis · Companies in Portugal

1 Introduction

In today's world, where technology is deeply integrated into business and society, Digital Transformation is changing how organizations manage their assets. This shift goes beyond adopting new technologies, requiring a complete rethinking of how assets are managed, maintained, and optimized. In Portugal, asset management has become a

© The Author(s), under exclusive license to Springer Nature Switzerland AG 2027
M. Calvete Gaspar et al. (Eds.): Regional Helix 2024, LNME, pp. 75–89, 2027.
https://doi.org/10.1007/978-3-032-36365-7_6

dynamic field, influenced by the demands of the digital era. Over recent years, digital transformation in asset management has grown, driven by the need to enhance efficiency, reduce costs, and improve decision-making processes.

Asset management is an increasingly relevant topic, encompassing buildings, machinery, manufacturing facilities, vehicles, software, and more. It is a strategic approach aimed at maximizing the value and performance of assets throughout their lifecycle [1]. All these assets have specific value when acquired, but they must be capitalized on over time. This implies defining policies and processes for developing, maintaining, monitoring, and replacing assets. Asset management aims to ensure that these assets are utilized in the best possible way to obtain the maximum benefit for the company.

Various aspects of organizational strategy are related to asset management. Implementing asset management systems based on ISO 55001 and the eventual strategic decision to certify this system bring significant advantages to organizations [2]. The asset management approach is particularly suitable for identifying and maximizing benefits in environments where the digitization of data, information, and management processes is predominant, ensuring that lifecycle costs and the risks of transformations underlying new ways of working are sustainable in the long term [3].

The digitization of industrial processes is a reality that aligns with new technologies and production methods, aiming to enhance production and workflow. Digitization in industry and various industrial processes brings new perspectives for improvement and development. With the emergence of Industry 4.0 technologies, asset management has become crucial to adequately integrate a higher level of automation and interconnectivity of assets in collaboration with human resources and organizational processes [3].

Integrating these new technologies revolutionizes asset management, providing a more efficient and precise approach to monitoring and optimizing resources through technologies such as IoT (Internet of Things), advanced data analytics, and specialized management software. Organizations can now obtain real-time information about asset performance, predict failures before they occur, and implement proactive maintenance strategies. Moreover, digitization allows for better resource allocation, reduced operational costs, and greater regulatory compliance, boosting efficiency and competitiveness across various industrial sectors.

This article aims to conduct a bibliometric analysis to understand the trends and existing publications related to asset management and digital transformation, using Bibliometrix R software and VOSviewer for data analysis.

2 Literature Review

2.1 Asset Management

Asset management's primary objective is to optimize a given asset's value and performance, reducing associated costs and maximizing its utilization [4].

Asset management focuses on good management practices and technical-financial application to decide which assets are necessary to meet the organization's objectives. To implement effective asset management, it is essential to consider the performance and risks associated with its management, ensuring sustainability in this process, which is crucial for the effective functioning of any organization or process [1].

ISO 55000 [5] defines asset management and is applied across various energy, transportation, infrastructure, and manufacturing sectors. It is used by companies aiming to improve operational efficiency, reduce costs, and mitigate risks associated with assets.

Asset management has been a common practice in the business world even before specific concepts and standards existed. In the 1980s, in the United Kingdom, the oil and gas sector recognized the need for effective management of physical assets to manage risks and achieve financial efficiency. This realization emerged after the Piper Alpha oil platform accident in the North Sea, which resulted in the death of 167 workers due to maintenance issues. This event and the drop in oil prices motivated the industry to adopt a more comprehensive approach to lifecycle asset management, significantly improving efficiency, safety, and productivity [6].

In 2004, with the development of this theme, the Publicly Available Specification 55 (PAS 55) was published by the British Standards Institute (BSI). This specification provides definitions, terminology, and requirements to assist organizations in implementing a management system. In 2008, more than 50 public and private organizations, spanning 10 countries and 15 sectors, contributed to the specification and user feedback [7].

Following the release of PAS 55 in 2014, the Institute of Asset Management (IAM) and the BSI produced international standards through the International Organization for Standardization (ISO), resulting in the three international standards ISO 55000, 55001, and 55002. These standards are now significant because they constitute a global consensus on asset management and its role in increasing the value generated by all organizations [7, 8].

2.2 Digital Transformation

Digitalization refers to the process by which companies adopt new technologies to improve all aspects of their operations, services, and processes, allowing them to adapt and remain competitive. Thus, companies must undergo a change that involves integrating new technologies across various areas [9].

The digitalization process includes increasing the use of robotics, automation, and computerization solutions, thereby reducing costs, improving efficiency and productivity, and allowing flexibility to adapt to changes [10].

Digital Transformation, an integral part of the latest industrial revolution, encompasses various concepts such as Artificial Intelligence (AI), the Internet of Things (IoT), Process Automation, and Augmented Reality. These technologies have improved operational efficiency, increased productivity, and enhanced organizational decision-making [11, 12]. Industry 4.0 is related to the integration of technologies and is the origin of the digital transformation process in the sector [9].

Digital transformation offers numerous opportunities to improve the quality of life in cities by optimizing urban infrastructure operations, reducing costs, and enhancing the quality of services. Developing a framework for this transformation aims to provide a clear structure for interested organizations and governments, identifying key issues and best practices to follow [13, 14].

2.3 Asset Management and Digital Transformation

Evans and Price [15] defined Information Asset Management (IAM) as fundamental to the success of Digital Transformation (DT) in organizations. It developed a holistic IAM model that emphasizes the need to treat data, information, and knowledge not merely as technological resources but as vital business assets essential for thriving digitally. The authors highlight that IAM is a critical component of DT, distinguishing it from Information Technology (IT) management by focusing on the effective creation and use of information.

Asset Management and Digital Transformation are explored concerning significant contributions from various authors. Drucker [16] highlights information assets as the most critical resource. Bontis et al. [17] consider these assets as drivers of business value. Laney [18] emphasizes the need to include information assets in financial reports. Roe [19] addresses information management as crucial for digital transformation. These authors help to understand that effective information asset management is essential for the success of digital transformation and for maintaining competitive advantages in today's market.

Digital Transformation can provide a framework that improves performance by collecting and integrating asset information into decision-making [20]. The technological resources embedded in assets enhance their performance through information management, essential in risk analysis and decision-making [21].

Using technology for asset management can improve asset performance by collecting and integrating information from reliable assets into the decision-making process. Although most companies use a variety of assets in their operations, many rely on inefficient methods, such as on-site inspections once or twice a year, instead of leveraging technology to manage their assets [22].

Intelligent asset management systems are becoming increasingly important. The Internet of Things (IoT) availability in the industrial environment incorporates new technologies, such as predictive maintenance analysis combined with Big Data and digital twin simulations. These new technologies are driving the growth of asset management platforms. However, using digital means in asset management can become a challenge, as it involves specific technological changes in the infrastructure and changes in the asset management process. The digital transformation of asset management must ensure that accurate business information and operational technology data are available at the right time, throughout the system, and during the entire asset lifecycle [23].

3 Research Methodology

This study adopted a bibliometric approach to examine the evolution of "Asset Management" and "Digital Transformation". The Scopus and Web of Science databases were utilized to gather relevant articles.

The data analysis was facilitated by visualizing the results through mapping with VOSviewer software, version 1.6.18 [24], and Bibliometrix R software [25], enabling the visualization of connections between studies and identifying trends. The VOSviewer visualization tool simplifies the interpretation of bibliometric maps, even when dealing with extensive datasets [26, 27]. According to Hood and Wilson [28], bibliometric analysis has become increasingly common and is now a standard practice in various fields of scientific research.

Given the significant expansion of literature with potential interest in the area, the first step was to conduct a bibliometric analysis to identify the most relevant, recent, and highly recommended authors and studies on the topics addressed. The main objective was establishing a correlation between Asset Management and Digital Transformation.

As mentioned earlier, the steps involved in the bibliometric analysis, adapted from Zupic and Cater [29], can be summarized as follows: (1) formulation of the study objectives, selection of the appropriate database, and application of filters to define the sample; (2) collection and processing of bibliometric data, including the application of predefined filters; (3) analysis of the data using specialized bibliometric analysis software; and (4) interpretation and description of the results.

Additionally, a methodological flowchart, following Donthu et al. [30] and the PRISMA-ScR framework developed by Tricco et al. [31], was used, as shown in Table 1.

Table 1. Bibliometric analysis procedure of the PRISMA-ScR flowchart.

Step 1: Definition of Objectives and Scope of the Bibliometric Study				
Comprehensive overview of research trends in asset management and digital transformation				
RQ1. How is scientific production in asset management and digital transformation advancing? RQ2. How is the research collaboration network in this field forming? RQ3. What is the research trend?				
Step 2: Selection of Bibliometric Analysis Techniques				
Scientific Databases: Web of Science and Scopus; **Bibliometric Mapping Software**: VOSviewer and Bibliometrix				
Step 3: Data Collection for Bibliometric Analysis				
Keywords (Web of Science)				
Search 1	((TS = ("Asset Management")) AND (TS = ("Digital Transformation")))			
Keywords (*Scopus*)				
Search 2	(TITLE-ABS-KEY ("Asset Management" AND "Digital Transformation"))			
Step 4: PRISMA-ScR Flow Diagram		**WoS**		**Scopus**
Identification	Recording of all fields obtained from the database search	47		134
	Total articles identified		181	
Screening	Articles after applying filters (English; conference paper; Article)	43		108
	Total articles after applying inclusion and exclusion criteria		151	

(*continued*)

Table 1. (*continued*)

Eligibility	Exclusion of duplicate articles		40	
	Total articles after applying inclusion and exclusion criteria	23		88
	Manual exclusion of articles by title, abstract, and keywords	4		2
Included	Included in the analysis	19		86
General	Final sample		**105**	
Step 5: Conducting the Bibliometric Analysis				
Performance Analysis	Summarizing the performance of prolific research indicators, such as the number of publications and citations; and the publication rate by journal, research area, author, affiliation, and country			
Scientific Mapping	Visualization of the intellectual structure through citation and co-authorship analysis, and of research themes and topics through co-citation analysis, bibliographic coupling, and co-occurrence analysis			
Step 6: Suggestions and Implications for Future Research				
Step 7: Conclusions				

4 Results

4.1 Bibliometric Analysis

This section provides a bibliometric analysis focused on Asset Management and Digital Transformation. It uses techniques such as co-citation maps, co-occurrence, and analysis of the most cited articles and authors to identify research trends. The following section discusses the bibliometric results from 105 selected articles, following the PRISMA-ScR methodology, highlighting the number of publications and the journals where they were published. These studies were evaluated for their relevance to the paper's research questions.

4.1.1 Global Publications by Year

Figure 1 shows the timeline of scientific publications on Asset Management and Digital Transformation. Significant growth between 2017 and 2019 reflected increased research activity in this field. The peak occurred in 2020, likely due to a rise in interest or investment in certain areas of study. However, after 2020, the number of publications dropped sharply, declining in 2021 and 2022. It's important to note that the graph only shows the number of publications, not their relevance or impact. Context, such as the potential influence of the COVID-19 pandemic in 2020, is missing and could help explain this trend.

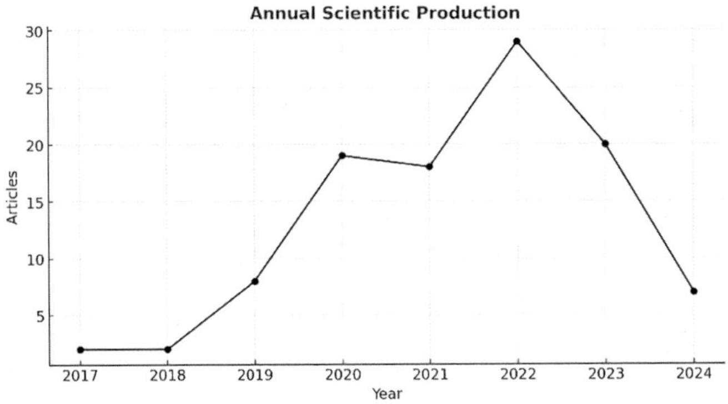

Fig. 1. Annual global scientific production obtained using Bibliometrix R software.

Publications by Year in Portugal Figure 2 shows a steady increase in scientific production in Portugal from 2020 to 2023, particularly in Asset Management and Digital Transformation. This consistent growth reflects a focus on these sectors, which are crucial for innovation, sustainability, and competitiveness. However, other important sectors seem to be less explored. Asset Management and Digital Transformation are essential for efficient operations and are globally recognized for their role in driving digital progress. Despite this, there are still gaps in the number of publications on these topics in Portugal.

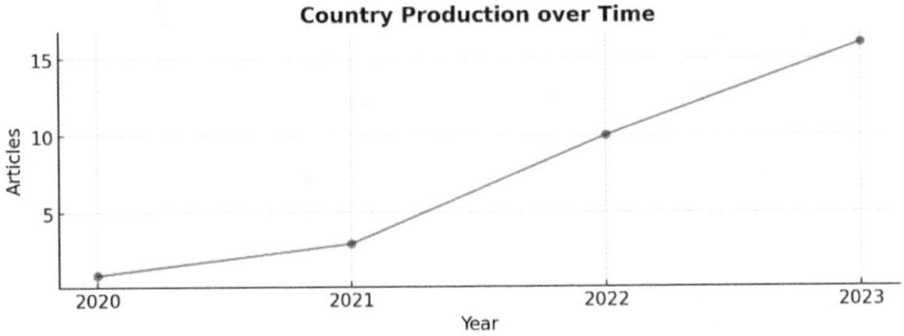

Fig. 2. Annual scientific production in Portugal obtained using Bibliometrix R Software.

4.1.2 Distribution of the Most Cited Publications by Country

This subsection analyzes the distribution of the most cited publications by country. Turkey stands out with many cited publications, followed by the United Kingdom, Australia, Italy, Finland, Canada, the Netherlands, India, Portugal, and others, as illustrated in Fig. 3.

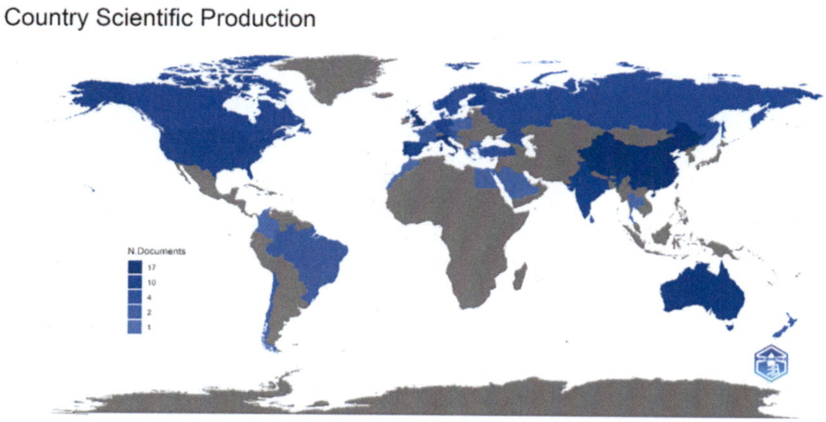

Fig. 3. Scientific production by country obtained using Bibliometrix R software.

4.1.3 Global Most Relevant Publications of Recent Years at the Global Level

This subsection identifies the most influential articles in the current literature, categorized through citation analysis (see Table 2). Each article was carefully reviewed to ensure relevance to the study's theme, resulting in a list of the 12 most significant recent articles.

Out of the 105 articles analyzed, only 6 were Portuguese, discussing both Asset Management and Digital Transformation. This suggests there is room for further research in Portugal. Vieira et al. [32] are notable as key contributors in this field, though not yet among the most cited authors.

4.1.4 Global Keyword Co-occurrence Analysis

Figure 4 shows the co-occurrence map created using VOSviewer and Bibliometrix R, representing the connections between keywords in academic research. From 758 frequent keywords found in titles, abstracts, and article keywords, 125 were selected for appearing together at least twice.

Each keyword is represented by a circle, with its size indicating frequency of use. The distance between circles shows the strength of their co-occurrence how often they appear together in the documents. Lines connect frequently paired words, and colors differentiate groups or related thematic areas.

This visual analysis helps detect research trends, thematic links, and gaps in the literature. Keywords like "asset management" stand out, highlighting it as a key topic, alongside related terms like "information management", "internet of things", "cybersecurity", and "network security".

Table 2. Most relevant publications of recent years at the global level.

Rank	Author	Title	Year	Citations
1	[33]	Development of an assessment model for industry 40: industry 4.0-MM	2017	455
2	[34]	Artificial intelligence based humancentric decision support framework an application to predictive maintenance in asset management under pandemic environments	2021	66
3	[15]	Development of a holistic model for the management of an enterprises information asset	2020	51
4	[35]	Geobim for built environment condition assessment supporting asset management decision making	2021	50
5	[36]	Digital transformation using blockchain technology in the construction industry	2020	41
6	[37]	Analysis of bim use for asset management in three public organizations in québec canada	2019	39
7	[38]	Intangible asset management and digital transformation evidence from intellectual property rights intensive industries	2022	32
8	[39]	Maintenance management through intelligent asset management platforms iamp emerging factors key impact areas and data models	2020	26
9	[12]	Digital twin development for airport management	2020	30
10	[40]	Digital transformation paradigm for asset management in water distribution networks	2021	16
11	[32]	Digital twins in asset management: Potential application use cases in rail and road infrastructures	2021	7
12	[14]	Managing urban infrastructure asset in the digital era challenges of municipal digital transformation	2024	4

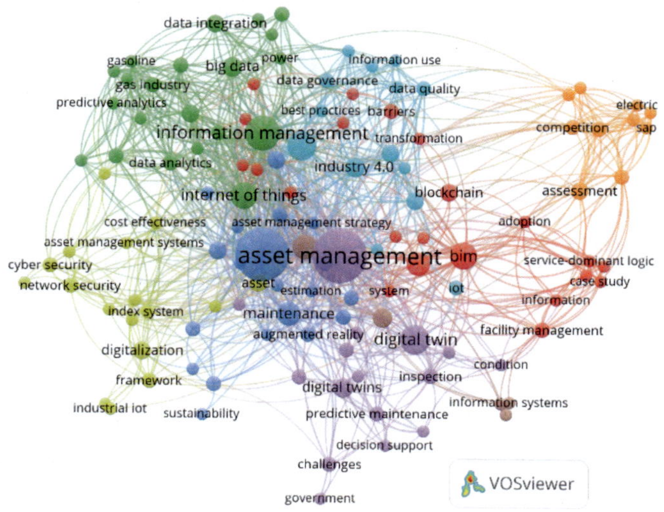

Fig. 4. Keyword map obtained using VOSviewer software.

This new co-occurrence map, produced using VOSviewer software, focuses on a specific subset of keywords within the original research, identifying 758 terms. In this case, a filter was applied to highlight only 14 keywords used at least five times in titles, abstracts, and article keywords, as illustrated in Fig. 5.

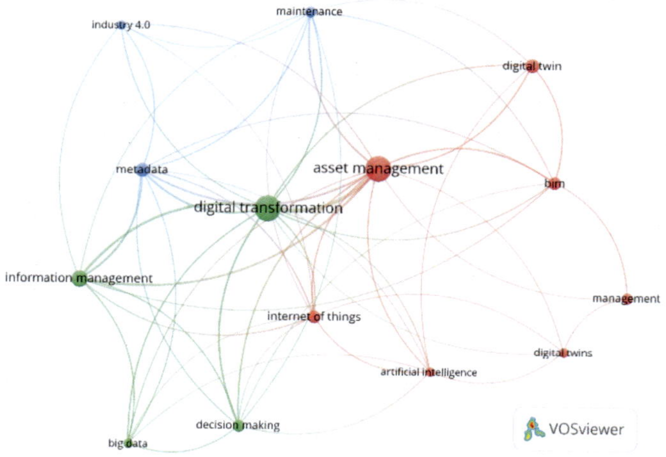

Fig. 5. Overlay of the keyword co-occurrence network obtained using VOSviewer software.

Keywords like "asset management" "digital transformation" and "information management" are central and frequently used indicating their importance in the research. Other relevant terms include "internet of things" "big data" "artificial intelligence" and "decision making".

Lines between keywords represent their co-occurrence, showing how often these topics are linked in studies. Thicker lines suggest stronger connections, such as between "digital transformation" and "industry 4.0".

Different colors indicate thematic clusters, with closely related keywords forming subdomains of the main study area. This layout helps visualize connections between concepts and identify dominant trends or potential areas for future research.

Analysis of the Keyword Cloud Map Figure 6 shows a word cloud created using Bibliometrix R software, where the size and color of words indicate their frequency in the dataset. More extensive terms like "asset management", "digital transformation", and "information management" highlight key themes. Smaller but still significant words include "metadata", "decision making", "maintenance", "internet of things", and "cybersecurity", pointing to technology, business, and digital asset management topics. This visual method helps identify research trends and thematic clusters within the literature, emphasizing connections between different areas of study.

Fig. 6. Keyword map obtained using Bibliometrix R software.

4.1.5 Analysis of Keyword Co-occurrences in Portugal

The co-occurrence map, created using VOSviewer, visually represents the connections between keywords used in academic research by Portuguese authors. It analyzed Scopus and Web of Science data, selecting 37 frequently co-occurring keywords from titles, abstracts, and article keywords.

In Fig. 7, "asset management" appears as the central theme, surrounded by interconnected nodes. To the left, keywords like "digital transformation" suggest a link between asset management and infrastructure modernization. On the right, terms like "cybersecurity" and "internal security" highlight a focus on security issues related to asset management.

The color gradient from light blue to yellow indicates the evolution of these themes from 2020 to 2023, mapping the emergence of different research topics over time.

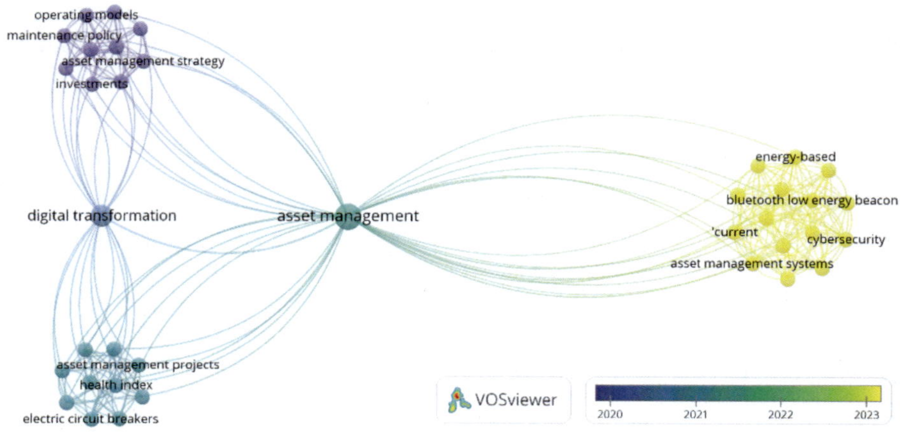

Fig. 7. Keyword co-occurrence map of Portuguese studies obtained using VOSviewer software.

4.1.6 Co-citation Analysis by Author in Portugal

Figure 8 presents a network map of co-citations between authors, generated using VOSviewer, a tool commonly used in bibliometric studies to show collaborations and citation patterns. The "nodes" represent authors, and the lines connecting them symbolize mutual citations. The frequency of these connections indicates the level of collaboration or citation frequency.

Larger nodes represent authors who are more frequently cited and considered central to the research area. Different colors highlight clusters of authors who tend to cite each other more frequently, forming distinct research communities. Out of 150 authors, only 25 are cited at least twice, suggesting these are the more influential figures within the network. The size and number of connections of each node help identify the most influential authors.

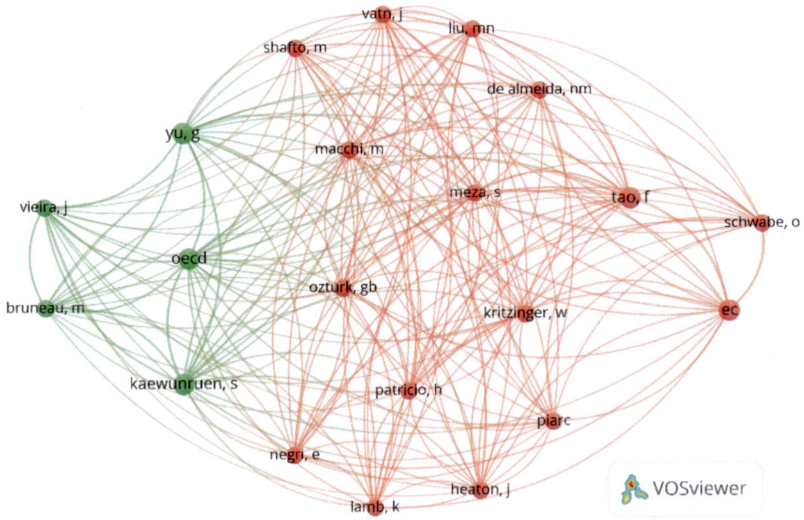

Fig. 8. Co-citation map of authors in Portuguese studies obtained using VOSviewer software.

5 Conclusion

With technological advances, new research topics such as digitalization have emerged, becoming increasingly integrated into people's daily lives and the industrial sector. Another currently well-addressed topic is asset management, which continues to evolve with new management and control methods.

This study revealed an upward trend in scientific production in Portugal between 2020 and 2023 related to emerging sectors such as Asset Management and Digital Transformation. The article also analyzes the distribution of the most cited publications by country. It was found that Turkey stands out with a high number of cited publications, followed by the United Kingdom, Australia, Italy, Finland, Canada, the Netherlands, India, and Portugal.

From this bibliographic study, it was noticeable that there is an abundance of publications on the topics researched in Portuguese. However, access to the majority of these publications was limited. In the context of digitalization/digital transformation, the availability of publications was poorer compared to asset management.

Of the 105 articles selected for this analysis, only six Portuguese documents address this topic, along with Asset Management and Digital Transformation. Among the publications on asset management and digitalization/digital transformation, only two articles with open access were selected, and these were the only ones relevant to the bibliographic study in question. This suggests a research field with room for growth in Portugal.

Acknowledgements. This work is supported by FCT – Fundação para a Ciência e a Tecnologia, I.P., within the scope of the project Ref. UIDB/05583/2020. Furthermore, we would like to thank the Digital Services Research Center (CISeD) and the Viseu Polytechnic Institute for their support.

References

1. Coelho, R.W.S.: Aplicação do conceito de gestão de ativos físicos numa estação elevatória de águas. Instituto Superior de Engenharia de Lisboa. (2015)
2. Figueiredo, J., Pinto, R., Braga, D., Farinha, J.T., Raposo, H.: Avaliação do impacto de um sistema informático de gestão da manutenção Para apoio à implementação de sistemas de gestão de ativos pela NORMA ISO 55001. RAE-Revista de Ativos de Engenharia. **1**(2), 71–85 (2023)
3. Almeida, N.: Fundamentos e perspetivas de inovação na gestão de ativos de engenharia. RAE-Revista de Ativos de Engenharia. **1**(1), 05–16 (2023)
4. Gaspar, D., Almeida, N., Sobral, J., Reguenga, D.: A manutenção na Gestão de ativos. In: Em Manual de Manutenção Em Edificações, Estudos, técnicas e aplicações, vol. 1–1, pp. 155–174. LEUD (2022)
5. ISO 55000: 2014: Asset Management—Overview, Principles and Terminology. ISO Geneva, Switzerland (2014)
6. Jones, M., Williams, W., Stillman, J.: The evolution of asset management in the water industry. J. AWWA. **106**(8), 140–148 (2014)
7. PAS 55: Part 1: Specification for the optimised management of physical assets. (2008)
8. Ma, Z., Zhou, L., Sheng, W.: Analysis of the new asset management standard ISO 55000 and PAS 55. In: 2014 China International Conference on Electricity Distribution (CICED), pp. 1668–1674. IEEE (2014)
9. Anifa, M., Ramakrishnan, S., Joghee, S., Kabiraj, S., Bishnoi, M.M.: Fintech innovations in the financial service industry. J. Risk Financ. Manag. **15**(7), 287 (2022)
10. Hole, G., Hole, A.S., McFalone-Shaw, I.: Digitalization in pharmaceutical industry: what to focus on under the digital implementation process? Int. J. Pharmaceut. X. **3**, 100095 (2021)
11. Kone, V.: Study on implementing smart construction with various applications using internet of things techniques. Int. J. Recent Technol. Eng. **7**, 188 (2019)
12. Oliveira, P.P.: Digital twin development for airport management. J. Airport Manag. **14**(3), 246–259 (2020)
13. Wang, M., Yin, X.: Construction and maintenance of urban underground infrastructure with digital technologies. Autom. Constr. **141**, 104464 (2022)
14. Lafioune, N., Desmarest, A., Poirier, É.A., St-Jacques, M.: Digital transformation in municipalities for the planning, delivery, use and management of infrastructure assets: strategic and organizational framework. Sustain. Futures. **6**, 100119 (2023)
15. Evans, N., Price, J.: Development of a holistic model for the management of an enterprise's information assets. Int. J. Inf. Manag. **54**, 102193 (2020)
16. Drucker, P.: Post-Capitalist Society. Routledge (2012)
17. Bontis, N., Dragonetti, N.C., Jacobsen, K., Roos, G.: The knowledge toolbox: a review of the tools available to measure and manage intangible resources. Eur. Manag. J. **17**(4), 391–402 (1999)
18. Laney, D.B.: Infonomics: How to Monetize, Manage, and Measure Information as an Asset for Competitive Advantage. Routledge (2017)
19. Roe, D.: Why Information Management Plays a Critical Role in Digital Transformation. CMS Wire (2019)
20. Attencia, G.F.: Transformação digital na gestão dos ativos: Análise das tecnologias adotadas e o processo de gestão. Revista Espacios. **41**(17) (2020)
21. Kinnunen, S.K., Hanski, J., Marttonen-Arola, S., Kärri, T.: A framework for creating value from fleet data at ecosystem level. Manag. Syst. Prod. Eng. **25**(3), 163–167 (2017)
22. Han, M., Kim, Y., Park, H., Kim, D.H.: A study on the asset smart management system based on IOT. Int. J. Eng. Technol. (UAE). **7**(2), 110–114 (2018)

23. Maletič, D., Maletič, M., Al-Najjar, B., Gomišček, B.: An analysis of physical asset management core practices and their influence on operational performance. Sustainability. **12**(21), 9097 (2020)
24. Van Eck, N.J., Waltman, L.: Manual for VOSviewer Version 1.6.8. CWTS Meaningful Metrics. Universiteit Leiden (2018)
25. Aria, M., Cuccurullo, C.: Bibliometrix: an R-tool for comprehensive science mapping analysis. J. Inform. **11**(4), 959–975 (2017)
26. Castillo-Vergara, M., Alvarez-Marin, A., Placencio-Hidalgo, D.: A bibliometric analysis of creativity in the field of business economics. J. Bus. Res. **85**, 1–9 (2018)
27. Mas-Tur, A., Roig-Tierno, N., Sarin, S., Haon, C., Sego, T., Belkhouja, M., et al.: Co-citation, bibliographic coupling and leading authors, institutions and countries in the 50 years of technological forecasting and social change. Technol. Forecast. Soc. Change. **165**, 120487 (2021)
28. Hood, W.W., Wilson, C.S.: The literature of bibliometrics, scientometrics, and informetrics. Scientometrics. **52**, 291–314 (2001)
29. Zupic, I., Čater, T.: Bibliometric methods in management and organization. Org. Res. Methods. **18**(3), 429–472 (2015)
30. Donthu, N., Kumar, S., Mukherjee, D., Pandey, N., Lim, W.M.: How to conduct a bibliometric analysis: an overview and guidelines. J. Bus. Res. **133**, 285–296 (2021)
31. Tricco, A.C., Lillie, E., Zarin, W., O'Brien, K.K., Colquhoun, H., Levac, D., et al.: PRISMA extension for scoping reviews (PRISMA-ScR): checklist and explanation. Ann. Internal Med. **169**(7), 467–473 (2018)
32. Vieira, J., Clara, J., Patrício, H., Almeida, N., Martins, J.P.: Digital twins in asset management: potential application use cases in rail and road infrastructures. In: World Congress on Engineering Asset Management, pp. 250–260. Springer International Publishing, Cham (2021)
33. Gökalp, E., Şener, U., Eren, P.E.: Development of an assessment model for industry 4.0: industry 4.0-MM. In: Software Process Improvement and Capability Determination: 17th International Conference, SPICE 2017, Palma de Mallorca, Spain, October 4–5, 2017, Proceedings, pp. 128–142. Springer International Publishing (2017)
34. Chen, J., Lim, C.P., Tan, K.H., Govindan, K., Kumar, A.: Artificial intelligence-based human-centric decision support framework: an application to predictive maintenance in asset management under pandemic environments. Ann. Oper. Res., 1–24 (2021)
35. Moretti, N., Ellul, C., Cecconi, F.R., Papapesios, N., Dejaco, M.C.: GeoBIM for built environment condition assessment supporting asset management decision making. Autom. Constr. **130**, 103859 (2021)
36. Prakash, A., Ambekar, S.: Digital transformation using blockchain technology in the construction industry. J. Inf. Technol. Case Appl. Res. **22**(4), 256–278 (2020)
37. Brunet, M., Motamedi, A., Guénette, L.M., Forgues, D.: Analysis of BIM use for asset management in three public organizations in Québec, Canada. Built Environ. Project Asset Manag. **9**(1), 153–167 (2019)
38. Trequattrini, R., Lardo, A., Cuozzo, B., Manfredi, S.: Intangible assets management and digital transformation: evidence from intellectual property rights-intensive industries. Meditari Acc. Res. **30**(4), 989–1006 (2022)
39. Crespo Marquez, A., Gomez Fernandez, J.F., Martínez-Galán Fernández, P., Guillen Lopez, A.: Maintenance management through intelligent asset management platforms (IAMP). Emerging factors, key impact areas and data models. Energies. **13**(15), 3762 (2020)
40. Ciliberti, F.G., Berardi, L., Laucelli, D.B., Giustolisi, O.: Digital transformation paradigm for asset management in water distribution networks. In: 2021 10th International Conference on Energy and Environment (CIEM), pp. 1–5. IEEE (2021)

Challenges and Opportunities for the Industry in Implementing Circular Economy Through Repair—Insights from the European Union Legislation

Berta Costa[1], Susana Rodrigues[2](✉), Susana Almeida[3], and Jacob C. Bazen[4]

[1] School of Tourism and Maritime Technology (ESTM), University of Leiria and Oeste, Santuário de Nossa Senhora dos Remédios, CITUR-Centre for Tourism Research, Development and Innovation, Campus 4, 2520-641 Peniche, Portugal

[2] School of Technology and Management (ESTG), University of Leiria and Oeste, Morro do Lena – Alto do Vieiro, CARME-Centre of Applied Research in Management and Economics, apartado 4163, 2411-901 Leiria, Portugal
susana.rodrigues@ipleiria.pt

[3] School of Technology and Management (ESTG), University of Leiria and Oeste, Morro do Lena – Alto do Vieiro, IJP-Portucalense Legal Institute (IJP/IPLeiria), apartado 4163, 2411-901 Leiria, Portugal

[4] School of Business, Building and Technology, Saxion University of Applied Sciences, M.H. Tromplaan 28, P.O. Box 70.000, 7500 KB Enschede, The Netherlands

Abstract. Circular economy has raised significant attention from researchers. It focusses on resource efficiency, waste reduction, and product longevity and is a promised solution to the established environmental and economic challenges. It lines up with sustainable development goals by dissociating economic growth from environmental degradation. It ensures a more equitable distribution of resources. Climate change, global resource scarcity, biodiversity loss, pollution, and land management have been identified as critical challenges of the modern era, which are threatening the planet's life-support systems and posing significant risks to well-being, prosperity, and economic competitiveness. The prevailing linear economic model, characterized by low-cost resource inputs and on product disposal, has reached its limits. Consumers discard products that are no longer desirable, overlooking their potential value. Repair enables an extension of products' lifespan, reduces waste, and creates new economic opportunities. It also involves fostering relationships with customers, understanding their needs, and offering tailored solutions. Nevertheless, this transition will require a collective effort from governments, businesses, and consumers to restructure production and consumption patterns, promote sustainable practices, and create a more equitable and prosperous society. The aim of this research is to understand the European Union legal and regulatory frameworks proposed, highlighting the key legislation and initiatives that guide this process and the challenges and opportunities it poses. It also explores good practices of some companies' Repair Circular Economy implementation. Literature was extracted from the Scopus and Web of Science databases considering the keywords defined to identify and better understand Circular Economy concepts. Legislation proposed by the European Union was considered and

© The Author(s), under exclusive license to Springer Nature Switzerland AG 2027
M. Calvete Gaspar et al. (Eds.): Regional Helix 2024, LNME, pp. 90–110, 2027.
https://doi.org/10.1007/978-3-032-36365-7_7

the industry websites were examined. The results showed that European Union legal and regulatory frameworks illustrated challenges as well as opportunities in the repair implementation process for companies. Companies' mandatory actions arising from the law were also identified. Examples of Repair in the industry as a regenerative practice were presented offering valuable insights for company's decision-makers.

Keywords: Circular economy · Repair · Products' life span · Waste industry

1 Introduction

The European Commission (EC) [1] identifies climate change as the main challenge of contemporary times, alongside global resource demand and scarcity, environmental degradation linked to biodiversity loss, pollution (air, soil, and water), and land management. These challenges threaten the life-support systems of planet Earth [2–6] and will presumably impose greater challenges on well-being and prosperity and, from a business perspective, on competitiveness and revenue. Consequently, there is an urgent necessity to shift towards more sustainable social and economic systems that remain within the environmental boundaries of the Earth [6–10]. In this context and given the constraints of the existing linear economic paradigm, based on the "extract-produce-utilise-discard" paradigm [11], supported by low-cost and accessible resource inputs that have made it possible to generate growth and stability [12], the effectiveness of the prevailing economic model is being questioned.

Consumers dispose of products that are no longer desirable, largely because they are broken, obsolete and no longer needed. However, a large number of these products still retain substantial value and, if they are maintained and improved through repair, upgrading or even re-manufacturing, they can be preserved and have economic value for a longer period of time. This reorganization would mean that companies no longer just sell products but keep them at their highest value for longer periods of time. This would also mean that customers would shift their focus from transactions to relationships, making adjustments or changes according to specific requirements [13].

Therefore, the adoption of a circular economic model that enables the transition towards sustainability seems to be the desirable way to separate economic growth from unsustainable resource use [14]. Thus, the motivation for the transition to a Circular Economy model has emerged. There is growing global recognition among academia, governments and stakeholders that the sustainable viability of planet Earth depends on this change, which will be the opportunity to restructure production and consumption models and obtain economic and social revenues, as well as opportunities to keep resources in effective use in the economy for as long as possible [14]. Additionally, consumers are becoming more conscious of environmental issues when making purchases and want to actively contribute to the green transition to a sustainable and circular economy [15].

In order to support and sustain this transition, the ability to monitor and measure circularity using monitoring processes, tools and the validation and evaluation of indicators is fundamental [16]. However, the development of indicators to measure the implementation of the Circular Economy is still in its early stages [17, 18].

The main goal of this study is to examine the legal and regulatory frameworks of the European Union by highlighting the key legislation and initiatives that shape this process, as well as to examine the challenges and opportunities they present.

1.1 The Concept of Circular Economy

Currently, humanity faces numerous environmental hazards and difficulties, ranging from global biodiversity loss and increasing pollution to climate change and the widening gap in income inequality. Industrial overexploitation of natural resources and excessive waste generation are also exacerbating these environmental, social, and economic crises. Addressing these challenges is essential for companies, stakeholders, governments, and ultimately communities, as building economic and environmental resilience has become more critical than ever.

The prevailing economic model is based on a linear paradigm [19], where resources are assumed to be abundant, easily accessible, consistently available, and inexpensive [20]. This model follows a "take, make, dispose" pattern [11], where natural resources are extracted, converted into goods, and discarded as waste after consumer use. In contrast, the Circular Economy model proposes a "make-use-reuse-remake-recycle" approach. This concept draws from various scientific and semi-scientific approaches, such as "circular flow of income," "cradle-to-cradle," "industrial ecology," "performance economy," "biomimicry," "eco-effectiveness and eco-efficiency," and "natural capitalism" [21]. It was first introduced to the scientific community by Pearce and Turner (1990) [22] as an attempt to address environmental constraints [23, 24]. CE identifies itself as a regenerative economic approach that reduces emissions, waste, and the use of raw materials. It also mitigates energy consumption by slowing down and closing energy loops, mimicking natural material cycles [6, 25–28].

According to the Ellen MacArthur Foundation (EMF) [29] and the European Commission (EC) [30], the CE model is "restorative and regenerative by intention and design," maintaining the value of products over extended periods while minimizing waste generation. The concept of "end-of-life" is replaced by product restoration. The model encourages the use of renewable energy sources and reduces the use of toxic substances that hinder recycling and reuse. Undoubtedly, better product design and manufacturing techniques will boost business performance and help them attain sustainable development, which entails creating environmental quality standards to guarantee the social justice and economic well-being of future generations [31].

By taking a preventive and regenerative approach, the Circular Economy has also been acknowledged as a trustworthy economic model that aids businesses in improving their operations. This model encourages a paradigm shift that fosters resilience and long-term adaptability while also reducing the negative externalities of the current linear economic model. It offers social and environmental advantages along with economic development opportunities.

Within the specific context of the European Union, the implementation of circular economy is increasingly centered on the "Right to Repair" as a primary industrial strategy. Recent legislative shifts, notably the EU's Circular Economy Action Plan and the Ecodesign for Sustainable Products Regulation (ESPR) [12, 13], represent a transition from voluntary initiatives to mandatory requirements for product reparability. These directives aim to address the systemic barriers that have historically favored replacement over repair, such as restricted access to spare parts and technical documentation [2]. By establishing a common framework for product durability and repair information, the EU aims to empower independent repairers and consumers, thereby decentralizing the maintenance of product value and reducing the environmental footprint of high-impact sectors like electronics and textiles [12, 13].

However, the industrial transition to repair-based models brings a complex range of challenges, particularly manufacturers accustomed to linear volume-based sales. Technical barriers, such as the use of adhesives and proprietary fasteners, often make products that are physically unrepairable, while economic constraints include high labour costs and the lack of efficient reverse logistics networks [2].

Furthermore, the industrial sector faces significant legal challenges as Intellectual Property (IP) rights may interfere with the disclosure of repair manuals. These obstacles demand a new "design for repair" philosophy and the development of specialized service-based business models, such as Product-as-a-Service (PaaS), to ensure that extending products' lives remain economically viable for the producer [5].

Despite these challenges, the shift towards a repair-oriented circular economy offers profound opportunities for industrial innovation and regional resilience. Leveraging Industry 4.0 technologies, such as additive manufacturing for on-demand spare parts and digital product passports (DPPs), can drastically reduce the costs associated with inventory and information asymmetry [5]. Beyond technical efficiency, a robust repair sector stimulates local economies by creating highly skilled service jobs that cannot be easily relocated, fostering a closer relationship between producers and consumers. Ultimately, the successful integration of repair within the EU legislative framework provides a blueprint for a resilient industrial ecosystem that decouples economic growth from resource depletion, transforming waste prevention into a competitive strategic advantage [2].

1.2 Circular Economy R-Principles

Over the past few decades, academia and stakeholders have employed various R-principles, though identifying the exact origins of these classifications has been difficult, with no single article serving as a definitive starting point [32, 33]. Research [34, 35] considers these R-principles as sequential stages within the Circular Economy framework, establishing them as core principles of CE.

In literature, the Circular Economy is commonly associated with the 3R-principles: Reduce, Reuse, and Recycle [25, 35–37, 39, 40]. Over time, additional R-principles have garnered attention. The European Union Waste Framework Directive [41] introduced the Recover principle, expanding the initial 3R framework to 4Rs. Various academic contributions have proposed alternative R frameworks, including the 6R framework [32], and even more comprehensive models like the 9R framework [42, 43].

The various R-principles should be organized hierarchically and prioritized according to their degree of circularity [31, 43]. They are viewed as strategies within circularity that help reduce the consumption of natural resources and materials while minimizing waste generation. These principles are applied at different stages of the value chain.

Within a Circular Economy framework, waste is regenerated, and the R-principles contribute to continuously cycling materials, reducing the need for raw materials by converting used ones into secondary raw materials. Potting et al. [43] present their interpretation of the 9R-principles and their prioritization based on circularity levels, as outlined in Fig. 1. However, this framework was expanded by introducing the Redesign, Remine, and Return R-principles, resulting in a comprehensive set of 13 R-principles. Specifically, Re-Educate (R2) and Redesign (R3) to the "Smarter product use and manufacture" category were added and Remine (R12) and Return (R13) were considered under the "Useful application of materials" category.

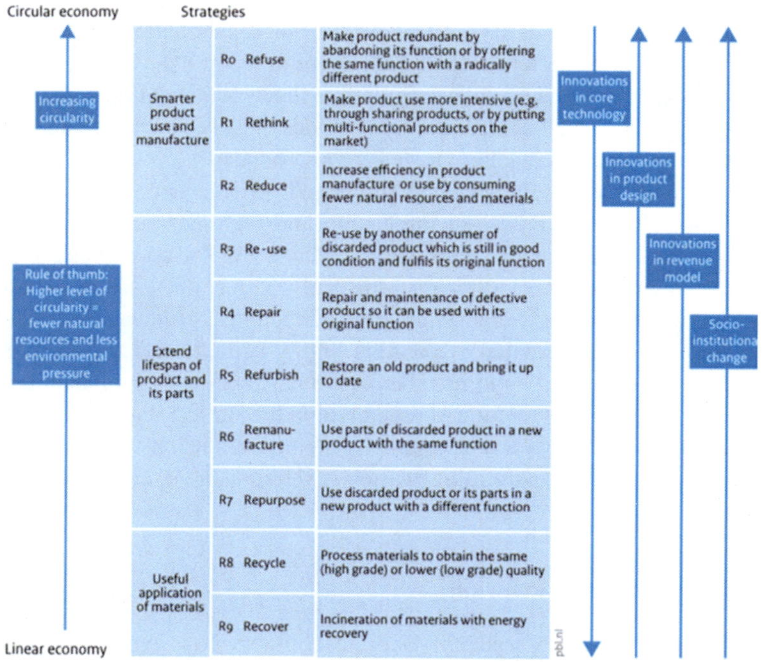

Fig. 1 Potting's representation of CE R-principles (2017:5).

Although as the CE R-Principles are presented and explained, this study highlights the repair R-principle. The R6-Repair principle focuses on prolonging a product's life by fixing its parts or making the required adjustments [44–46]. Lacy and Rutqvist [13] suggest that prioritizing this R-principle within a Circular Economy model can lead to substantial economic benefits. By ensuring that goods maintain their maximum usefulness and worth for extended periods of time, these guidelines contribute to the environment's inherent ability to regenerate [47, 48].

The implementation of the repair principle within the European industrial landscape is increasingly determined by evolving legislative frameworks that attempt to transform product design and consumer rights. In this sense the "Right to Repair" is a legal instrument which aims to dismantle the technical and economic barriers established by linear business models, such as intentional obsolescence and the restriction of spare parts to authorized networks. By requiring that manufacturers provide access to technical documentation and components for a minimum period, the legislation seeks to shift the market toward a "repair-first" hierarchy, thereby decoupling economic growth from material throughput [49].

However, the transition to a repair-based economy presents both transformative opportunities and structural challenges for the European industry. On one hand, repair offers a pathway to localized job creation and the development of high-value service-oriented business models, such as "Product-as-a-Service" (PaaS), which align manufacturer incentives with product longevity [5]. On the other hand, technical design barriers continue to make products physically unrepairable, while high labor costs relative to the price of new goods often discourages consumers from choosing repair [50]. Bridging this gap would allow the repair principle to become an affordable and transparent strategy of the European circular transition [51].

1.3 The Repair R-Principle: A Cornerstone of Circular Economy

The Repair R-Principle is a critical factor of the circular economy, emphasizing the importance of extending product lifespans through repair rather than disposal. By repairing products, we can reduce waste, conserve resources, and mitigate environmental impact.

While repair has historically been undervalued, recent years have seen a renewed interest in its potential to drive sustainable consumption and production patterns. This shift is driven by various factors, including growing environmental concerns, increasing re-source scarcity, and a desire for more sustainable lifestyles.

Independent repair, carried out by individuals and small businesses, plays a significant role in promoting circular economy principles. By repairing products outside of traditional manufacturer-controlled channels, independent repairers can challenge corporate control and offer affordable repair options to consumers. However, they face challenges such as limited access to spare parts and repair information.

Consumer attitudes and behaviors are crucial in shaping the success of repair initiatives. Recent studies have shown a growing willingness among consumers to repair products, particularly for certain categories like electronics and household appliances. However, factors such as product design, availability of spare parts, and repair skills can influence consumer choices.

To encourage repair and support a circular economy, policymakers and regulators ought to play a more crucial role. Implementing policies that promote right-to-repair legislation, standardized repair information, and access to spare parts can create a more favorable environment for repair.

By understanding the multifaceted role of repair and addressing the challenges and opportunities associated with it, we can accelerate the transition to a more sustainable and resilient future.

The Repair R-Principle is a cornerstone of the Circular Economy. By emphasizing the importance of repair, it promotes the extension product lifespans, thereby reducing waste and conserving resources. This principle encourages the prioritization of repair as a sustainable practice, moving away from a disposable culture towards a more circular model of consumption.

Beyond environmental benefits, the Repair R-Principle also generates important socio-economic value. Repair activities support local economies by creating skilled jobs, strengthening small and medium-sized enterprises, and fostering community-based initiatives such as repair cafés. These localized repair networks reduce dependence on global supply chains and increase economic resilience, particularly in times of resource shortages or supply disruptions [9, 10, 52].

Furthermore, product design plays a decisive role in enabling or preventing repair. Design choices such as modular components, standardized fasteners, and accessible repair documentation can significantly improve product repairability and longevity. Policies such as eco-design requirements and repairability scoring systems encourage manufacturers to integrate repair into the product development phase, thereby aligning industrial innovation with circular economy objectives [52].

1.4 The Importance of *Repair* in Circular Economy

Repairing products is a more resource-efficient strategy than recycling, as it conserves both materials and energy [33]. By extending the lifespan of existing items, repair minimizes the need for new resource extraction and reduces the energy demands associated with manufacturing. This shift towards a more sustainable approach is reflected in the growing emphasis on repair initiatives within European legislation [33]. Community repair movements, such as repair cafés, not only empower individuals to engage in sustainable practices but also foster valuable social connections [53]. By enabling individuals to repair their own belongings, these initiatives challenge the dominance of planned obsolescence and advocate for consumer rights to repair and modify products, ultimately fostering a more equitable and sustainable consumption model [53].

Existing literature examines repair as an entirely means of extending product lifespans, arguing that it can inadvertently perpetuate consumerism and industrial capitalism [54]. This perspective emphasizes that a broader understanding of repair must include its social and political dimensions, recognizing its potential to facilitate transformative changes in consumption patterns [54]. While the Repair R-Principle promotes sustainability, it is crucial to acknowledge the risk of inadvertently reinforcing consumerist behaviors if not critically examined. This highlights the need for a balanced perspective on repair within the Circular Economy framework.

Repair also plays a crucial role in waste prevention, which sits at the top of the waste hierarchy. By delaying or avoiding end-of-life treatment, repair reduces the volume of products entering recycling or disposal streams, where material losses and environmental impacts still occur. Studies have shown that even high-quality recycling processes cannot fully recover all embedded materials and energy, making repair a more effective strategy for preserving product value and reducing environmental footprints [55, 56].

Moreover, repair can act as an entry point for broader behavioral change towards sustainable consumption. Engaging in repair practices can increase consumers' awareness of product lifecycles, labor inputs, and material value, potentially leading to more conscious purchasing decisions and longer-term use patterns. This experiential learning aspect of repair has been identified as a catalyst for sufficiency-oriented lifestyles, which prioritize reduced consumption and greater care for existing goods [57].

Finally, the comprehensive integration of repair into circular economy strategies requires alignment across policy, industry, and society. While CE strategies play an important role, their impact remains limited without supportive regulatory frameworks and business models that prioritize durability and service-based value creation. To achieve a truly circular economy, repair must be institutionalized within governance structures and corporate models. This ensures that repair initiatives do not simply compensate for the negative impacts of overproduction but instead serve as a pillar for a resilient industrial system designed for longevity and resource efficiency [56, 57].

1.5 Repair and the EU Perspective

The growing focus on the circular economy, particularly within European academia and policy circles, has motivated increased research interest in repair [8, 34, 58]. As a core element of circularity, repair extends the lifespan of products, contributing to a more regenerative economic system [59]. By restoring, reconfiguring, and remediating, repair helps mitigate the environmental damage caused by the linear economy [3, 60].

This study investigates the role of independent repair, which involves individuals, non-profit organizations, and small businesses engaging in product repair without formal authorization from brand owners. Independent repair often utilizes both original and non-original spare parts and tools.

To provide context and direction for this research, we conducted a literature review on repair and its regulation. Our research approach combined semi-structured interviews and thematic analysis to delve into the experiences and perspectives of key stakeholders. The results of our study are presented in the following sections, along with a discussion of repair as a regenerative practice and the potential role of regulation in supporting independent repair.

At the policy level, the European Union has increasingly recognized repair as a key factor for achieving circular economy objectives. Initiatives such as the Circular Economy Action Plan and revisions to the Ecodesign Directive emphasize product durability, repairability, and access to spare parts and repair information as key regulatory priorities [12, 13]. These measures aim to shift responsibility upstream to producers, encouraging design choices that facilitate repair and reduce premature product obsolescence across multiple product categories.

More recently, right-to-repair debates have gained prominence within the EU, highlighting tensions between intellectual property protection, market competition, and sustainability goals. While regulatory progress has improved conditions for authorized repair networks, independent repairers often remain constrained by limited access to technical documentation, software locks, and proprietary components. Understanding how EU regulation can better balance innovation, consumer rights, and environmental objectives is therefore essential to unlocking the full potential of repair as a regenerative and inclusive practice within the European circular economy framework [12, 13].

2 Repair Policies and Regulations

2.1 The European Framework for Consumer Protection and Empowerment for the Green Transition: Main Lines

The new "right to repair" policy (R2R) created by the EU emerges as a fundamental step in the roadmap outlined in the new Circular Economy Action Plan (published in March 2020) and within the context of the European Green Deal (published in December 2019) to achieve climate neutrality in the EU by 2050 and to accelerate the transition to a circular economy model.[1] It also stands as a key initiative within the framework of the New Consumer Agenda, aiming for sustainable consumption through product repair, thereby preventing a systemic waste of resources [52].

In November 2020, the European Commission defined the EU consumer policy for the 2020–2025 period in the New Consumer Agenda, identifying the green transition as a key priority among the five priority areas.[2]

In this area, the Commission seeks to empower consumers to take an active role in the ecological transition. This can only be achieved by ensuring that sustainable products are available in the EU market and that consumers have access to the information they need to make informed and sustainable choices, thus preventing greenwashing and planned obsolescence [52].

In this context, and within the framework of the initiatives outlined in the New Consumer Agenda and the Circular Economy Action Plan, emerges, on the supply side and at the production stage, the Regulation (EU) 2024/1781 of the European Parliament and of the Council of 13th June 2024. This Regulation, as highlighted in recital 6, "should provide for the setting of new eco-design requirements to improve product durability, reliability, repairability, upgradability, reusability and recyclability, improve possibilities for the refurbishment and maintenance of products, address the presence of hazardous chemicals in products, increase the energy and resource efficiency of products, including with regard to the possibility of recovery of strategic and critical raw materials, reduce their expected generation of waste and increase the recycled content in products", as well as addresses practices associated with premature obsolescence [52].

Furthermore, the EU drafted in March 2022 and concluded in February 2024 the Directive (EU) 2024/825 of the European Parliament and Council. This Directive sets out the demand-side requirements, focusing primarily on the point of contracting or purchasing decisions. It seeks to empower consumers for the ecological transition through better protection against unfair commercial practices and better information on durability and reparability of goods at the point of sale. This Directive aims to amend Directive 2005/29/EC on unfair commercial practices (UCPD) and Directive 2011/83/EU on consumer rights (CRD), with the overall objectives of contributing "to a circular, clean and green EU economy by enabling consumers to take informed purchasing decisions

---

[1] (COM(2020) 98 final, https://eur-lex.europa.eu/resource.html?uri=cellar:9903b325-6388-11ea-b735-01aa75ed71a1.0022.02/DOC_1&format=PDF, p. 6).

[2] (COM/2020/696 final, New Consumer Agenda, https://eur-lex.europa.eu/legal-con-tent/EN/TXT/PDF/?uri=CELEX:52020DC0696, p. 5).

and therefore contribute to more sustainable consumption" (EUROPEAN COMMISSION, "Proposal for a Directive of the European Parliament and of the Council amending Directives 2005/29/EC and 2011/83/EU as regards empowering consumers for the green transition through better protection against unfair commercial practices and better in-formation," COM (2022) 143 final – Explanatory Memorandum.[3]

In March 2023, the European Commission presented a proposal for a Directive on common rules promoting the repair of goods, thus establishing a new (R2R) both within and beyond the legal warranty period. This initiative aims to encourage and facilitate the repair and reuse of products, in line with the objectives set out in the European Green Deal (EGD). This process ultimately led to Directive (EU) 2024/1799 of the European Parliament and of the Council, dated June 13, 2024 [52].

2.2 Key Aspects of the Right to Repair Directive (2024/1799) and the Paradigm Shift

The European Union's Right to Repair Directive (2024/1799) represents a significant shift towards sustainable consumption and waste reduction. By empowering consumers and pressing manufacturers, this legislation aims to extend the lifespan of products and promote a circular economy.

The main key aspects of this Directive are to propose the Manufacturer's Obligation to Repair, i.e. manufacturers are now required to offer repair services for products beyond the standard legal guarantee period. This obligation covers common household items such as washing machines, vacuum cleaners, and smartphones, provided they are technically repairable under EU law. Repairs must be conducted at a reasonable price and within a reasonable timeframe, ensuring consumers have continued access to functional products. It also indicates that there should be provided access to spare parts and repair information to facilitate repairs. Manufacturers must also provide spare parts, tools, and necessary repair information to both consumers and independent repairers at reasonable prices. This provision aims to eliminate barriers that previously discouraged repair, such as high costs or lack of information, thereby supporting a more robust repair market. The European Repair Information Form is also brought forward. It indicates that a standardized form will be introduced to enhance transparency in repair services. Consumers can request this form from repairers to obtain clear information regarding the nature of the defect, repair costs, and duration. This initiative enables consumers to make informed decisions when seeking repair services. Furthermore, this directive presents the European Online Repair Platform, which is a centralized online platform which will be established to connect consumers with local repair services and sellers of refurbished goods. This platform aims to simplify the process of finding reliable repair options, thereby encouraging consumers to opt for repairs over replacements. Incentives for repair are also presented, i.e. member states are encouraged to implement measures that promote repair activities. These may include repair vouchers, funds to subsidize repair costs, information campaigns, and support for community-led repair initiatives like repair cafés. Such incentives are designed to make repairs more appealing and cost-effective for consumers.

[3] https://eur-lex.europa.eu/legal-content/EN/TXT/?uri=celex:52022PC0143).

2.2.1 Paradigm Shift Brought by the Directive

The Right to Repair Directive signifies a transformative change in consumer culture and manufacturing practices within the EU. It promotes Empowered Consumers by ensuring access to affordable and transparent repair options, consumers are empowered to make sustainable choices, reducing the need for premature product replacements.

It also assists sustainable consumption, as it encourages a shift from a disposable culture to one that values longevity and sustainability, aligning with broader environmental goals to reduce waste and conserve resources. Economic Opportunities are also stimulated as the emphasis on repair services is expected to encourage growth in the repair sector, creating jobs and fostering local economies centered around sustainable practices. In addition, *Manufacturer Accountability* will also be promoted as by mandating repair obligations, manufacturers are held accountable for the durability and repairability of their products, potentially influencing design practices towards more sustainable and long-lasting products.

So, the Right to Repair Directive (2024/1799) not only enhances consumer rights but also promotes environmental sustainability and economic resilience, marking a pivotal shift towards a more circular economy within the European Union [52].

2.3 Benefits and Challenges of the Right to Repair Directive

2.3.1 Benefits Posed by the Right to Repair Directive

The European Union's Right to Repair Directive (Directive 2024/1799) aims to promote sustainable consumption by facilitating the repair of goods, thereby extending product lifespans and reducing waste. It offers the following benefits:

1. *Environmental Sustainability*: by encouraging the repair of goods, the directive aims to reduce waste and promote a lifecycle that includes reuse and refurbishment, contributing to environmental conservation.
2. *Consumer Empowerment*: the directive mandates that manufacturers and sellers facilitate easier, more affordable repairs, thereby extending product lifespans and reducing waste.
3. *Economic Opportunities*: the emphasis on repair services is expected to stimulate growth in the repair sector, creating jobs and fostering local economies centered around sustainable practices.
4. *Extension of Legal Guarantees*: the directive amends the existing Sale of Goods Directive by granting consumers an additional year of legal guarantee if they choose to repair a product instead of replacing it under the legal guarantee. This encourages repair and enhances consumer protection.
5. *Promotion of Circular Economy*: by shifting the focus from product replacement to repair, the directive fosters a circular economy. This approach emphasizes resource efficiency and waste reduction, contributing to environmental sustainability.
6. *Incentives for Repairability*: the directive encourages the reform of Extended Producer Responsibility (EPR) schemes to promote repairability. This reform incentivizes manufacturers to design products that are easier to repair, aligning with sustainability goals.

These benefits underscore the directive's role in advancing consumer rights, environmental sustainability, and the transition towards a circular economy within the European Union [52].

2.3.2 Challenges Posed by the Right to Repair Directive

While the directive offers several benefits, it also presents certain challenges.

1. *Intellectual Property Concerns*: the directive's provisions allowing independent repairers to carry out repairs and use refurbished goods may raise legal questions regarding existing industrial property rights, potentially leading to conflicts between repair objectives and intellectual property protections.
2. *Increased Operational Costs for Businesses*: manufacturers and sellers are required to provide repair services beyond standard warranty periods, which could result in additional operational expenses, including maintaining spare parts inventories and offering extended customer support.
3. *Consumer Awareness and Behavioural Shift:* encouraging consumers to opt for repairs over replacements requires significant changes in consumer behaviour and awareness, as the prevailing "throw-away" culture may hinder the directive's objectives.
4. *Technological and Security Concerns*: modern products, especially electronics, often contain proprietary software and complex technologies. Ensuring that independent repairs do not compromise product security or violate software licenses presents a substantial challenge.
5. *Standardization and Compatibility Issues*: the lack of standardized components and repair procedures across different manufacturers can complicate repair processes, making it difficult for independent repairers to service various products effectively.

Addressing these challenges is crucial for the successful implementation of the Right to Repair Directive, aiming to promote sustainability and consumer empowerment within the European Union. Based on the literature and the European Union legislation reviewed, the following research questions were defined:

RQ1: How does European Union law guide the repair principle implementation by companies?

RQ1.1. What are the challenges for companies in implementing the repair principle?

RQ1.2 What are the opportunities for companies in implementing the repair principle?

RQ2: What are the companies repair principle practices?

The following section addresses data collection process to answer these research questions.

3 Methodology

The research methodology was designed to offer a comprehensive exploration of how circular economy and repair initiatives can foster industrial entrepreneurial sustainable practices. To respond to the research questions, the study combined multiple data sources and analytical approaches to build a robust understanding of both the theoretical and practical dimensions of this subject.

Relevant academic literature was sourced from the Scopus and Web of Science databases. This review provided insights into the theoretical background of the circular economy concept and repair as a regenerative practice.

In parallel, the research involved a detailed analysis of the European Union's legislative framework. Key directives, such as the Directive on common rules promoting the re-pair of goods (R2R Directive), were examined to understand how EU policy is designed to encourage repair, extend product lifecycles, and promote sustainability.

Furthermore, a range of industry websites were reviewed to identify real-world examples of how companies are integrating repair initiatives into their business models. These examples provided practical context to complement the legislative and theoretical frameworks. The collected information was then synthesized using content analysis, which enabled the identification of challenges, opportunities, and repair practices. This structured approach ensured that both qualitative data (such as policy objectives and consumer behavior) and quantitative data (including sales figures and repair metrics) were coherently integrated.

3.1 Examples of the Industry

Table 1 provides some examples of Repair initiatives implemented.

Different sources of data aimed to gather comprehensive insights into the challenges, opportunities, and practices related to the implementation of the repair circular economy principle among various companies.

3.2 Repair Services

3.2.1 Legislative Framework and Obligations

The Directive on Repair of Goods (R2R Directive) establishes several measures aimed at promoting repair as part of sustainable consumption. Manufacturers are obliged to offer repair services within a "reasonable time" and at a "reasonable price." This includes:

1. *Obligation to repair*: manufacturers must repair goods that are technically repairable, ensuring that repair services are accessible even beyond the legal guarantee.
2. *Transparency and information*: manufacturers are required to provide clear information about repair services and indicative repair costs through freely accessible platforms.
3. *Extended legal guarantee*: consumers choosing repair over replacement may benefit from an extension of the product's liability period [52].

3.2.2 Digital Solutions for Repair Services

1. *European online repair platform*: a new digital platform will be established as an extension of the "Your Europe" portal. This platform is designed to facilitate the search for a repairer, to provide a unified online interface that links national repair service offerings, and to enhance transparency and accessibility for consumers looking for repair solutions.

Table 1 Companies repair initiatives

Example	Key features/initiatives	Notes
Miele	Family-owned with a long history (since 1899); Emphasis on durable design, repairability, and sustainability; Operates a central spare parts warehouse ensuring availability for up to 15 years post-production	Distinguished for its holistic sustainability approach, including CO_2 neutrality and long-term product testing
Philips (Online Repair Support)	Provides an online troubleshooting guide and repair request system; Offers step-by-step diagnostics and repair service requests via their website	Demonstrates a consumer-friendly digital approach to facilitating repairs
Servicemanuals.net	An online resource offering repair manuals and technical documentation for various products	Serves as a reference platform to support repair professionals and DIY repair enthusiasts
Electrical appliance repair training course	A specialized training course (e.g., 7-day course costing around £795+ VAT) aimed at enhancing repair and servicing skills in the consumer electrical appliance sector	Highlights the importance of skill development in supporting a sustainable repair ecosystem
Nudie Jeans	Offers lifetime repair on every pair of jeans. https://www.nudiejeans.com/	Aligned with the movement: Fixing Fashing. https://fixing.fashion/
Biochef	Designs appliances to last, with some having a lifetime warranty Replacement parts are also available to reduce waste and ensure continued use of working parts	Appliances are made in a way that makes repair and service easy
Go Green	Each Go Green school lunch box comes with a reusable gel freezer pack, all of which are leached and lead free. https://www.biome.com.au/	They offer replacement parts to increase the lunchbox's lifetime

(*continued*)

Table 1 (*continued*)

Example	Key features/initiatives	Notes
Patagonia	It will repair all of its gear covered by its Ironclad Guarantee free of charge. If not covered by this guarantee, it will still repair damage due to wear and tear for a reasonable amount	It has Do-It-Yourself Repair Tutorials that provide step-by-step instructions for repairing it yourself
Citizen wolf	They only make what they sell. They have zero inventory, which means zero waste They will repair anything they make for free, for life. The longer customers wear the clothes they already own, the better it is for the planet. Customers only pay for shipping	They also recycle customers' old clothes into new fabrics. It's a radically simple idea to transform the wildly destructive fashion industry https://www.citizenwolf.com/
Moulinex Repair Centre	They will repair appliances free of charge if the good is under warranty. They will provide a list of authorized repairers. If customers prefer to repair the products themselves, they will provide online help	

Source: Own data research

2. *European Repair information form*: this form allows repair service providers to present standardized information regarding their repair services. Key aspects include the detailed breakdown of repair processes, pricing, and estimated time frames, and the guarantee that information remains valid for 30 days, ensuring consumers can make informed comparisons.

3.2.3 Benefits and Challenges

Benefits These benefits collectively contribute to a more sustainable and consumer-friendly re-pair ecosystem. By extending product lifecycles and reducing waste, repairability supports environmental goals while also offering economic advantages for consumers. Enhanced consumer rights and improved access to information foster greater trust in repair services, encouraging more people to opt for repairs instead of replacements. Additionally, the integration of digital platforms and standardized repair information enhances accessibility and efficiency, making the repair process more transparent, convenient, and widely adopted.

Challenges These challenges highlight the complexities of implementing effective repair initiatives within the circular economy framework. Ensuring enforcement and compliance across a diverse range of products requires regulatory oversight and industry cooperation, which can be difficult to achieve. Additionally, manufacturers may face higher costs when redesigning products to meet repairability standards, potentially impacting pricing and business models. Lastly, consumer awareness and education play a crucial role in maximizing the benefits of repair services. Without sufficient knowledge and engagement, even the most well-structured policies and platforms may struggle to drive meaningful changes in consumer behaviour.

3.3 Offering Repair Courses: Acquiring Competences and Skills

3.3.1 Specialized Training Programmes

Repair courses are available to enhance technical skills and competences in servicing and repairing consumer products.

Example: Electrical Appliance Repair Training Course.

This is a structured 7-day training program designed to teach essential repair skills and is focused on practical knowledge and hands-on experience.

3.3.2 Importance of Repair Education

Repair education plays a vital role in supporting the circular economy by equipping individuals with the skills needed to extend product lifecycles and reduce waste. Training programs help address the growing shortage of skilled repair technicians across various industries, ensuring that repair services remain accessible and efficient. By acquiring repair knowledge, both professional service providers and DIY enthusiasts can actively contribute to sustainable consumption practices. Additionally, online platforms and technical resources, such as repair manuals and instructional guides, further enhance learning opportunities, enabling continuous skill development and fostering a more resilient repair ecosystem.

3.3.3 Role of Online and Technical Resources

Online and technical resources play a crucial role in supporting repair education and skill development. Platforms like Servicemanuals.net offer access to repair manuals and technical documentation for a wide range of products, making valuable information readily available to both professionals and DIY repair enthusiasts. These resources are essential for self-learning, allowing individuals to develop and refine their repair skills at their own pace. By facilitating continuous skill development, online platforms contribute to a more knowledgeable and capable repair workforce, ultimately strengthening the circular economy and promoting sustainable consumption practices.

3.3.4 Repair in One of the Most Polluting Industries: Clothing

The clothing and textile industry is widely recognized as one of the most environmentally damaging sectors due to its intensive resource consumption, pollution, and waste generation. Fast fashion - defined by rapid production cycles, short product lifespans, and low-cost clothing - intensifies these environmental issues, contributing significantly to global carbon emissions and water pollution [61]. In recent years, the European Union has demanded to address these challenges through legislative and policy measures that encourage sustainability, with repair emerging as a key strategy for extending garment lifecycles and reducing waste [62].

Repair initiatives within the clothing industry can disrupt the prevailing linear model by extending the use phase of garments, reducing overall waste, and promoting a more circular economic model [8]. Historically, clothing repair has been undervalued as consumer habits and low-cost production methods led to frequent disposal of items rather than their maintenance. However, current EU policies - such as proposals under the Circular Economy Action Plan and emerging guidelines within the Textile Strategy for Sustainable and Circular Textiles - aim to encourage the design for durability and reparability. These initiatives are intended to promote extended producer responsibility and encourage the integration of repair services within business models [61].

Despite these promising initiatives, several challenges prevent the widespread adoption of repair practices in the fashion sector. Many brands continue to rely on business models that favour continuous production and rapid consumption, making it difficult to justify investments in repair infrastructure. Consumer perceptions also play a role; repairing garments is sometimes viewed as inconvenient or less desirable than purchasing new products. Moreover, there is a notable decline in the availability of skilled labour trained in textile repair, further preventing the scalability of repair services.

Nonetheless, opportunities are emerging as both industry leaders and policymakers begin to reframe the narrative around repair. Prominent companies like Patagonia and Nudie Jeans have pioneered repair programs - offering free or low-cost mending services to extend product lifespans and reduce environmental impacts. Additionally, digital platforms and localized tailoring services are making it easier for consumers to access repair solutions, thereby supporting a shift towards a more sustainable consumption model [63]. With continued support from regulatory frameworks and consumer advocacy, the integration of repair within the clothing industry could serve as a critical component in the transition to a circular economy.

4 Conclusions

This study emphasizes that incorporating repair as a fundamental element of the circular economy not only addresses environmental and economic challenges but also creates new opportunities for industry transformation. By analysing the European Union's legal and regulatory frameworks, the research demonstrates that initiatives such as the Right to Repair Directive are crucial for reshaping consumer culture and manufacturing practices. These initiatives promote product durability by instructing repairability and ensuring that manufacturers provide access to spare parts and essential repair information, which, in turn, extends the lifespan of products and supports the transition away from a disposable economic model.

Furthermore, the study reveals that achieving a sustainable, repair-oriented circular economy demands a holistic approach involving governments, businesses, and consumers. The integration of digital tools, like the European Online Repair Platform and standardized repair information forms, enhances transparency and accessibility, making repair services more viable and appealing. Additionally, the focus on repair education and specialized training programs highlights the importance of building a skilled workforce capable of supporting sustainable consumption practices. While challenges such as increased operational costs, intellectual property issues, and the need for a significant shift in consumer behaviour remain, the research concludes that these obstacles can be overcome with coordinated efforts. This would not only reduce environmental degradation and resource consumption but also foster long-term economic and social resilience.

The current research provides some examples of companies' repair initiatives. More examples are desirable to be considered for benchmarking best practices.

Future research should concentrate on evaluating the long-term impacts of EU legislative initiatives on industry practices, consumer behaviour, and environmental outcomes. It would be beneficial to conduct longitudinal studies that track the effectiveness of policies like the Right to Repair Directive over time, providing insights into how these measures influence repair practices and product lifespans. Additionally, further investigation is needed into how repair initiatives vary across different sectors such as electronics, household appliances, and automotive industries. This approach would help identify the unique challenges and benefits faced by each sector and contribute to more targeted policy recommendations. Another aspect of the repair initiative that deserves attention in future research is the paradox that in some cases it is better to not repair appliances, but instead replace them with newer, more energy-efficient models, so that the overall environmental impact measured over a couple of years will be lower. To tackle this apparent paradox, it is helpful to focus in further research on the different aspects of the life cycle assessment (LCA) of products, and to calculate the environmental impact of production, energy use during its lifetime and impact of end-of-life of products.

Research should also explore the potential of emerging technologies, including artificial intelligence, the Internet of Things, and blockchain, so as to enhance repair services. These technologies could improve traceability, transparency, and security in the repair process, offering innovative solutions to current operational challenges. Moreover, there is a need to understand better the barriers faced by independent repairers and small businesses, such as issues related to intellectual property and access to spare parts. In addition, future studies could delve deeper into consumer behaviour and cultural shifts, analysing the factors that drive the decision to repair rather than replace products. This deeper understanding could inform educational campaigns and incentive structures, ultimately accelerating the shift towards a more sustainable, circular economy.

References

1. Andersen, M.S.: An introductory note on the environmental economics of the circular economy. Sustain. Sci. **2**, 133–140 (2007)
2. Balkenende, R., Bocken, N., Bakker, C.: Design for the circular economy. In: Egenhoefer, R.B. (ed.) The Routledge Handbook of Sustainable Design, pp. 1–19. Routledge, London, UK (2017)
3. Bastein, O.: The role of biodiversity in supporting ecosystem services in Natura 2000 sites. Ecol. Indic. **24**, 12–22 (2013)

4. Bradley, K., Persson, O.: Community Repair in the Circular Economy—Fixing More Than Stuff, Local Environment, vol. 27(10–11), pp. 1321–1337. Taylor & Francis Online (2022)
5. Cayzer, S., Griffiths, P., Beghetto, V.: Design of indicators for measuring product performance in the circular economy. Int. J. Sustain. Eng. **10**, 289–229 (2017)
6. Circularity Gap Report: The Circle Economy. Retrieved from Circularity Gap Report (2019)
7. Costa, B., Rodrigues, S., Moreno, P.: Sustainable tourism and the circular economy—a theoretical overview. In: Mapping, Managing, and Crafting Sustainable Business Strategies for the Circular Economy. IGI Publications, Publisher of Timely Knowledge (2020)
8. Diddi, S., Yan, R.-N.: Consumer perceptions related to clothing repair and community mending events: a circular economy perspective. Sustainability. **11**, 5306 (2019)
9. Ellen MacArthur Foundation: Towards the Circular Economy. Vol. 1: Economic and Business Rationale for a Circular Economy. Ellen MacArthur Foundation (2012)
10. Ellen MacArthur Foundation: Towards the Circular Economy. Vol. 2: Opportunities for the Consumer Goods Sector. Ellen MacArthur Foundation (2013)
11. European Commission: The European Union Waste Framework Directive (2008). https://eur-lex.europa.eu/legal-content/EN/TXT/?uri=celex%3A32008L0098
12. European Commission: Towards a circular economy: a zero waste programme for Europe. In: Communication from the Commission to the European Parliament. The Council, the European Economic and Social Committee and the Committee of the Regions, Brussels (2014)
13. European Commission: Sustainable development in the European Union. 2015 Monitoring Report of the EU Sustainable Development Strategy (2015)
14. Feng, Z., Yan, N.: Putting a circular economy into practice in China. Sustain. Sci. **2**, 95–101 (2007)
15. Geissdoerfer, M., Savaget, P., Bocken, N., Hultink, E.: The circular economy—a new sustainability paradigm? J. Clean. Prod. **143**, 757–768 (2017)
16. Geng, Y., Doberstein, B.: Developing the circular economy in China: challenges and opportunities for achieving "leapfrog development". Int. J. Sustain. Dev. World Ecol. **15**, 231–239 (2008)
17. Giurco, D., Littleboy, A., Boyle, T., Fyfe, J., White, S.: Circular economy: questions for responsible minerals, additive manufacturing and recycling of metals. Resources. **3**, 432–453 (2014)
18. Jackson, T.: Prosperity Without Growth. Economics for a Finite Planet. Earthscan, London, New York (2009)
19. King, A., Burgess, S., Ijomah, W., Mamahon, C., King, A.: Reducing waste: repair, recondition, remanufacture or recycle? Sustain. Dev. **14**, 257–267 (2006)
20. Kirchherr, J., Reike, D., Hekkert, M.: Conceptualizing the circular economy: an analysis of 114 definitions. Resour. Conserv. **127**, 221–232 (2017)
21. Lacy, P., Long, J., Spindler, W.: The Circular Economy Handbook. Realizing the Circular Advantage. Palgrave MacMillan, England (2020)
22. Lacy, P., Rutqvist, L.: Waste and Wealth—The Circular Economy Advantage. Palgrave MacMillan, England (2015)
23. Lett, A.: Las amenazas globales, el reciclage de resíduos y el concepto de economia circular. Revista Argentina de Microbiologia. **46**(1), 1–2 (2014)
24. McLaren, P.: In a broken world: towards an ethics of repair in the Anthropocene. Anthropoc. Rev. **5**(2), 136–154 (2018)
25. Markard, J., Raven, R., Truffer, B.: Sustainability transitions: an emerging field of research and its prospects. Res. Policy. **41**, 955–967 (2012)
26. Meadows, H., Randers, J., Meadows, L.: The Limits to Growth: The 30-Year Update. Routledge, London (2004)
27. Mesa, J., Esparragoza, I., Maury, H.: Developing a set of sustainability indicators for product based on the circular economy model. J. Clean. Prod. **196**, 1429–1442 (2018)

28. Ness, D.: Sustainable urban infrastructure in China: towards a factor 10 improvement in resource productivity through integrated infrastructure system. Int. J. Sustain. Dev. World Ecol. **15**, 228–301 (2008)
29. Pearce, D., Turner, R.: Economics of Natural Resources and the Environment. The John Hopkins University Press, Baltimore (1990)
30. Potting, J., Hekkert, M., Worrel, E., Hanemaaaijer, A. Circular Economy: Measuring Innovation in the Product Chain. (2017)
31. Preston, F.: A Global Redesign: Shaping the Circular Economy. London, Chattam House: The Royal Institute of International Affairs (2012)
32. Reh, L.: Process engineering in circular economy. Particuology. **11**(2), 119–133 (2013)
33. Ren, Y.: The circular economy in China. J. Mater Cycles Waste Manag. **9**, 121–129 (2007)
34. Riisgaard, H., Mosgaard, M., Zacho, K.O.: Local circles in a circular economy—the case of smartphone repair in Denmark. EJSD. **5**, 109 (2016)
35. Rockström, J., Steffen, W., Noone, K., Persson, A., Chapin, F., Lambin, E., Lenton, T., Scheffer, M., Folke, C., Schellnhuber, H., Nykvist, B., De Wit, C.A., Hughes, T., van der Leeuw, S., Rodhe, H., Sörlin, S., Snyder, P., Costanza, R., Svedin, U., Falke, M.: Planetary boundaries: exploring the safe operating space for humanity. Ecol. Soc. **14** (2009)
36. Rodríguez, C., Florido, C., Jacob, M.: Circular economy contributions to the tourism sector. Sustainability. **12**, 4338 (2020)
37. Saidani, M., Kendall, A., Yannou, B., Leroy, Y., Cluzel, F.: Closing the loop on platinum from catalytic converters: contributions from material flow analysis and circularity indicators. J. Ind. Ecol. 1143–1158 (2019)
38. Sakai, S., Yoshida, H., Hirai, Y., Asari, M., Takigami, H., Takahashi, S., Tomoda, K., Peeler, V., Wejchert, J., Schmid-Unterseh, T., Douvan,.R., Hathaway, R., Hylander, D., Fischer, C., Oh, J., Jinhui, L., Chi, K. International comparative study of 3R and waste management policy developments. J. Mater. Cycles Waste Manag. 13:86–102 (2011)
39. Seiffert, M., Loch, C.: Systemic thinking in environmental management: support for sustainable development. J. Clean. Prod. **13**(12), 1197–1202 (2005)
40. Sihvonen, S., Ritola, T.: Conceptualizing ReX for aggregating end-of-life strategies in product development. Proc. CIRP. **29**, 639–644 (2015)
41. Stahel, W.: The Performance Economy, 2nd edn. Palgrave Macmillan, Basingstoke, New York, USA (2010)
42. Su, B. Heshmati, A., Geng, Y., Yu, X. A review of the circular economy in China: moving from rhetoric to implementation. J. Clean. Prod. 42 (0), 215–227 (2013)
43. Thierry, M., Salomon, M., Van Nunen, J., Van Wassenhove, L.: Strategic issues in product recovery management. Calif. Manag. Rev. **37**(2), 114–135 (1995)
44. Van Buren, N., Demmers, M., van der Heijden, R., Witlox, F.: Towards a circular economy: the role of Dutch logistics industries and governments. Sustainability. **8**, 647 (2016)
45. World Business Council for Sustainable Development. Eco-efficiency learning module (1992). www.wbcsd.org
46. World Wild Fund for Nature (WWFN): Living Planet Report. Global Footprint Network (2014) https://www.footprintnetwork.org/content/images/article_uploads/Living_Planet_Report_2014.pdf
47. Yan, J., Wu, N.: Technology supporting system of circular economy of mining cities. In: 2011 Asia-Pacific Power and Energy Engineering Conference, pp. 1–5. IEEE (2011)
48. Yang, S., Feng, N.: A case study of industrial symbiosis: Nanning sugar co., ltd in China. Resour. Conserv. Recycl. **52**, 813–820 (2008)
49. Svensson-Hoglund, S., Richter, J.L., Maitre-Ekern, E., Russell, J.D., Pihlajarinne, T., Dalhammar, C.: Barriers, enablers and market governance: a review of the policy landscape for repair of consumer electronics in the EU and the U.S. J. Clean. Prod. **276**, 124228 (2020)

50. Maitre-Ekern, E., Dalhammar, C.: Towards a hierarchy of consumption strategies in the circular economy. Maastricht J. Eur. Compar. Law. **26**(3), 394–420 (2019)
51. Nascimento, D.L., Alencastro, V., Quelhas, O.L., Caiado, R.G., Garza-Reyes, J.A., Rocha-Lona, L., Tortorella, G.: Exploring industry 4.0 technologies to enable circular economy practices in a manufacturing context. J. Manuf. Technol. Manag. **30**(3), 607–627 (2019)
52. European Commission: Proposal for a Directive on Common Rules Promoting the Repair of Goods, p. 155 final. COM(2023) (2023)
53. Zhu, Q., Geng, Y., Lai, K.: Circular economy practices among Chinese manufacturers varying in environmental-oriented supply chain cooperation and the performance implications. J. Environ. Manag. **91**(6), 1324–1331 (2010)
54. Zhu, L., Zhou, J., Cui, Z., Liu, L.: A method for controlling enterprises access to an eco-industrial park. Sci. Total Environ. **408**(20), 4817–4825 (2010)
55. Ghisellini, P., Cialani, C., Ulgiati, S.: A review on circular economy: the expected transition to a balanced interplay of environmental and economic systems. J. Clean. Prod. **114**, 11–32 (2016). https://doi.org/10.1016/j.jclepro.2015.09.007
56. Stahel, W.R.: The circular economy. Nature. **531**(7595), 435–438 (2016). https://doi.org/10.1038/531435a
57. Jaeger-Erben, M., Jensen, C., Hofmann, F., Zwiers, J.: There is no sustainable circular economy without a circular society. Resour. Conserv. Recycl. **168**, 105476 (2021). https://doi.org/10.1016/j.resconrec.2021.105476
58. Türkeli, S., Huang, B., Stasik, A., Kemp, R.: Circular economy as a global business activity: mobile phone repair in the Netherlands, Poland and China. Energies. **12**(3), 498 (2019)
59. Van der Velden, M.: Community repair and a sustainable circular economy. J. Cleaner Prod. **304**, 127151 (2021). https://doi.org/10.1016/j.jclepro.2021.127151
60. Steffen, W., Sanderson, A., Tyson, P., Jäger, J., Matson, P., Moore, B., Oldfield, F., Richardson, K., Schellnhuber, H., Turner, B., Wasson, R.: Global Change and the Earth System. A Planet Under Pressure. Springer, Berlin Heidelberg New York (2004)
61. Ellen MacArthur Foundation: Fashion and the Circular Economy. Ellen MacArthur Foundation (2017)
62. European Commission. (2021). https://ec.europa.eu/info/strategy/priorities-2019-2024/european-green-deal/delivering-european-green-deal_en
63. Bocken, N.M., De Pauw, I., Bakker, C., Van Der Grinten, B.: Product design and business model strategies for a circular economy. J. Ind. Prod. Eng. **33**(5), 308–320 (2016)

Proportional Cox Model and Its Potential Application to Predictive Maintenance in Industry 5.0

Jorge Siopa[1,2](✉), Pedro Custódio[1,2], Joana F. A. Valente[2], João Marcelino[2], Diogo Montalvão[3], and João Pombo[4]

[1] School of Technology and Management, Polytechnic of Leiria, Leiria, Portugal
jorge.siopa@ipleiria.pt
[2] Centre of Rapid and Sustainable Product Development, Polytechnic of Leiria, Leiria, Marinha Grande, Portugal
[3] Department of Design and Engineering, Bournemouth University, Poole, UK
[4] Institute of Railway Research, University of Huddersfield, Huddersfield, UK

Abstract. Industry 5.0, also known as the Fifth Industrial Revolution, integrates advanced technology and AI, aiming beyond efficiency and productivity, to reinforce the role and the contribution of industry to society. This study explores the implementation and benefits of predictive maintenance within the framework of Industry 5.0. Utilising multivariate reliability models, particularly Cox Proportional Hazard Models, the time to failure of equipment is evaluated and the impact of several explanatory variables on survival probability is analysed. The results demonstrate significant improvements in operational efficiency and equipment reliability. This work outlines the methodology and key findings, providing a comprehensive overview of predictive maintenance as a relevant tool to increase the life-cycle of the assets, minimise in-service failures and reduce maintenance costs, supporting the development of Industry 5.0.

Keywords: Predictive maintenance · Industry 5.0 · Cox proportional hazard models · Reliability · Operational efficiency

1 Introduction

Industry 5.0 is a broad concept that provides a vision of industry that places the wellbeing of workers at the centre of the production process and uses new technologies to provide to whole society, while respecting the environment and the production limits of the planet [1]. It marks a paradigm shift toward enhanced collaboration between humans and machines, aiming to foster smarter and more efficient production environments. It goes beyond the automation and connectivity of Industry 4.0 by emphasizing the integration of human intelligence, creativity, and decision-making with advanced technologies such as artificial intelligence (AI), the Internet of Things (IoT), and robotics. This shift aims to create production systems that are not only highly efficient and customizable but also sustainable and human-centric. Predictive maintenance plays a pivotal role in this new

© The Author(s), under exclusive license to Springer Nature Switzerland AG 2027
M. Calvete Gaspar et al. (Eds.): Regional Helix 2024, LNME, pp. 111–124, 2027.
https://doi.org/10.1007/978-3-032-36365-7_8

industrial revolution by leveraging advanced technologies to forecast equipment failures, optimize maintenance processes, and reduce downtime—ultimately enhancing overall productivity and cost-efficiency.

The goal of maintenance practices aligned with Industry 5.0 is not only to improve the accuracy of failure prediction but also to reduce the environmental impact of maintenance activities themselves. Unplanned repairs, for instance, often require urgent interventions, expedited delivery of parts, and potentially resource-intensive fixes, all of which contribute to higher emissions and waste. By adopting more predictive and condition-based maintenance strategies, organizations can significantly reduce unnecessary interventions and associated resource consumption. Furthermore, sustainable maintenance practices support the broader goals of Industry 5.0 by promoting the circular economy through extended equipment lifespans and by prioritizing repairs or replacements only when genuinely necessary.

Within this context, the development and implementation of effective maintenance strategies become essential. One of the most widely used preventive maintenance policies is the age replacement strategy, which involves replacing a system either upon failure or at a predetermined age, whichever occurs first. This approach is intended to prevent unexpected failures and mitigate the risks associated with unplanned downtime. However, most current age-based replacement policies fail to consider actual operating conditions, which can lead to suboptimal maintenance schedules and increased costs. Factors such as environmental conditions, usage patterns, wear and tear, and operator skill levels can significantly influence a system's actual lifetime. Consequently, traditional age-based policies may not be adequate for dynamically changing production environments.

This research focuses on enhancing the effectiveness of preventive maintenance by optimising the timing of such interventions. Specifically, we aim to estimate the average cost per unit of time associated with performing preventive maintenance at a given time point. The primary objective is to determine the optimal replacement time that minimizes the expected maintenance cost while accounting for the various factors that affect system performance and reliability.

The Cox proportional hazards model, a well-established statistical tool in survival analysis, is particularly suitable in this context. It enables the estimation of how different covariates impact the hazard rate, which reflects the instantaneous risk of failure. By incorporating variables related to operating conditions—such as environmental factors, usage intensity, and other relevant parameters—the Cox model provides a more accurate assessment of a system's failure risk over time. This allows for the development of more refined and context-aware maintenance policies tailored to the specific conditions in which a system operates.

This approach supports the shift toward more dynamic and adaptive maintenance strategies in line with Industry 5.0 principles, where human insight and advanced predictive technologies collaborate to improve the reliability and efficiency of production systems. The integration of data-driven models into maintenance decision-making processes not only helps to reduce costs but also extends equipment lifespans, reduces waste, and promotes sustainable manufacturing practices. Ultimately, the goal is to optimize preventive maintenance by determining the optimal intervention time that minimizes expected maintenance costs while considering real-world operating conditions and system-specific characteristics. This strategy not only improves the efficiency and effectiveness of maintenance operations but also aligns with the overarching vision of Industry 5.0, where the synergy between humans, machines, and data drives innovation and sustainable growth.

This convergence of technological advancement and human oversight represents a significant evolution in industrial practices, marking a shift from purely efficiency-driven production to a model that emphasizes sustainability, adaptability, and ethical considerations as central elements of manufacturing. By placing equal importance on human factors and technological progress, Industry 5.0 offers a framework that transcends production outputs to generate a holistic impact on societal progress, environmental stewardship, and the evolving role of the workforce in shaping the industry's future. Through intelligent maintenance policies, data-rich systems, and adaptive scheduling guided by real-time analytics, Industry 5.0 is poised to redefine industrial standards for efficiency, sustainability, and resilience.

System failures can have substantial consequences for society. Due to technical constraints, ensuring system reliability during the design phase is often difficult. Maintenance plays a critical role in preserving reliability, with the two most common types being preventive and corrective maintenance. However, performing preventive maintenance either too frequently or too sporadically can be inefficient. Consequently, various maintenance strategies have been developed, including age replacement policies and condition-based maintenance.

In 1960, Barlow and Hunter [2] introduced and examined fundamental age replacement policies known as Policy I and Policy II. In Policy I, preventive maintenance is performed after T hours of continuous operation, provided the system has not failed beforehand. If a failure occurs before T, an emergency replacement is conducted, and the scheduled maintenance is deferred. In Policy II, preventive maintenance takes place once the system accumulates a total of T operating hours, regardless of how many failures have occurred in the meantime. Under this policy, only minimal repairs are carried out after each failure, maintaining the system's original failure rate. The authors derived the optimal maintenance interval, T^*, to maximize system availability. Building on these models, Makabe and Monirmura [5] proposed Policy III, which is considered simpler and more practical than Policy II. In this model, the system is replaced after the k-th failure, with only minimal repairs conducted for the first $k-1$ failures. The optimal age replacement policy was analysed in terms of efficiency and maintenance cost rate.

In recent years, several prominent researchers have further refined age replacement policies. In 1975, Tahara and Nishida [3] introduced a maintenance policy in which a unit is replaced either after the first failure occurring after t_0 hours of operation or when the total operating time reaches T_0 $(0 \leq t_0 \leq T_0)$, whichever happens first. Failures occurring within the interval $]0, t_0[$ are addressed with minimal repairs. Notably, if $t_0 = 0$, this model simplifies to the basic age replacement policy.

Later, Segawa, Ohnishi, and Ibaraki (1992) [4] research an optimal age replacement policy in which a system is replaced either at time T or after N failures, whichever happens first, with minimal repairs performed between replacements. In this policy, the decision variables are T and N. If $N = 1$, the policy simplifies to the standard age replacement policy.

With advancements in technology and data analytics, maintenance strategies have undergone significant evolution in recent years. Condition-based maintenance, for example, has gained traction as it allows for more accurate predictions of system failures by continuously monitoring the health of equipment. By using sensors and data analysis, systems can be assessed in real time, and maintenance can be scheduled based on actual conditions rather than just time or failure-based criteria. This approach not only improves system reliability but also reduces unnecessary maintenance, leading to cost savings and improved operational efficiency. Moreover, the integration of artificial intelligence (AI) and machine learning (ML) into maintenance decision-making has opened new avenues for optimising maintenance strategies. AI algorithms can analyse large datasets to predict failure patterns, optimise maintenance schedules, and even recommend corrective actions. This shift from traditional maintenance strategies to AI-driven predictive maintenance is transforming industries, making maintenance operations more proactive and reducing the likelihood of unexpected system failures, which can be costly and disruptive.

2 Methodology

In this study, a predictive maintenance methodology based on Cox Proportional Hazard Models and adaptative age replacement, [5–7] and [8], will be developed. It was analysed a dataset comprising condition monitoring data for various equipment, including operating times until failure or suspension and successive measurements of covariates over time. The Cox Models were used to assess how these covariates influence the survival probability of the equipment, which has a consequence delay or anticipate the optimal moment of preventive maintenance action.

In the Age Based replacement model taken into account in [5], the following assumptions were made. A preventive replacement performs after T hours of operation without failure, and the cost of preventive replacement is C_m. Failure of a system can be known immediately. If the system fails before T, an emergency replacement performs instantaneously with cost C_e, where $C_e > C_m$, and the preventive maintenance is rescheduled at the time. The time of replacement can be neglected, and the system is restored to a new state condition after each replacement. The lifetime distribution of a system is denoted by $F(t)$.

We denote the expected cost rate per unit time by $C(T)$:

$$C(T) = \frac{C_m\overline{F}(T) + C_e F(T)}{\int_0^T \overline{F}(t)dt} \tag{1}$$

where $\overline{F}(T) = 1 - F(t) = R(t)$. Moreover, note that $C_m\overline{F}(t) + C_e F(t)$ and $\int_0^T \overline{F}(t)dt$ are respectively the average cost and the average time for one replacement cycle.

Assuming the lifetime of a system follows a Weibull distribution, the reliability function, denoted by $R(t)$, is given as follows:

$$R(t) = \overline{F}(t) = e^{-\left(\frac{t}{\lambda}\right)^{\rho}} \tag{2}$$

In this case, the expected cost rate per unit time is given by

$$C(T) = \frac{C_m e^{-\left(\frac{t}{\lambda}\right)^{\rho}} + C_e\left(1 - e^{-\left(\frac{t}{\lambda}\right)^{\rho}}\right)}{\int_0^T e^{-\left(\frac{t}{\lambda}\right)^{\rho}} dt} \tag{3}$$

Given a certain data related to a machine functionality, we aim to minimise the function $C(T)$ and consequently find the optimal time instant to execute the predictive maintenance action.

Based on John Fox [7] and Jessica Ko [8], that studied the cox proportional-hazard regression for survival data, given a vector $X \in R^d$ of d features and a parameter $\beta \in R^d$, the hazard function is defined as:

$$h(t) = h_0(t) \times e^{\sum_{j=1}^{d} \beta_j X_j} = h_0(t) \times k \tag{4}$$

where β is the vector with the adjustment coefficients, X the set of variables and k is the proportional risk coefficient.

Doing $R(t) = e^{-\int_0^t h(t)dt}$ and applying the above equation we obtain the following result:

$$R(t) = e^{\int_0^t h_0(t) \times K dt} = R_0(t)^K \tag{5}$$

Denote as $C_K(T)$ the expected cost rate per unit time with proportional risk coefficient K, we are going to simply denote $C(T)$ when $K = 1$. From Eq. (5) we get the follow:

$$C_K(T) = \frac{C_m\left(e^{-\left(\frac{t}{\lambda}\right)^{\rho}}\right)^K + C_e\left(1 - \left(e^{-\left(\frac{t}{\lambda}\right)^{\rho}}\right)^K\right)}{\int_0^T \left(e^{-\left(\frac{t}{\lambda}\right)^{\rho}}\right)^K dt} \tag{6}$$

3 Results

In this section the functions $C_K(T)$ will be minimized and then analysed for some distinct values of the proportional risk coefficient K. Then, the results will be compared with the obtained ones with function $C(T)$.

Assume without loss of generality the `data` variable is a matrix with columns `duration`, `event`, `var1`, `var2` and `var3`. Note that duration is the column given in a specific data set`data`. In Fig. 1 it is possible to analyse the data structure of the data used on the described procedures during this paper.

	ID	stop	var1	var2	var3	event	start
0	2	252.0	0.19	0.18	0.71	False	0
1	2	522.0	0.09	0.13	0.19	False	252
2	2	751.0	0.08	0.01	0.45	False	522
3	2	989.0	0.18	0.03	0.23	False	751
4	2	1269.0	0.08	0.01	0.39	False	989
...	...	...	...	...	...	...	...
2246	299	5520.0	0.29	0.15	0.57	False	5235
2247	299	5773.0	0.3	0.05	0.82	False	5520
2248	299	5997.0	0.34	0.08	0.84	False	5773
2249	299	6250.0	0.25	0.18	0.92	False	5997
2250	299	6306.0	0.2	0.18	0.81	False	6250

[2251 rows x 7 columns]

Fig. 1. Data set `data`.

Now, using the data format previously described, it is going to be estimate the parameters λ and ρ from the Weibull distribution using `WeibullFitter` and `COXPHFITTER` from library `lifelines` [9]. The `WeibullFitter` class implements a Weibull model for univariate data. The model has the parameterised form:

$$R(t) = \exp(-(t/\lambda) \wedge \rho), \lambda > 0, \rho > 0 \tag{7}$$

The λ scale parameter represents the time when 63.2% of the population has died. The greater the value of λ sooner the failures tend to occur. The ρ shape parameter controls if the cumulative hazard is convex or concave, representing accelerating or decelerating hazards.

The cumulative hazard rate is

$$H(t) = \left(\frac{t}{\lambda}\right)^{\rho} \tag{8}$$

And the hazard rate is

$$h(t) = \frac{\rho}{\lambda}\left(\frac{t}{\lambda}\right)^{\rho-1} \tag{9}$$

After calling the `.fit` method, we have access to properties like `cumulative_hazard_`, `survival_function`, `lambda_` and `rho_`. A summary of the fit is available with the method `print_summary()`, which output for the used data can be seen in the following Fig. 2.

For the collected data it was obtained $\lambda = 11692$ and $\rho = 4$. The reliability of Weibull distribution for those parameters can be algebraically represented as follows:

$$R(t) = \exp\left(-\left(\frac{t}{11692}\right)^{4}\right) \tag{10}$$

The Weibull distribution density function for same parameters can be algebraically represented as follows:

$$f(t) = -\frac{\partial R(t)}{\partial t} = \frac{4}{11692}\left(\frac{t}{11692}\right)^{3} \exp\left(-\left(\frac{t}{11692}\right)^{4}\right) \tag{11}$$

```
                 coef   se(coef)  coef lower 95%  coef upper 95%
lambda_  11691.749290 464.664953    10781.022717    12602.475864
rho_         4.007592   0.452566        3.120579        4.894606

                 z              p   -log2(p)
lambda_  25.159525  1.111699e-139 461.595239
rho_      6.645642   3.018987e-11  34.947145
```

Fig. 2. Output Weibull estimation.

In Fig. 3, it is possible to observe its graphical representation.

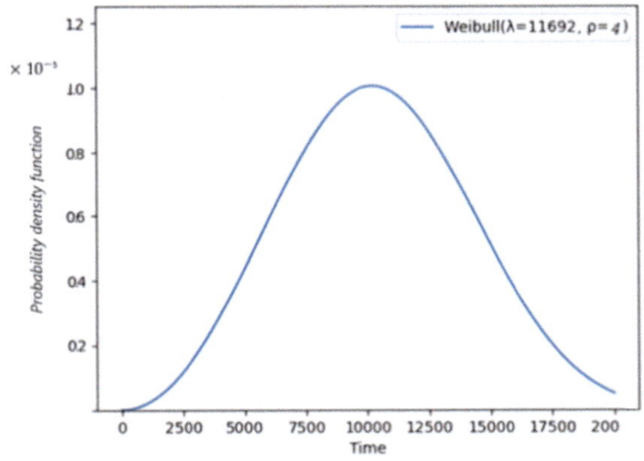

Fig. 3. Output Weibull distribution with $\lambda = 11692$ and $\rho = 4$.

Afterwards, it was applied the `CoxTimeVaryingFitter` class, which fits the Cox's time varying proportional hazard model:

$$h(t|x(t)) = h_0(t) \exp\big((x(t) - \overline{x})'\beta\big) \tag{12}$$

where β is the level satisfying the confidence intervals.

In the following Fig. 4 we can observe the output from implementation of `fit` and `print_summary` functions from the `CoxTimeVaryingFitter` class.

It might be to select the proportional risk coefficient as 0.8 and 1.2 as we are going to see in the following section.

Assuming that $C_m = 1$ and $C_e = 10$, applying the values calculated previously into the expected cost rate per unit time function and doing $K = 1$, we obtain the following equation that algebraically defines the function $C(T)$.

$$C(T) = \frac{e^{-\left(\frac{t}{11692}\right)^4} + 10\left(1 - e^{-\left(\frac{t}{11692}\right)^4}\right)}{\int_0^T e^{-\left(\frac{t}{11692}\right)^4} dt} \tag{13}$$

Thus, the graphical representation of the dimensionless expected cost rate per unit time, $C(T)$ is as follows (Fig. 5).

model	lifelines.CoxTimeVaryingFitter
event col	'event'
penalizer	0.1
number of subjects	70
number of periods	2251
number of events	41
partial log-likelihood	-118.51
time fit was run	2024-07-23 09:32:12 UTC

	coef	exp(coef)	se(coef)	coef lower 95%	coef upper 95%	exp(coef) lower 95%	exp(coef) upper 95%	cmp to	z	p	-log2(p)
var1	0.18	1.20	0.22	-0.24	0.61	0.79	1.84	0.00	0.85	0.40	1.33
var2	0.23	1.25	0.22	-0.20	0.65	0.82	1.92	0.00	1.03	0.30	1.73
var3	0.03	1.03	0.13	-0.23	0.29	0.79	1.34	0.00	0.21	0.84	0.26

Partial AIC	243.02
log-likelihood ratio test	1.82 on 3 df
-log2(p) of ll-ratio test	0.71

Fig. 4. Output of *CoxTimeVaryingFitter.*

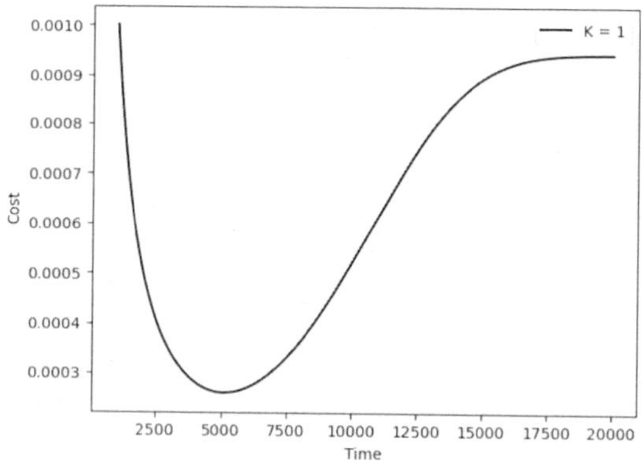

Fig. 5. Function $C(T)$.

3.1 Preventive Maintenance Time Optimisation and Variability Analysis

In this section it is going to be found the time that minimises the expected cost rate per unit time $C(T)$, which is the cost divided by the working time.

Using the mathematical bisection method of John Mo, Sherman C. P. Cheung and Raj Das [6] it was obtained $T^* = 5127$. Therefore, for this particular data set and taking into account the assumptions already done, if one pretend to minimise the average cost per unit time of preventive maintenance, it must be taken into account action at time $T^* = 5127$.

As mentioned before, it is useful to analyse the behaviour of expected cost rate per unit time when the proportional risk coefficient K changes. In Fig. 6, one can observe the functions $C(T)$, $C_{0,5}(T)$ and $C_{1,5}(T)$.

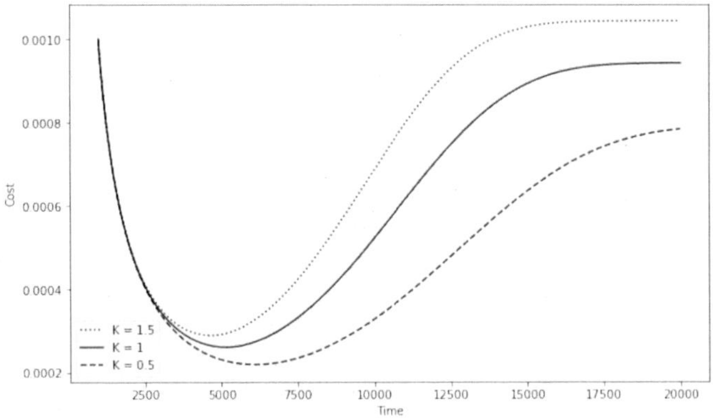

Fig. 6. Functions $C_K(T)$.

In Fig. 7 one can also observe several other functions $C_K(T)$ for some tested values K.

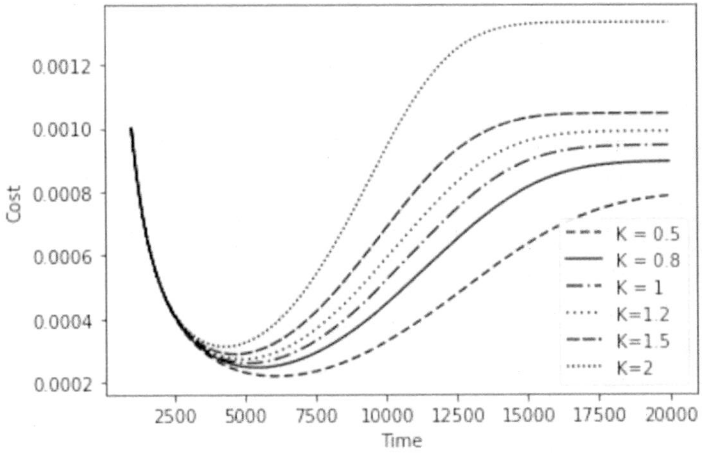

Fig. 7. Functions $C_K(T)$.

In Table 1, one can observe the optimal values obtained for each proportional risk coefficient for the average cost function per unit time. It can also be observed the Run to Failure value denoted as $C_K(\infty) = \lim_{T\to\infty} C_K(T)$, which means the expected cost rate per unit time when no preventive maintenance strategy is applied (Run to Failure). Moreover, the mean time between failures (MTBF) is denoted as $\lim_{T\to+\infty} \int_0^T R(t)dt$.

In Table 2, it can be observed a percentage comparation between the minimums, minimisers and Run to Failure value in the functions $C_K(T)$ for the considered values of K.

Table 1. Variability analysis.

K	0.5	0.8	1	1.2	1.5	2
T^*	6102	5422	5127	4901	4634	4273
$C_K(T^*)$	2.19e-4	2.46e-4	2.60e-4	2.72e-4	2.88e-4	3.15e-4
$C_K(\infty)$	7.93e-4	8.92e-4	9.43e-4	9.87e-4	1.04e-3	1.33e-3
MTBF	12610	11211	10604	10132	9615	7519

Table 2. Variability analysis in percentage.

K	0.5 (%)	0.8 (%)	1 (%)	1.2 (%)	1.5 (%)	2 (%)
T^*	119.0	105.8	100.0	95.6	90.4	83.3
$C(T^*)$	84.2	94.6	100.0	104.6	110.8	121.2
$C(\infty)$	305.0	343.1	362.7	379.6	400.0	511.5
MTBF	246.0	218.7	206.8	197.6	187.5	146.7

Analysing Tables 1 And 2 one can state that the higher the value of K then the higher its minimum $C(T^*)$ and the lower its minimiser T^*. Conversely, the lower the value of K then the lower its minimum $C(T^*)$ and the higher its minimiser T^*. As expected, these results accurately represent the scenario, since the higher the proportional risk coefficient of a piece of equipment, the sooner it will require maintenance. The other consequence is that the Run to Failure strategy cost $C(\infty)$ is in the interval [305%, 511%] of the median optimal cost $C_1(T^*)$ and *MTBF* is in the interval [146%, 246%] of T^*.

3.2 Individual Analysis of Equipment

In this section its going to be analysed the expected cost rate per unit time and the value K of equipment 11, 149 and 237. It was chosen three equipments between the 70 equipments with distinct k values to better understand what happens in each case. In Fig. 8 we can observe that this is a young piece of equipment that failed at the end of 17 inspections, with an age of 4030 and an increasing K with an average value of 1.0609. Thus, it serves as an example of one of the two pieces of equipment that cannot be preventively suspended either by Age Based Replacement or by applying Optimal Risk Replacement with Cox Proportional Hazard Models.

In Fig. 9 one can observe that it is a piece of equipment that failed at an Age of 12777, with an extremely increasing K at the end of its life but with an average value of 0.9942. Thus, it serves as an example of the 48 pieces of equipment that were preventively suspended both by Age-Based Replacement and by applying Optimal Risk Replacement with Cox Proportional Hazard Models.

In Fig. 10, one can observe that the equipment 237 has the highest K, with an average value of 1.1181, yet it has not failed after 47 inspections with an Age of 11191. Thus,

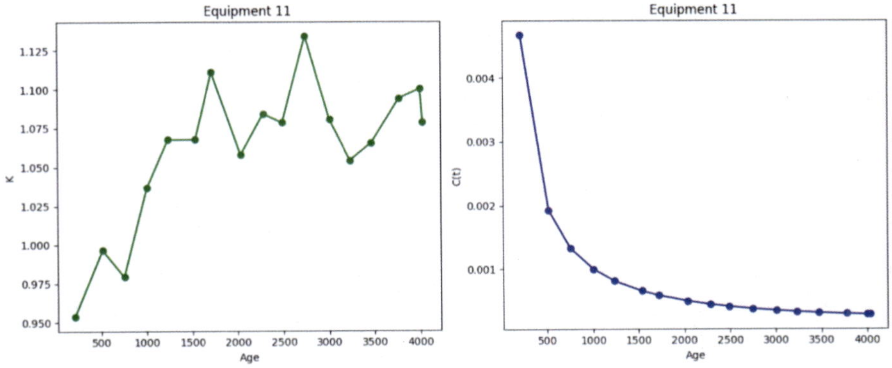

Fig. 8. Analysis of equipment 11, k value and cost time function.

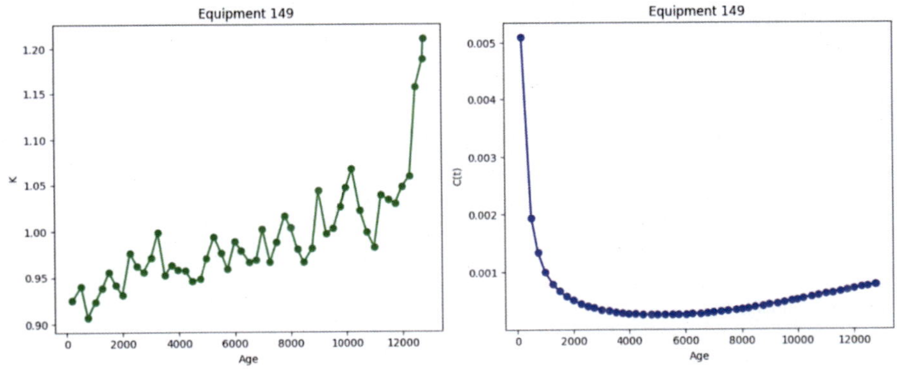

Fig. 9. Analysis of equipment 149, k value and cost time function.

it serves as an example of equipment that is preventively suspended both by Age-Based Replacement, at an age of 5127, and by applying Optimal Risk Replacement with Cox Proportional Hazard Models, at the 21st inspection with an age of 5266. However, it can be used for a much longer period.

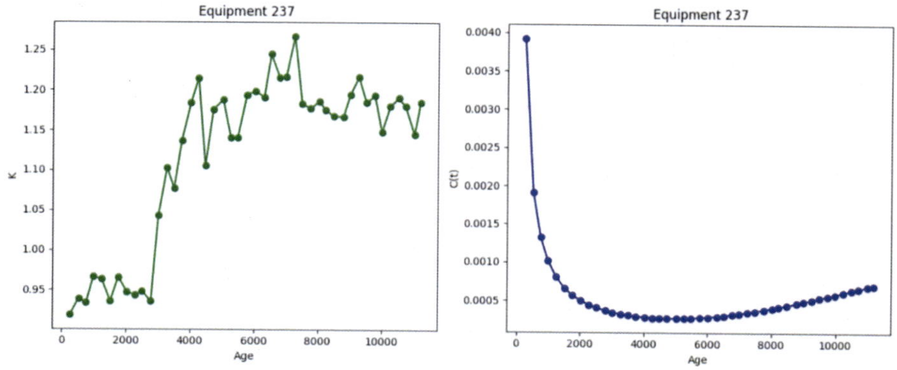

Fig. 10. Analysis of equipment 237, k value and cost time function.

3.3 Summary of Global Data Set Values

In this section it is going to be analysed the expected cost rate per unit time of the example data set in three maintenance policies: Run to Failure, Age Based Replacement (AGE) and Optimal Risk Replacement with Cox Proportional Hazard Models (PHM).

Table 3. Maintenance policy *(optimal replacement).*

	Run to Failure	AGE	PHM
Event type (number)	ES = 29	ES = 20	ES = 20
	EF = 41	OAF = 2	OPHMF = 2
		OAS = 48	OPHMS = 48
Work time	554 606	310 085	310 802
Total cost	439	88	88
	7.92E-04	2.92E-04	2.83E-04

In Table 3, one can observe several data values, including total cost, work time, and expected cost rate per unit time for the entire dataset. Based on this information, one should consider an age-based replacement policy because, if one disregard the inspection costs, the age-based cost (AGE = PHM = 88) is equal and the work time is slightly lower. Thus, the expected cost rate per unit time is higher under the age-based replacement policy.

Regarding the Run to Failure policy, it reveals itself to be an inefficient maintenance policy, as its expected cost rate per unit time is 171% higher than the optimal rate. That is, a 79% increase in work time corresponds to a 399% increase in total cost.

The reason that both the Age Based Replacement and the Optimal Risk Replacement using the Cox Proportional Hazard Model are quite similar is due to the fact that both involve 20 suspensions of young individuals (ES = 20), 48 suspensions due to optimal replacement (OAS = OPHMS = 48), and do not prevent 2 failures (OAF = OPHMF = 2), resulting in a total cost of $88 = (20 + 48) \times 1 + 2 \times 10$. An opposite scenario is the

Run to Failure maintenance policy, where only 29 of the 70 pieces of equipment did not fail, resulting in a total cost of $439 = 29 \times 1 + 41 \times 10$.

4 Conclusions

The integration of predictive maintenance techniques in Industry 5.0 represents a substantial upgrade in industrial operations management. Using sophisticated survival analysis models like Cox Proportional Hazard Models, companies can improve equipment reliability and operational efficiency. This research highlights the practical advantages and feasibility of predictive maintenance, positioning it as a valuable strategy for the future of smart manufacturing. The proposed analysis revealed that the application of Cox Proportional Hazard Models allows for accurate predictions of equipment failures. This predictive capability enables proactive maintenance interventions, significantly reducing faults, unplanned downtime and maintenance costs. The findings of this work highlight the importance of integrating predictive maintenance into Industry 5.0 strategies to enhance operational efficiency and reliability.

Beyond cost savings and downtime reduction, predictive maintenance aligns with the broader objectives of Industry 5.0 by fostering more sustainable practices. Reduce waste from unnecessary repairs and extend equipment lifespans contribute to lower resource consumption and energy use, aligning industrial operations with global sustainability goals. This transition also creates opportunities for workforce development, as maintenance strategies evolve from reactive troubleshooting (corrective maintenance) to data-driven decision-making and continuous optimisation of machine performance. Workers are empowered to take on more analytical roles, enhancing skillsets, and job satisfaction and wellbeing.

In conclusion, predictive maintenance not only serves as a tool for operational improvement, but also as a pathway to realise the full potential of Industry 5.0. By embracing these advanced maintenance practices, organisations can achieve greater productivity, sustainability, and workforce empowerment. The findings of this re-search provide a foundation for further exploration and application of predictive models in maintenance, underscoring their transformative impact on modern manufacturing and their essential role in shaping the future of the industrial landscape.

Acknowledgements. Prepared within the scope of the Packing of the Future ®, Green Agenda for Business Innovation, investment project n.º 59, funded by the PRR, NextGenerationEU under the PPS16 - Predictive and Preventive Maintenance Tower.

References

1. Strauka, O., et al.: ERA Industrial Technologies Roadmap on Human-Centric Research and Innovation for the manufacturing sector. In: European Commission, Directorate-General for Research and Innovation (Brussels), 1st edn (2024)
2. Barlow, R., Hunter, L.: Optimum preventive maintenance policies. Oper. Res. **8**(1), 90–100 (1960)

3. Tahara, A., Nishida, T.: Optimal replacement policy for minimal repair model. Eng. Math. (1975)
4. Segawa, Y., Ohnishi, M., Ibaraki, T.: Optimal minimal-repair and replacement problem with age dependent cost structure. Eng. Math. Comput. Sci. (1992)
5. Jin, L., Yamamoto, W.: Adaptive age replacement using on-line monitoring. Procedia Eng. **174**, 117–125 (2017)
6. Mo, J., Cheung, S.C.P., Das, R.: Chapter 2 Basic numerical techniques. In: Demystifying Numerical Models, pp. 5–31. Butterworth – Heinemann (2018)
7. Fox, J.: Cox proportional-hazards regression for survival data. Appendix to *An R* and *S-plus* Companion to Applied Regression (2018)
8. Ko, J.: Chapter 2 Cox proportional hazards model. In: Solving the Cox Proportional Hazards Model and its Applications, pp. 3–5. Electrical Engineering and Computer Sciences University of California at Berkeley (2017)
9. Davidson-Pilon: Lifelines: survival analysis in Python. J. Open Source Softw. **4**(40), 1317 (2019). https://doi.org/10.21105/joss.01317

Implementing SMED in the Tool Change of Glass Inspection Machines: A Case Study

Marcelo Cardoso[1], M. L. Alves[1,2](✉), and M. Jacometti[3]

[1] School of Technology and Management, Polytechnic University of Leiria, 2411-901 Leiria, Portugal
leopoldina.alves@ipleiria.pt

[2] School of Technology and Management, INESC Coimbra-Institute for Systems Engineering and Computers at Coimbra, Polytechnic Institute of Leiria, Polytechnic of Leiria, 2411-901 Leiria, Portugal

[3] Universidade Tecnológica Federal do Paraná, Paraná, Brasil

Abstract. Manufacturing industries face growing pressure to enhance efficiency and optimize operations in an increasingly globalised and technologically driven economy. To sustain competitiveness, these companies must address dedicated challenges, thus requiring higher process productivity and minimised downtime. Reducing setup time for high-volume production processes can increase productivity and efficiency. The lean methodologies play a significant role in identifying opportunities for improvement that can lead to waste reduction and increased value in operations. The main goal of this current research is to implement the Single Minute Exchange of Die (SMED) approach within the specific context of tool change in automatic inspection machines used within the glass industry. Consequently, this initiative yielded a significant global decrease of 18% in the total average of tool change. As a result of the proposed research, the reduction in overall setup time was considered highly significant in a well-established production process. Also, this improvement enabled the automated glass selection machines to maintain the same percentage of non-rejected samples, assuring the process quality. This study confirmed the resulting enhancement in efficiency related to the tool change on automated inspection machines used in the glass industry and the upholding of the quality of the changes, which remained uncompromised by the proposed setup sequence.

Keywords: Glass industry · Lean manufacturing · Automatic inspection machines · Continuous improvement · Standardization · SMED

1 Introduction

In today's highly competitive industrial landscape, achieving consistent and rapid process execution is a key driver for companies and industries. A higher production rate often translates into increased revenue, making efficiency the focus of business success.

The project centres on the fine-tuning section of a multinational glass company whose primary function is to change the tools in the automatic inspection machines

© The Author(s), under exclusive license to Springer Nature Switzerland AG 2027
M. Calvete Gaspar et al. (Eds.): Regional Helix 2024, LNME, pp. 125–135, 2027.
https://doi.org/10.1007/978-3-032-36365-7_9

(Mias) whenever there is a change in the product model to be produced (usually bottles), reducing them by 20% [1–3].

This work arose from the need to reduce the tool changeover times on these machines, always bearing in mind that this improvement could not compromise the performance of the factory. As such, it was necessary to ensure that the quality of the tuning was not being jeopardised. This project is significant for the company, as it wants to make changes to the product model produced as quickly as possible, improving tool change times and fine-tuning the bottle defect inspection machines as soon as they arrive in the respective factory area.

To achieve these objectives, Single Minute Exchange of Die (SMED) was implemented to reduce tool change times [4, 5].

Before a glass container comes into the hands of consumers, it goes through a long process to transform the raw material into the containers we daily use. In the company under study, this process can be divided into four main stages: composition and melting; hot zone; cold zone I; and cold zone II.

Composition and Melting

The glass production process begins by grinding a set of raw materials, in particular silicon dioxide (SiO_2), which comes from sand, calcium oxide (CaO), magnesium oxide (MgO), aluminium oxide (Al_2O_3), sodium oxide (NaO) and sodium sulphate (Na_2SO_4). The company uses sand, limestone, soda and 48% recycled glass as raw materials, which can be present in different percentages. Once the raw materials have been mixed, they are taken to the furnace, homogenized and melted at a temperature of 1500 °C. The mixture is placed in the furnace at the same speed as the glass is molded in the individual section machines (IS Machines) used in the production of glass containers, such as bottles and jars, which allows the amount of glass in the furnace to be constant [1].

Hot Zone

Once the raw material has been melted, it is sent to the IS machines as large drops of molten glass. These are formed by cutting the molten glass with scissors as directed towards the IS machines, where they arrive at approximately 1200 °C. IS machines allow for an increase in production speed, as they can mold from 2 to 4 bottles per cycle [1]. Once the drop has been created, various molding processes can be used, and the company, in this case, uses the Blow and Blow and Narrow Neck Press and Blow processes.

Once the final shape of the bottle has been obtained, a hot coating is carried out in which a layer of tin chloride or titanium is placed on the surface of the bottle, which aims to prevent defects due to the bottles sliding on the surfaces of the equipment and increase the mechanical strength of the bottle [1].

After applying the hot coating, the bottles are sent to the annealing oven, where they remain for approximately 45 min. When the bottles are in the oven, they are heated and then gradually cooled to room temperature. The annealing oven reduces the internal stresses in the glass caused during molding and makes the bottles more stable [1].

Cold Zone I

When the bottles leave the annealing oven, they are cold coated by spraying them with polyethylene wax. The cold coating lubricates the bottles to glide better between each other and makes them easier to transport.

The bottles are then sent for checking by automatic inspection machines (Mias), which analyze all the bottles produced. The two types of automatic inspection machines that are the focus of the SMED implementation will be identified as Machine A and Machine B.

There are three types of Mias: (i) flat inspection Mias, which check for defects in the bottle wall; (ii) rotary inspection Mias, which check the thickness and existence of cracks in the bottle wall and check the geometry of the neck, (iii) and Mias for inspecting the mouth and bottom of the bottle.

After being checked by the Mias, these bottles are sent for palletizing and labelling. If the bottles are rejected during the inspection, they are sent to the cask production (recycled material) [1].

Cold Zone II

The bottles are palletised in this area, with some bottles undergoing prior decoration processes through serigraphy and labelling. The palletizing process is followed by labelling and placing an RFID (Radio Frequency Identification) chip in the bottles. The final stage in Cold Zone II is packaging and storage in the company's finished product warehouses for later transportation to the customer. Distribution to the customer and transportation between warehouses is carried out using the company's trucks or, in exceptional cases, by subcontracting transportation services [1].

This study focuses on the Cold Zone section, specifically the tuning sub-section, whose primary function is to carry out tool changes on the Mias, which occur whenever a change in the bottle model is produced. These changes occur daily and sometimes on several production lines, so the time spent on this task should be minimised. According to the internal objectives set by the company, the aim is to reduce these times by around 20%. Still, this improvement must not compromise the performance of the factory or the quality of the Mias tuning. To achieve these objectives, it was necessary to use Single Minute Exchange of Die (SMED) to reduce tool change times.

To change tools and fine-tune the Mias, the fine-tuners have downtime to work when bottles from the previous model are no longer in production and bottles from the next model start arriving.

2 Theoretical Background

Single-Minute Exchange of Die (SMED) was developed to reduce tool change times (in this case, tool change times on Mias) to less than ten minutes. This tool aims to reduce setup times, the time between the last good bottle produced in the current model and the first good bottle made in the next model, to less than ten minutes [6].

SMED was developed in Japan by Shigeo Shingo in 1950, and he spent around 19 years experimenting to improve the tool into what it is known for today. This period of experimentation can be divided into three major phases, which correspond to Shigeo

Shingo's work at Mazda Toyo Kogyo, Mitsubishi Heavy Industries and Toyota Motors Company [6–9].

Shigeo Shingo divided the SMED implementation process into four stages: Preliminary Stage, Stage 1, Stage 2 and Stage 3.

Preliminary Stage

After mapping all the activities carried out in the process, there is no distinction between Internal Setup and External Setup activities in the first phase of SMED implementation. During tool changes at Mias, operators often carry out tasks with the machine stopped (Internal Setup) that could be carried out with the machine running (External Setup). Therefore, the SMED implementation process begins by checking how the tool changes occur. This analysis involves observing the setup process, identifying all the steps and recording the time it takes to complete each task.

In addition to this analysis, communication with operators is essential to motivate them to get involved in the SMED implementation process. This communication with the operators will also make it possible to identify the main difficulties in the process and the proposals for improvement they suggest to reduce setup times [6].

Two methods can be used to check how the setup is carried out: ➢ Time the execution of the tasks carried out and list them; ➢ Use film cameras, which make it possible to analyse how the tasks are carried out and their duration in greater detail.

Stage 1

In this second phase of SMED implementation, the task list obtained in the Preliminary Stage is used, and a differentiation is made between internal setup tasks and external setup tasks. This stage is essential in implementing SMED since carrying out as many External Setup operations as possible will reduce the time needed for Internal Setup by 30 to 50%. To ensure that the External Setup activities are carried out while the machine is operating, checklists must be created to ensure that the operators prepare the material and carry out tasks before the machine stops [6].

Stage 2

At this stage in the implementation of SMED, the Internal Setup and External Setup classifications made in Stage 1 should be reviewed to check that no steps have been incorrectly classified as Internal Setup and could be assigned to External Setup. At this stage, it is necessary to standardize how all the operators carry out the various tasks to reduce their execution times while ensuring that the factory's performance and the quality of the Mias' tuning are not compromised [6].

Stage 3

The last stage of SMED implementation can be carried out at the same time as Stage 2 since this consists of analyzing the various tasks, both Internal Setup and External Setup, and optimizing them. This can be done by using tasks in parallel, improving material transportation to the equipment, and changing from manual to automatic tools, among other factors that can help operators optimize their functions [1].

The SMED methodology enables a significant reduction in tool changeover times, which is associated with a series of advantages that result from this implementation, namely [7–9]:

- Equipment downtime is reduced, which increases its availability and, consequently, the company's productivity;
- Provides greater production flexibility and product variety;
- It allows smaller batches to be produced, which is reflected in a more effective response to customer requests and provides stock reductions, which lowers the costs associated with the process;
- Tool change processes are simplified, which reduces the risk of set-up errors and work accidents when tasks are carried out.

3 Methodology

To achieve the aimed objectives, it was necessary to outline a set of steps to fulfil such goals. The methodology used can be seen in the flowchart in Fig. 1.

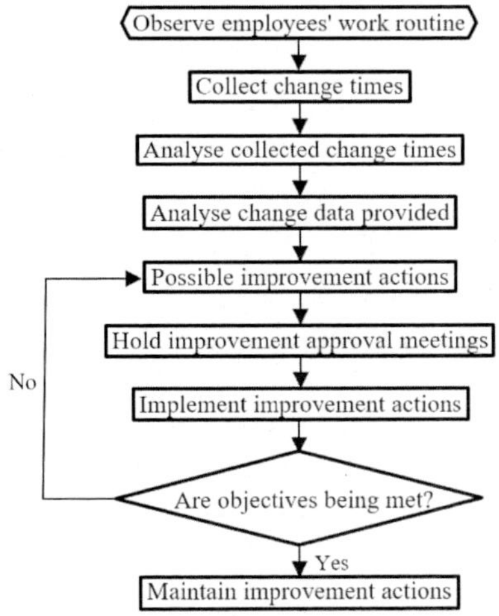

Fig. 1. Flowchart of project steps.

Verifying the current state of the tool change process was essential to propose improvements and subsequently analyse the progress made with the implementation of the project. The analysis was carried out in two phases: (a) analysing the data collected during tool change observations (January to June 2023); (b) analysing the data previously recorded by the company on tool changes (July to November 2022).

Data collection began by mapping the Cold Zone, from the moment the bottles leave the respective annealing cabinet until they reach the palletizers, followed by collecting the tool change times (which include the preparation, mechanical tuning and software problem tuning phases) carried out by the different collaborators involved in the process.

The data obtained corresponds to a period in which there were no significant changes between the type of bottles that were produced.

Based on these analyses, opportunities for improving the tool change process were identified, some of which were implemented, and the results of their implementation were analysed.

The collected times were analysed for both types of machines (Machine A and Machine B) using the same approach, which was carried out using a Gantt Chart, which allowed a better perception of the impact of each task on the total changeover time. Following the typical SMED application steps, Internal Setup activities were separated from External Setup activities; it was analysed which activities could be converted from Internal Setup to External Setup activities, and, finally, it was checked which activities could be optimised. This analysis was fundamental for defining the improvement actions implemented later [4, 5].

4 Results

The Mias tool changes are divided into four categories, applicable to both machine types (A and B). The categories of tool changes are as follows:

1. Tool change of marshes.
2. Tool change due to a change between similar products, with only positioning adjustments.
3. Substitution of tools.
4. Tool changes between different products.

The collected times were analysed for both types of machines (Machine A and Machine B), allowing to identify the impact of each task on the total changeover time for all types.

Analysing the data previously recorded by the company enabled the identification of the types of changeover that occurred most frequently for each of the machines and verified that the non-standardisation of the tool changeover method was reflected in a pronounced difference in results between operators. Based on the data provided, it was also possible to determine the average changeover time for each of the machines before implementing the improvement actions that were to be implemented.

Table 1 shows an example of the results obtained for machine A, where the tasks involved in type 3 tool changing are identified.

Following the typical SMED application steps, Internal Setup activities were separated from External Setup activities; it was analysed which activities could be converted from Internal Setup to External Setup activities and, finally, it was checked which activities could be optimised. This analysis was fundamental for defining the improvement actions implemented later [4, 5].

To identify proposals for improving the process, an analysis was carried out of the wastes relating to Machine A and Machine B, and it was determined that there was an opportunity to improve tool changes on both machines if the pure waste tasks could be eliminated and some of the necessary waste tasks could be minimised.

The improvement proposals that resulted from analysing the data collected, the information provided by the company and the waste analysis can be divided into three groups. The first group corresponds to improvements aimed at SMED (13 proposals), the second group relates to improvements in work standardisation and the training of new operators (2 proposals), and the third group corresponds to other actions that have been proposed and which make it possible to improve the company's functioning, but which do not fit into the first two groups mentioned (2 proposals).

The improvement actions identified followed different timings and, at the time of this study, some of the actions had been fully implemented, others were awaiting material for their implementation, and others were still in the process of being decided by the company, or there are no conditions/interest for their implementation (Table 2).

After implementing the improvement actions, it was possible to check the overall average changeover time (without maintenance) before and after implementing the improvement actions, as shown in Fig. 2 (average obtained between machines A and B). The changeover time values for the year 2022 were 88 (machine A) and 79 (machine B) respectively. For 2023, the monthly values for each machine were January: 80 (A) and 84 (B); February: 53 (A) and 54 (B); March: 73 (A) and 68 (B); April: 63 (A) and 74 (B); May: 75 (A) and 84 (B) and June: 68 (A) and 70 (B).

The lower figures for February result from the fact that there were only type 1 and 2 changeovers, which are less time-consuming.

The graph in Fig. 2 shows that in the last month of the project (June 2023), it was possible to achieve a reduction in tool changeover time of approximately 18%, a very positive result given that this company is a reference for the other companies in the group in terms of tool changeovers.

Quantitatively, the number of bottles and tons of glass saved with these improvements can be estimated. To make this estimate, a production speed of 300 bottles per minute was considered, and each bottle weighed 200 g. This estimate was necessary because the number of bottles produced per minute and their weight vary from model to model. Thus, with this data, it is estimated that the overall saving per tool change would be 4,500 bottles, equivalent to 0.9 tons of glass.

Since this project to improve changeover times was aimed at Machine A and Machine B, it is essential to check the percentage of time saved in carrying out tool changes on each of them. Concerning Machine A, a reduction of approximately 23% was achieved in the overall changeover time on this machine. Machine B, on the other hand, only attained a decrease of roughly 11%. This smaller reduction in Machine B can be explained by the fact that there is less potential for improvement on this machine due to the greater frequency of more complex changes that were taking place at the beginning of the project.

On a quantitative level, it is estimated that Machine A would save 6000 bottles per tool change, equivalent to 1.2 tons of glass, while Machine B would save 2700 bottles, equivalent to 0.54 tons of glass.

Although the aim is to reduce changeover time on both machines, it is positive that the best reduction was made on Machine A because, in the future, Machine B will be replaced by Machine A, making these machines the priority for the company.

To analyse the impact of each of the improvement actions implemented, tables were created for each machine where tool changes took place, identifying the improvement

Table 1. Tasks inherent to type 3 of tool change performed on Machine A.

Description of the tasks	Setup classification	Main steps	Time (s)
Going to get a tool bag	External	Tuning preparation	1
Fill in the tool material control sheet	External	Tuning preparation	1
Moving to the machine	External	Tuning preparation	2
Picking up the bottles (samples)	External	Tuning preparation	3
Waiting for time off	External	Tuning preparation	2
Cleaning the machine with an air hose	Internal	Mechanical tuning	1
Check the height of the teeth is good	Internal	Mechanical tuning	2
Removing the ring	Internal	Mechanical tuning	6
Removing a part of the star	Internal	Mechanical tuning	4
Go to the section to get a key and remove all the parts of the plate	Internal	Mechanical tuning	19
Assembling the star	Internal	Mechanical tuning	2
Put the ring on	Internal	Mechanical tuning	2
Adjusting guides and rotators	Internal	Mechanical tuning	4
Delete the work of the incoming model	Internal	Tuning software defects	1
Copy work that is working	Internal	Tuning software defects	1
Upload work that is running	Internal	Tuning software defects	1
Change genie options	Internal	Tuning software defects	2
Adjust genie height	Internal	Tuning software defects	1
Clearing the genie	Internal	Tuning software defects	7
Fine-tuning the genius	Internal	Tuning software defects	1
Fine-tuning the thickness	Internal	Tuning software defects	5
Finish tuning the genius	Internal	Tuning software defects	2
Fine-tuning the endless	Internal	Tuning software defects	2
Thinning horizontal silk	Internal	Tuning software defects	4
Thinning vertical silks	Internal	Tuning software defects	2
Passing the bottles around to see if they were all rejected	Internal	Tuning software defects	1
Going for a snack	Internal	Tuning software defects	20

(*continued*)

Table 1. (*continued*)

Description of the tasks	Setup classification	Main steps	Time (s)
Sample visual defects and make adjustments	Internal	Tuning software defects	9
Cleaning and fine-tuning the mold reader	Internal	Tuning software defects	4
		Total time (s)	112

Table 2. Current status of implementation of proposed improvements.

improvement action	Implementation status
Eliminate snack breaks	Completed
Purchase of material	Completed
Training of employees	Continuous implementation
Re-label machine cables	Completed
Purchase and replacement of Genie covers	Waiting for the material
Testing and implementation of sidewall flap opening system	Waiting for the material
Next day tool preparation audits	Continuous Implementation
Reduction of the number of samples	Under review
Define and post changeover times and number of non-rejected samples	Continuous Implementation
Standardisation of machinery	Under implementation
Positioning of machine components before changeover	Completed
Preparing the base plates of Machine B in the workshop before the changeover	Under review
Attach a vice to the trolleys that are taken to the changes	Under review
Working instructions for tool changes	Completed
Silk tuning instructions	Completed
Carry out 5S audits in the tuners' workshop	Continuous implementation
Pass "good" bottles at the end of tuning the machine	Under review
Eliminate snack breaks	Completed

action that was implemented, the estimated time saved on a tool change where this improvement action was applied, the percentage of time saved taking into account the overall average of changes on that machine, the estimated rate of changes in which the improvement action was applied and, finally, the percentage value that the improvement action was estimated to have influenced in reducing the overall time of tool changes.

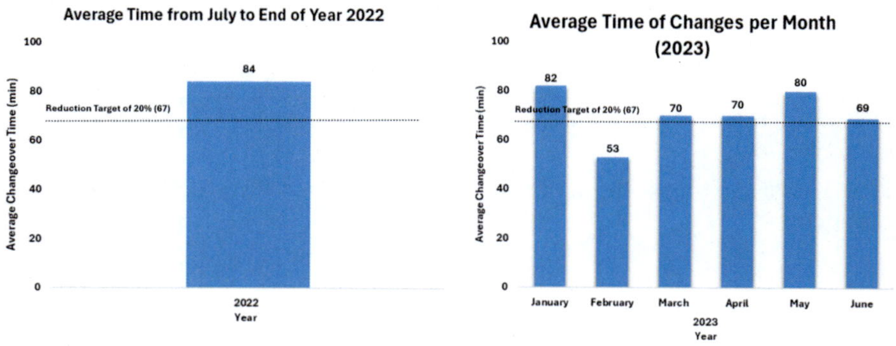

Fig. 2. Average value of changeover time before (left) and after (right) implementing the improvement actions.

5 Conclusion

The SMED implementation of the tool change process on two types of machines in a multinational glass industry company resulted in an average reduction of 15 min in the setting time of the defect inspection equipment located on a glass bottle production line, which is equivalent to 18% of the average time of all tool changes carried out. This reduction in time results in an overall saving per tool change of 4,500 bottles, equivalent to 0.9 tonnes of glass, assuming a speed production of 300 bottles per minute weighing 200 g each. When analysing the overall time savings per machine, it was found that Machine A could reduce the overall time by 23%. In comparison, Machine B was only able to reduce it by 11%, which turns out to be positive since Machine B will be successively replaced by Machine A, making these machines the priority for the company.

The company to which this SMED implementation was applied is part of a group of companies in the glass industry with similar production processes. Throughout the group, this case study was presented as an example of good practice to be implemented in the various production units, with the aim of replicating the reduction in time and costs associated with the tool change operation.

It was also found that the quality of the changes made was maintained, and this factor was monitored over time by analysing the number of samples not rejected in each tool change.

Acknowledgements. This work was funded by national funds through FCT/MCTES (PIDDAC) and FCT Pluriannual Funding UID/308: Institute for Systems Engineering and Computers at Coimbra-INESC Coimbra.

References

1. Manoel, J.: Análise de Processos Fabris na BA Vidro. MsC Thesis of Faculdade de Engenharia da Universidade do Porto, Porto (2010)

2. Jan, F.: The single minute exchange of die methodology in a high-mix processing line. J. Compet. **8**(2), 59–69 2016. ISSN 1804–171X (Print). ISSN 1804–1728 (On-line) https://doi.org/10.7441/joc.2016.02.05
3. Cardoso, M.: Implementação de SMED em máquinas de inspeção automáticas da indústria do vidro. MsC Thesis of Polytechnic Institute of Leiria, Portugal (2023)
4. Dillon, A.P., Shingo, S.: A Revolution in Manufacturing: The SMED System. Taylor & Francis (1985)
5. Carrizo Moreira, A., Campos Silva Pais, G.: Single minute exchange of die: a case study implementation. J. Technol. Manag. Innov. **6**(1), 129–146 (2011). https://doi.org/10.4067/S0718-27242011000100011
6. Shingo, S.: A Revolution in Manufacturing: The SMED System (1985)
7. Bidarra, T., Godina, R., Matias, J., Azevedo, S.: SMED methodology implementation in an automotive industry using a case study method. Int. J. Indus. Eng. Manag. **9** (2018). https://doi.org/10.24867/IJIEM-2018-1-101
8. Juárez-Vite, A., Corona Armenta, J., Rivera-Gomez, H., Oscar, M., Medina, J.: Application of the SMED methodology through folding references for a bus manufacturing company. Int. J. Indus. Eng. Manag. **14**. 232–243 (2023). https://doi.org/10.24867/IJIEM-2023-3-335
9. Emekdar Karaman, E., Açıkgöz Tufan, H., Sahin, U., Bahadir, S., Tuluk, B., Şimşek, A.: Process improvement and efficiency study using single-minute exchange of dies method applied to Set-Up and operation of screen printing machines. Color. Technol. **139** (2023). https://doi.org/10.1111/cote.12676

The Integration of Digital Transformation in Asset Management in Portuguese Companies

Samuel Messias[1], André Guimarães[2,3](✉), Hugo Raposo[4], and Daniel Gaspar[2]

[1] ISEC-Higher Institute of Engineering of Coimbra-Polytechnic Institute of Coimbra, Coimbra, Portugal

[2] CISeD-Research Centre for Digital Services-Polytechnic Institute of Viseu, Viseu, Portugal
aguimaraes@estgv.ipv.pt

[3] CISE-Electomechatronic Systems Research Centre-University of Beira Interior, Covilhã, Portugal

[4] RCM2+-Research Centre for Asset Management and Systems Engineering, Polytechnic Institute of Coimbra, Coimbra Institute of Engineering, Coimbra, Portugal

Abstract. This article explores the importance of integrating emerging technologies in Asset Management in Portuguese companies, drawing on data from a questionnaire administered to 50 professionals across various sectors. The digital transformation of asset management, propelled by the adoption of advanced technologies such as the Internet of Things (IoT), Artificial Intelligence (AI), Big Data, and specialized Asset Management software, has brought significant advancements in optimizing operations, reducing costs, and improving overall efficiency. The study delves into the digital maturity of these companies. It evaluates the extent to which they have implemented ISO 55001 certification, providing insights into how digitalization reshapes asset management practices. Despite the increasing recognition of the benefits of digital transformation, challenges remain, including resistance to change, a shortage of specialized skills, and limited awareness of advanced Asset Management systems. These obstacles highlight the need for more significant efforts in aligning technology adoption with strategic asset management practices to unlock the full potential of digital tools. The findings underscore the importance of continued investment in Digital Transformation and formal certification to enhance the efficiency and competitiveness of Portuguese companies in an increasingly digitalized market.

Keywords: Asset management · Digital transformation · Digital maturity · Asset management maturity

1 Introduction

Digital technologies have emerged rapidly, playing a crucial role in modernizing and optimizing business processes. Adopting these solutions is significantly transforming asset management, particularly in the context of Portuguese companies, where this area has undergone substantial changes driven by the increasing use of emerging digital tools

© The Author(s), under exclusive license to Springer Nature Switzerland AG 2027
M. Calvete Gaspar et al. (Eds.): Regional Helix 2024, LNME, pp. 136–151, 2027.
https://doi.org/10.1007/978-3-032-36365-7_10

[1]. The maturity level in Asset Management and Digital Transformation within companies is assessed based on the adaptation of maturity models, including the verification of ISO 55001 certification and the presence of an integrated asset management system [2, 3].

Studies on Digital Transformation and Asset Management Maturity aim to assess the impact of adopting emerging technologies such as the Internet of Things (IoT), Artificial Intelligence (AI), Big Data, and asset management software on optimizing business operations [4, 5]. This research explores how these technologies can improve processes, reduce costs and increase overall asset management efficiency. A questionnaire developed by Messias et al. [6] was applied to professionals from 50 Portuguese companies, revealing how these technologies are being integrated and highlighting trends, challenges, and opportunities in the specific context of asset management in Portugal.

2 Literature Review

2.1 Asset Management

Asset Management is considered an essential area in a company with a significant number of assets or high-value assets, as effective management can result in improved operational performance of the investment, more substantial cost reduction, and increased efficiency [7].

Asset management involves managing a company's financial, physical, and intangible resources. Its main goal is to optimize the value and performance of a given investment, reducing associated costs and maximizing its utilization. This concept emerged from optimizing costs, reducing risks, and maximizing performance throughout the asset's lifecycle [8, 9].

ISO 55000 [10] defines an asset as "An item, thing, or entity that has potential or actual value to an organization" and asset management as the "Coordinated activity of an organization to realize the value of assets". However, the same standard defines Physical Assets as "Items such as facilities, machinery, buildings, roads, vehicles, railways, airplanes, pipelines, cables, communication equipment, and other infrastructure".

Assets can include buildings, machinery, manufacturing facilities, vehicles, software, contracts, e-books, infrastructure, information technologies, knowledge, and investments [11]. Assets can be grouped into different categories, as demonstrated by [12]:

- **Physical Assets**—These add value to the organization and its shareholders. Physical assets include machinery, facilities, roads, vehicles, buildings, etc.
- **Human Assets**—These are human beings endowed with unique characteristics and knowledge, capable of performing many tasks.
- **Financial Assets**—These are the investments and financial resources necessary to acquire, for example, other types of assets and perform specific tasks.
- **Information Assets**—This refers to the quality and quantity of information for asset management. These are valuable resources related to information and data within an organization. They can include any information or digital data essential for the company's operations and strategies.

- **Intangible Assets**—These are all assets that are not tangible, meaning they have no physical form or substance, such as software, licenses, franchises, copyrights, intellectual capital, and more.

2.2 Digital Transformation

Digitalization involves companies adopting new technologies to enhance various aspects of their operations, services, and processes. This transformation allows businesses to remain competitive by integrating innovative tools across multiple areas, driving efficiency and adaptation in a rapidly evolving market [13].

The need for complex and repetitive physical operations has been reduced, increasing the demand for highly skilled labor [14]. Additionally, there have been changes in recruitment patterns and work organization [15, 16]. Industry 4.0 addresses interoperability, decentralization of information, real-time data collection, and greater flexibility, representing the critical aspects of the fourth industrial revolution. It began with mechanization in the first industrial revolution, followed by the introduction of electricity in the second, and automation with Information Technology in the third. The fourth industrial revolution is characterized by the interconnection of new technological processes, focusing on digital development over physical [17, 18], as illustrated in Fig. 1.

Fig. 1. Industrial revolutions [18].

Key Enabling Technologies (KETs), such as advanced sensors, Big Data, Machine Learning (ML), Mechatronics, Advanced Robotics, Cloud Computing, Cybersecurity, Additive Manufacturing, Digital Twin, and Machine-to-Machine (M2M) communication, have further driven these benefits, promoting substantial growth and development in organizations [15, 17, 19].

2.3 Asset Management and Digital Transformation

Asset management involves various processes that require intensive data utilization, as well as a constant need for data collection and analysis to support decision-making. With access to digital tools, this need is driving the transition to intelligent and digital infrastructure systems. Using smaller, more powerful, and affordable sensors, along with technological advancements in computing, data transfer, storage, and management solutions, can provide new decision-making capabilities and opportunities to improve efficiency in infrastructure networks [20].

Evans and Price [11] argue that IAM should be viewed as a discipline independent of IT, which is crucial for driving business performance. Effective management of information assets is critical to the success of DT. It is one of the fundamental pillars of any transformation or organizational change project, making companies more competitive and profitable.

Digital solutions focused on data utilization advance the operations and maintenance of assets. Information from sensors, cloud computing, and mobility drives demand in asset management through the standardized adoption and use of various technologies. Many organizations responsible for asset management are exploring these technologies as a new approach to gathering data supported by monitoring systems [21, 22].

2.4 Maturity Models

Portuguese industrial culture is going through some challenges and difficulties in the new phase of digitalization noting that not all companies have the same pace and ease of adopting technologies [23]. In this sense, it is essential to understand the maturity level of these companies both at a digital and asset management level.

A maturity model involves a sequence of discrete maturity levels for a specific process class that spans one or more business domains [24]. Maturity models are widely used to evaluate business processes and organizational behavior. The main advantage of using maturity models is identifying the organization's strengths and weaknesses about its objectives, establishing procedures, and efficiently prioritizing actions [9].

3 Research Methodology

The research methodology used in this study was designed to ensure reliable and valid results. A quantitative approach assessed how emerging technologies are integrated into Asset Management in Portuguese companies. The methods provided a representative and diverse sample while maintaining data consistency. Next, it describes how the data is collected, defines the sample, and the general research process.

The methodology included collecting primary data from 50 Portuguese companies through a questionnaire administered in 2024, validated by experts through an adaptation of the Delphi Method [6], to evaluate the level of integration of new technologies in Asset Management. The questionnaire was created on the Typeform platform to make it more engaging and more accessible to complete.

The questions cover various topics and areas, assessing whether the company has an Asset Management System (SGA), whether it is certified by NP EN 55001, what levels of digital maturity are in Asset Management, which technologies are considered most relevant, and which expectations for the company's future are related to Asset Management and Digital Transformation.

From the sample under study, two main sectors of activity were identified: 10 companies belong to the automotive industry (20%) and 18 are in the manufacturing sector (36%), with the remaining 21 companies (44%) representing various other sectors. The methodology structure for this investigation followed a specific sequence of procedures, as illustrated in Fig. 2.

Fig. 2. Survey methodology steps.

4 Results and Discussion

This section analyzes the role of technology in Asset Management, focusing on how Portuguese companies incorporate digital tools to enhance efficiency and decision-making. The study highlights key technologies such as the Internet of Things (IoT), Artificial Intelligence (AI), and Big Data, which reshape asset management practices. The findings provide insights into the level of technological adoption and its impact on operational performance.

4.1 Asset Management System

This subsection explores the adoption and use of software solutions for Asset Management, evaluating their functionalities and benefits in optimizing asset lifecycle management. It examines the extent to which companies utilize specialized Asset Management software and how these digital tools contribute to improved monitoring, maintenance strategies, and compliance with ISO 55001 standards. Figure 3 provides a clear overview of the surveyed companies' adoption of asset management systems and ISO 55001 certification.

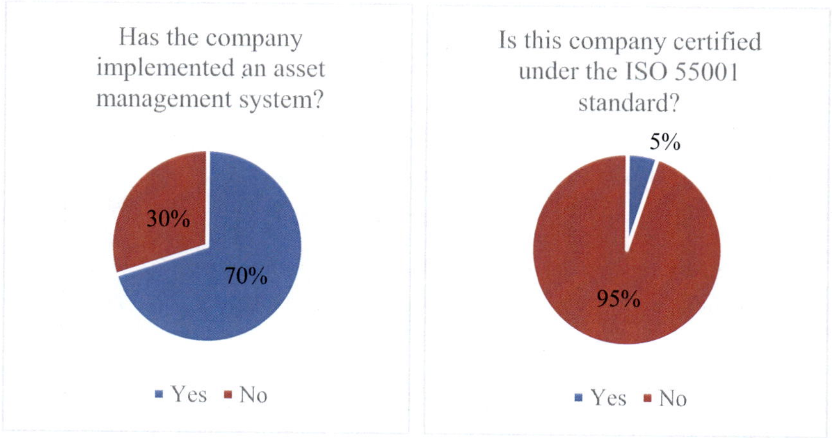

Fig. 3. Percentage of companies with an implemented Asset Management System (on the left) and ISO 55001 certification (on the right).

By analyzing the graphs in Fig. 3, it is evident that a significant proportion of the surveyed companies already have an asset management system in place. According to the data, 70% of these companies have adopted some form of asset management system. However, it is important to note that only 5% of the companies are certified under the ISO 55001 standard, highlighting a considerable gap between the adoption of asset management practices and their formalization and certification according to international standards.

This discrepancy suggests that, while many companies are aware of the importance of asset management and have implemented internal systems, there is still resistance or difficulties in obtaining certification, which may involve additional costs, bureaucratic processes, or a lack of technical readiness. The absence of ISO 55001 certification in companies could be related to the perception that asset management systems, without certification, are already sufficient to meet operational needs, without taking into account the additional benefits that certification can bring, such as increased stakeholder confidence and process optimization.

The following figures show the distribution of ISO 55001 certifications across two different sectors: the **Automotive Industry** and the **Manufacturing Industry**.

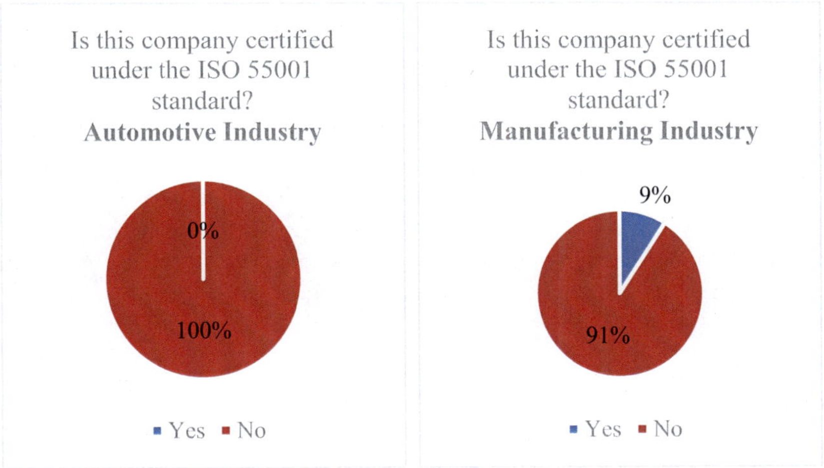

Fig. 4. Distribution of ISO 55001 certification in the automotive and manufacturing industries.

As illustrated in Fig. 4, the Automotive Industry shows that none of the companies in this sector are ISO 55001 certified, with 0% of affirmative responses. This may indicate that asset management certification is not a priority or that there is less awareness of its benefits, such as improved efficiency and asset management. In contrast, the Manufacturing Industry presents a slightly more favorable situation, with 9% of companies certified under ISO 55001, although the majority (91%) have not yet implemented this certification. These data suggest that, despite some efforts, there is still significant room for improvement in the sector, particularly regarding adopting best asset management practices that could enhance efficiency and competitiveness.

4.2 Maturity Models

Figure 5 compares the maturity levels of Asset Management (AM) and Digital Transformation (DT) across different companies.

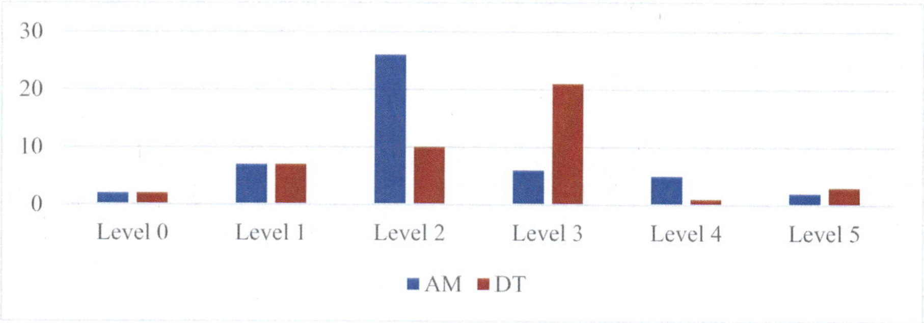

Fig. 5. Comparison of maturity levels in asset management and digital transformation.

By observing Fig. 5, it is noticeable that most Asset Management (AM) responses are concentrated at level 2, suggesting that many companies are still developing their asset management systems. On the other hand, in Digital Transformation (DT), most responses are concentrated at level 3, indicating that companies are more advanced and better prepared to adopt digital technologies in this area. This difference may reflect a prioritization of digital technologies over asset management.

4.3 Use of Technology in Asset Management

This subsection explores the technologies that companies consider most relevant to Asset Management. The aim is to identify which digital tools and systems are gaining prominence in Industry 4.0.

4.3.1 Use of Software for Asset Management

Figure 6 is part of the analysis regarding using technological tools in asset management within Portuguese companies.

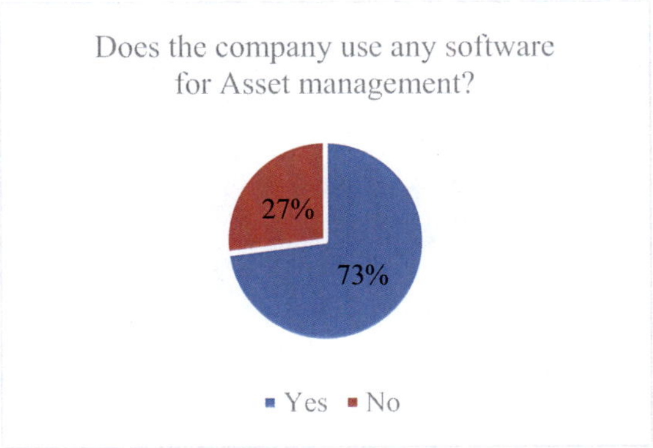

Fig. 6. Distribution of companies using software for asset management.

Figure 6 shows that 73% of companies use software for asset management, while 27% have not yet adopted these technological tools. Using specific software for asset management is considered essential, as it facilitates more accurate decision-making, minimizes human error, and optimizes process control. The high adoption rate of this technology highlights the importance that companies place on digitalization and automation, aiming for more efficient and competitive asset management in the market.

4.3.2 Impact of Data on Asset Management

Figure 7 aims to explore the perception of Portuguese business owners regarding the importance of data in asset management. The impact of data has become increasingly relevant in the efficient management of business assets, as it enables monitoring, forecasting, and more informed decision-making, ensuring optimal asset performance throughout their lifecycle.

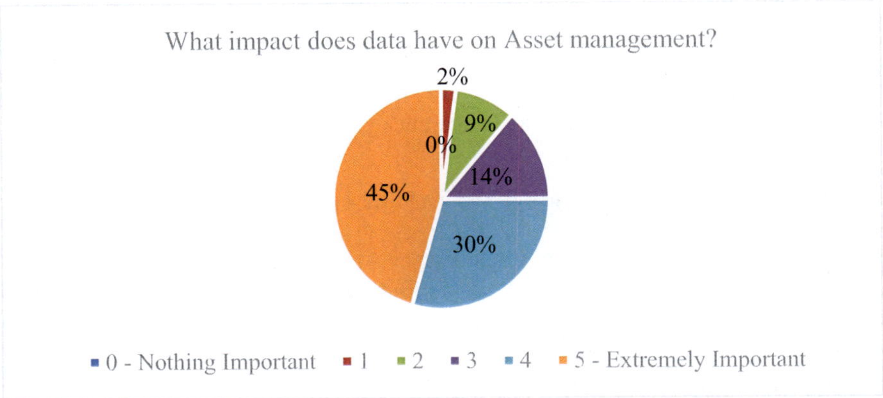

Fig. 7. Impact of data utilization in asset management.

Analyzing the results in Fig. 7 shows that using data for asset management is widely considered necessary. Most professionals gave high ratings to the significance of data usage, with 45% classifying it as "Extremely Important" and 30% as "Important". These results reinforce the perception that data utilization, through specialized software, facilitates decision-making and optimizes asset management processes, making it a crucial element for organizational efficiency.

4.3.3 Information System in Asset Lifecycle Management

This subsection analyzes the role of information systems in asset lifecycle management, focusing on their adoption and effectiveness in Portuguese companies. It explores how these systems support asset tracking, maintenance planning, and decision-making processes. Figure 8 provides an overview of the current state of information system usage, highlighting key trends and areas for improvement.

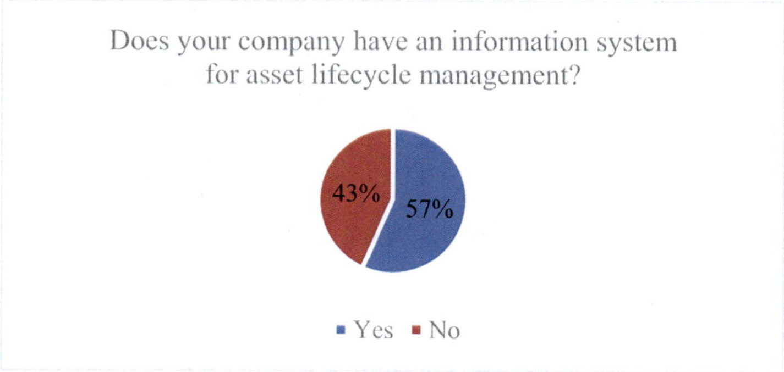

Fig. 8. Proportion of companies using an information system in asset lifecycle management.

According to the data obtained, 57% of companies are already using information systems to optimize the management of their assets throughout their lifecycle, while 43% have not yet implemented these tools. Although most companies have adopted digitalization in this area, a significant portion is still not leveraging the available technological advantages for more efficient asset management.

4.4 Relevant Aspects of the Digital Area About Asset Management

Figure 9 identifies the most relevant digital aspects by the surveyed Portuguese companies regarding adopting digital technologies to improve efficiency and asset management within their organizations. Several critical areas of digital transformation were addressed, including software, sensorization of equipment and production lines, digitalization of administrative processes, the application of new technologies (such as augmented reality and digital twin), and the use of artificial intelligence (AI) tools. These results provide insight into the technological priorities of companies on their digital transformation journey.

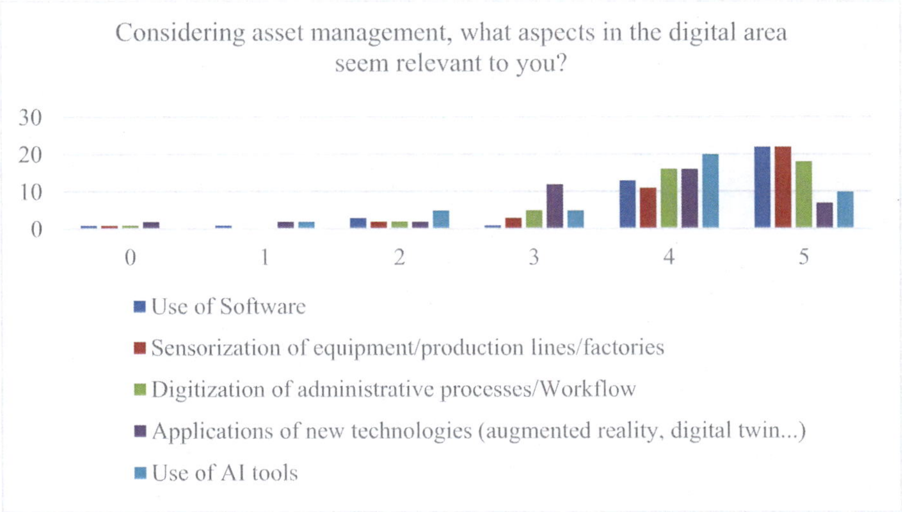

Fig. 9. Relevance of digital technologies in asset management in Portuguese companies.

Figure 9 reveals the relevance of digital aspects in asset management, as indicated by the surveyed Portuguese companies. The topics "Use of Software" and "Sensorization of Equipment/Production Lines" stood out as the most relevant, with 22 responses at level 5—"Extremely Relevant". The "Digitalization of Administrative Processes/Workflow" was also considered necessary, with 18 responses at the same level.

On the other hand, "Applications of New Technologies" and "Use of AI Tools" were primarily rated with a level 4 relevance, receiving 16 and 20 votes, respectively, indicating that while these technologies are not yet as widely implemented, they are considered highly relevant for the future of asset management.

Figure 10 presents a boxplot comparing the relevance attributed to different digital aspects related to asset management by the surveyed Portuguese companies, revealing the variation in responses provided by professionals regarding the impact of technologies such as the use of software, sensors, process digitalization, applications of new technologies (such as augmented reality), and AI tools. The chart displays the distribution of responses and the median for each factor, offering a clear visualization of the relative importance assigned to each technological option.

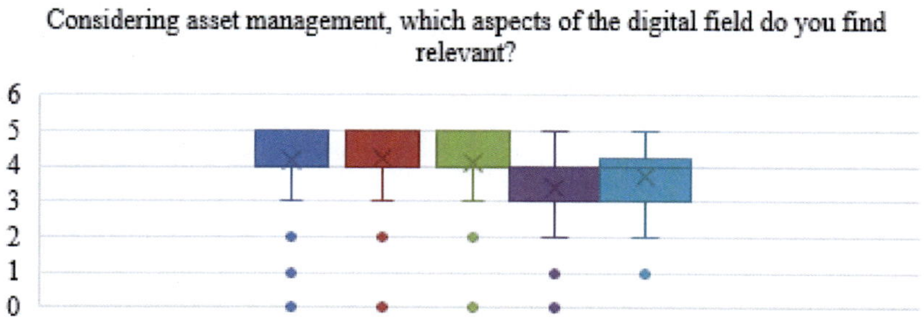

Fig. 10. Trend of the relevance of digital aspects in asset management.

Figure 10 shows the average and standard deviation of the responses to the question about which digital aspects are most relevant in asset management. The average responses for the first three options were close to 4, indicating they are essential. The last two questions, related to new technologies and AI tools, had an average between levels 3 and 4, with a stronger tendency towards level 4 for the latter. Professionals assign high relevance to digitalization, particularly concerning software, sensors, and process digitalization.

4.5 General Future Perspective of Companies

Figure 11 presents the future perspectives of Portuguese companies regarding Digital Transformation (DT) and Asset Management (AM). Professionals were surveyed about their development plans in these areas, with four main options: planned investment, continuation of the current system, ongoing implementation, and topic exploration. The responses provide an overview of the following steps companies are considering in adopting new technologies and improving asset management.

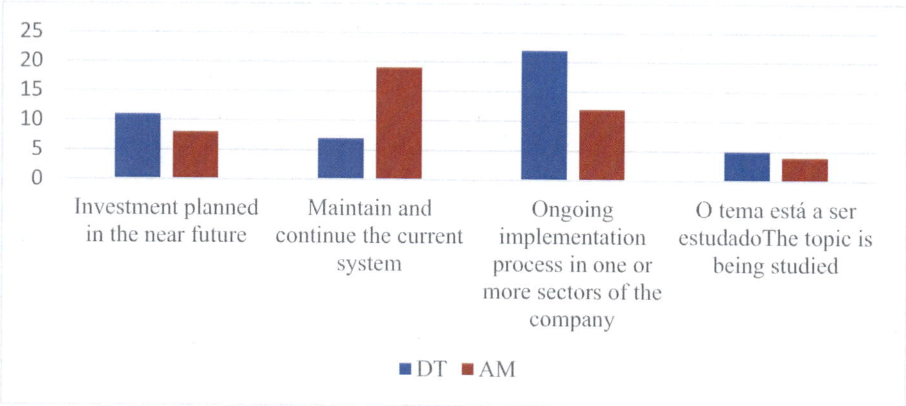

Fig. 11. General future perspective of companies in digital transformation and asset management.

Figure 11 reveals that, in terms of Digital Transformation (DT), most companies are implementing in one or more sectors, with a significant number of responses in this category. Additionally, many companies foresee new investments shortly, demonstrating a continued interest in digitalizing and modernizing their operations.

Regarding Asset Management (AM), a relevant proportion of companies intend to maintain and continue their current systems, suggesting that many already have established processes in this area. Nevertheless, a significant portion is also implementing new asset management systems, indicating room for improvement and development in this field.

In both areas, some are still studying the subject. Still, this rep in both areas represents the smallest percentage, suggesting that most companies are already committed to digital transformation and optimizing asset management.

4.5.1 Future Perspective in the Automotive Industry

Figure 12 refers to the future perspectives of companies in the automotive sector regarding Digital Transformation (DT) and Asset Management (AM). This data helps understand the strategic planning of automotive companies in Portugal concerning digitalization and asset management optimization.

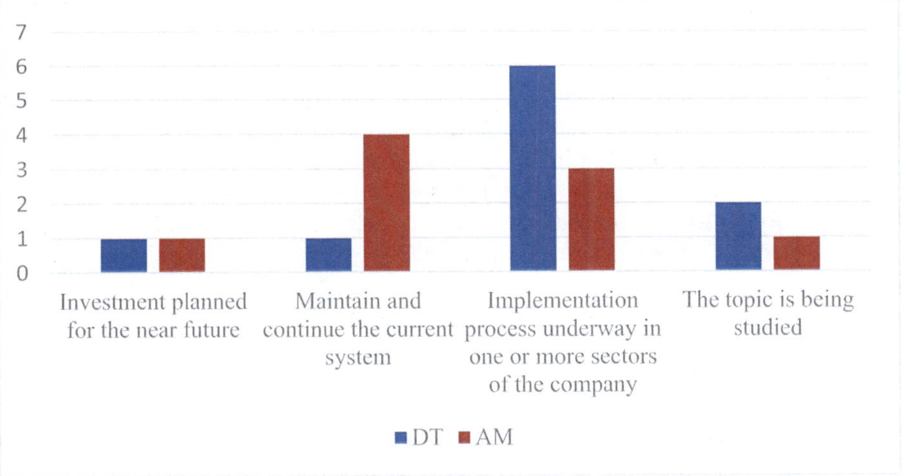

Fig. 12. Future perspectives on digital transformation and asset management in the automotive industry.

Most companies in the automotive sector indicated that they are implementing Digital Transformation (DT), while in the area of Asset Management (AM), the primary focus is on maintaining and continuing existing systems. A smaller percentage of companies mentioned having investment plans or studying the subject in DT and AM, suggesting that these areas are not currently a priority compared to the established practices already in place.

4.5.2 Future Perspective in the Manufacturing Industry

Figure 13 depicts the future perspectives of companies in the Manufacturing Industry regarding Digital Transformation (DT) and Asset Management (AM).

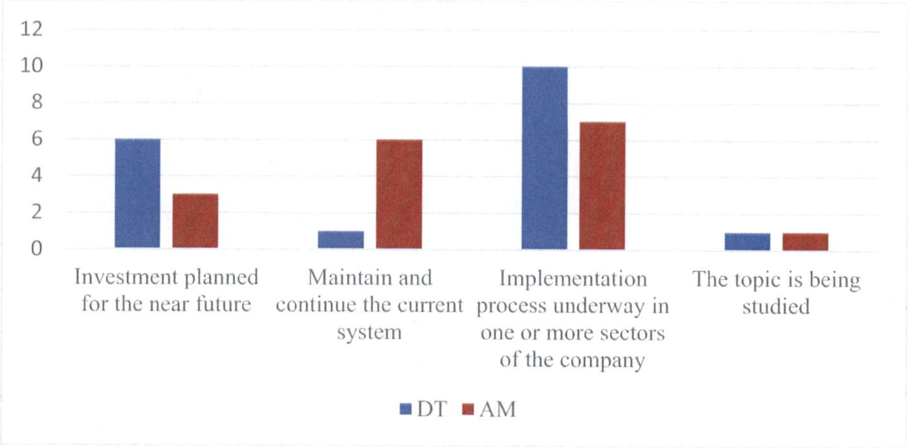

Fig. 13. Future perspectives on digital transformation and asset management in the manufacturing industry.

Professionals in the manufacturing industry highlight the "Process of implementing digital transformation" as the central future perspective, with ten responses, followed by "Planned investment shortly", which received six responses.

Regarding Asset Management, the responses are more divided: 7 refer to the ongoing implementation process, 6 indicate the continuation of the current system, and 3 point to new investments. These data suggest a balance between maintaining existing systems and moving forward with the implementation of new solutions.

5 Conclusion

The analysis of the data collected in the study reveals valuable insights into the current state of Digital Transformation (DT) and Asset Management (AM) in Portuguese companies. Most of the surveyed companies have already implemented some asset management systems, demonstrating a growing awareness of the importance of this area for operational efficiency. However, adherence to ISO 55001 certification remains limited, especially outside the manufacturing industry, indicating a significant gap between adopting asset management practices and their formalization according to international standards.

Regarding maturity, the results suggest that both Asset Management and Digital Transformation in Portuguese companies are at intermediate stages of development, with many organizations positioned at levels 2 and 3 of the maturity models. This reality shows that, although there is an apparent effort to implement digital and asset management practices, there is still considerable progress to be made before these practices are fully optimized and integrated into business processes.

The study further highlights the growing importance of adopting new technologies, such as specialized software, sensors, and process digitalization, to improve the efficiency of Asset Management. The use of data has become a critical factor in informed decision-making, enabling more effective and efficient management of assets throughout their lifecycle. However, challenges remain, including resistance to change, a lack of specific technical skills, and the perception that ISO 55001 certification may not be an immediate priority.

Despite significant progress in digitalization, the low rate of ISO 55001 certification suggests that many companies are not yet fully capitalizing on the advantages of aligning their Asset Management practices with international standards, which could enhance stakeholder confidence and further optimize internal processes. The lack of certification may be associated with the perception that existing systems are sufficient to meet immediate operational needs without considering the additional benefits formal certification can offer.

For the future, the study suggests that companies should continue to invest in Digital Transformation and the formalization of their Asset Management systems to reach higher levels of digital maturity and alignment with international standards. It is recommended that future research further explore how companies can overcome barriers to certification and how the synergy between Digital Transformation and Asset Management can contribute to long-term organizational success.

Acknowledgements. This work is supported by FCT—Fundação para a Ciência e a Tecnologia, I.P., within the scope of the project Ref. UIDB/05583/2020. Furthermore, we would like to thank the Digital Services Research Center (CISeD) and the Viseu Polytechnic Institute for their support.

List of Acronyms and Abbreviations

IoT	Internet of Things.
IA	Artificial Intelligence.
Big Data	Large Volumes of Data.
ISO	International Organization for Standardization.
SGA	Asset Management System.
NP EN	Portuguese European Standard.
DT	Digital Transformation.
AM	Asset Management.
KETs	Key Enabling Technologies.
ML	Machine Learning.
M2M	Machine-to-Machine.
IAM	Information Asset Management.

References

1. Koumou, K.O., Isafiade, O.: Asset Management Trends in Diverse Settings Involving Immersive Technology: A Systematic Literature Review. IEEE Access (2024)
2. IAM.: Asset Management—An Anatomy. https://theiam.org/ (2015)
3. Aras, A., Büyüközkan, G.: Digital transformation journey guidance: a holistic digital maturity model based on a systematic literature review. Systems **11**(4), 213 (2023)
4. Haryanti, T., Rakhmawati, N.A., Subriadi, A.P.: The extended digital maturity model. Big Data Cogn. Comput. **7**(1), 17 (2023)
5. Tripathi, A., Dadi, G.B., Nassereddine, H., Sturgill, R.E., Mitchell, A.: Assessing technology implementation success for highway construction and asset management. Sensors **23**(7), 3671 (2023)
6. Messias, S., Guimarães, A., Raposo, H., Gaspar, D.: Método Delphi Modificado para Abordar a Transformação Digital na Gestão de Ativos. RAE—Revista de Ativos de Engenharia (2024)
7. Cóppola, G.: Gestão de ativos: O que é e como fazer na sua empresa? Rabbot | Automatize sua Gestão de Frotas. https://rabbot.co/blog/gestao-de-ativos/ (2020)
8. Gaspar, D., Almeida, N., Sobral, J., Reguenga, D.: A manutenção na Gestão de ativos. Em Manual de Manutenção em Edificações, Estudos, técnicas e aplicações, 1a, 1–1, pp. 155–174. LEUD (2022)
9. Sobral, J., Tancredo, F.: Desenvolvimento e aplicação de metodologia de avaliação da maturidade em gestão de ativos. RAE-Revista de Ativos de Engenharia **1**(2), 01–15 (2023)
10. ISO 55000: 2014: Asset Management–Overview, Principles and Terminology. ISO Geneva, Switzerland (2014)
11. Evans, N., Price, J.: Development of a holistic model for the management of an enterprise's information assets. Int. J. Inf. Manage. **54**, 102193 (2020)
12. Pereira, J., Raposo, H., Farinha, J.T., de-Almeida-e-Pais, J.E.: Physical assets life cycle analysis in a food industry company-case study. Arch. Food Nutr. Sci. **7**(1), 012–029 (2023)
13. Green, S.: A digital start-up project–carm tool as an innovative approach to digital government transformation. Comput. Syst. Sci. Eng. **35**(4), 257–269 (2020)
14. Pylaeva, I.S., Podshivalova, M.V., Alola, A.A., Podshivalov, D.V., Demin, A.A.: A new approach to identifying high-tech manufacturing SMEs with sustainable technological development: empirical evidence. J. Clean. Prod. **363**, 132322 (2022)
15. Branca, T.A., Fornai, B., Colla, V., Murri, M.M., Streppa, E., Schröder, A.J.: The challenge of digitalization in the steel sector. Metals **10**(2), 288 (2020)
16. Wang, M., Yin, X.: Construction and maintenance of urban underground infrastructure with digital technologies. Autom. Constr. **141**, 104464 (2022)
17. Matyi, H., Veres, P., Bányai, T., Tamás, P., Demin, V.: Digitalization in industry 4.0: The role of mobile devices. J. Produc. Eng. 75–78 (2020)
18. Jan, Z., Ahamed, F., Mayer, W., Patel, N., Grossmann, G., Stumptner, M., Kuusk, A.: Artificial intelligence for industry 4.0: Systematic review of applications, challenges, and opportunities. Expert. Syst. Appl. **216**, 119456 (2023)
19. Okorie, O., Russell, J., Cherrington, R., Fisher, O., Charnley, F.: Digital transformation and the circular economy: creating a competitive advantage from the transition towards net zero manufacturing. Resour. Conserv. Recycl. **189**, 106756 (2023)
20. Vieira, J., Clara, J., Patrício, H., Almeida, N., Martins, J.P.: Digital twins in asset management: potential application use cases in rail and road infrastructures. In: World Congress on Engineering Asset Management, pp. 250–260. Springer International Publishing, Cham (2021)
21. Hua, L., Junguo, Z., Fantao, L.: Internet of things technology and its applications in smart grid. TELKOMNIKA Indones. J. Electr. Eng. **12**(2) (2014)

22. O'Donovan, P., Leahy, K., Bruton, K., O'Sullivan, D.T.: An industrial big data pipeline for data-driven analytics maintenance applications in large-scale smart manufacturing facilities. J. Big Data **2**, 1–26 (2015)
23. Guimarães, A., Reis, P., Charrua-Santos, F.: Industry 4.0-assessment of digital readiness of manufacturing companies in Portugal. In: 2023 IEEE International Conference on Industrial Engineering and Engineering Management (IEEM), pp. 0557–0561. IEEE (2023)
24. Tarhan, A., Turetken, O., Reijers, H.A.: Business process maturity models: a systematic literature review. Inf. Softw. Technol. **75**, 122–134 (2016)

Enhancing G-SIB Evaluation with the PSI-WISP Method: A Proposed Framework

Felipe Fortuna Lucas[1,4(✉)], Carlos Francisco Simões Gomes[1], Marcos dos Santos[1], Solange Maria Fortuna Lucas[2], Fernanda Amélia Ferreira[3], and Maria Teresa Pereira[4]

[1] Fluminense Federal University (UFF), Rua Passo da Pátria, 156, Niterói, RJ, Brazil
felipefortuna@id.uff.br, cfsg1@bol.com.br

[2] EST, Polytechnic of Castelo Branco, Av. do Empresário, 6000-767 Castelo Branco, Portugal
solangefortunalucas@gmail.com

[3] UNIAG, ESHT, Polytechnic of Porto, rua D. Sancho I, N.º 981, 4480-876 Vila do Conde, Portugal
faf@esht.ipp.pt

[4] LAETA-INEGI, ISEP, Polytechnic of Porto. Dr. António Bernardino de Almeida, 431. 4249-015, Porto, Portugal
{mtp,fflas}@isep.ipp.pt

Abstract. Financial market regulation has been crucial since the first Basel Accord in 1988. However, current methods for assessing global systemically important banks (G-SIBs) require improvement to provide a more precise evaluation of systemic risk. The Basel Committee on Banking Supervision (BCBS) assesses G-SIBs based on size, interconnectivity, substitutability, complexity, and transnationality. This study proposes an enhanced evaluation approach using multicriteria decision aid (MCDA) tools. By applying the PSI-WISP method to the 2022 BCBS dataset, we compare the rankings generated by our model with those of the committee. Our results indicate significant differences in classifications, suggesting that PSI-WISP can offer a more accurate and detailed assessment of systemic risk. The findings have practical implications for regulators and policymakers, as they provide an alternative methodology that enhances transparency and robustness in G-SIB rankings, ultimately contributing to a more resilient global financial system.

Keyword: Multicriteria decision aid · Banking regulation · Basel III

1 Introduction

The Basel Committee was created in 1974, a group created by the 10 main central banks in the world, to circumvent various disturbances in the international currency and banking markets.

After the Latin American debt crisis of the 1980s, the members of the Basel committee studied an optimal capital composition to overcome the crisis and optimize the

© The Author(s), under exclusive license to Springer Nature Switzerland AG 2027
M. Calvete Gaspar et al. (Eds.): Regional Helix 2024, LNME, pp. 152–170, 2027.
https://doi.org/10.1007/978-3-032-36365-7_11

risk management of the main banking institutions involved. With this, in 1988, the first Basel Accord (Basel I) was born [5].

Model-driven capital adequacy assessment presents a promising approach to addressing a growing regulatory challenge. Current standards governing capital requirements for commercial bank loans trace their origins to the 1988 Basel Accord [18].

To modernize these guidelines, the Basel Committee introduced a comprehensive revision in June 1999, establishing a three-pillar framework known as Basel II. This overhaul aimed to refine risk sensitivity and institutional accountability through the following core components:

- **Minimum Capital Standards**: Enhancement of the original accord's standardized rules, incorporating advanced methodologies for quantifying credit and operational risks.
- **Supervisory Oversight**: Mandatory regulatory evaluation of banks' internal capital adequacy assessments and risk management frameworks to ensure alignment with financial stability goals.
- **Market Discipline**: Strategic disclosure requirements to promote transparency, enabling market participants to incentivize prudent banking practices through informed scrutiny.

In response to the 2007–09 crisis, documents were created that reviewed aspects of the agreement and outlined new guidelines, over the year's new adjustments were made, until in 2017 the Basel III agreement was made, which remains in force.

Basel III reinforced the financial resilience of European banking institutions through elevated capital thresholds and rigorous qualitative standards for capital composition. This paradigm incentivizes banks to sustain or enhance their capital buffers, ensuring compliance not only with regulatory mandates but also investor confidence and market demands [8].

Concurrently, regulatory authorities acknowledged the systemic risks posed by large-scale financial entities. The potential collapse of a major institution, they concluded, could destabilize the broader financial ecosystem. Consequently, capital frameworks were recalibrated to reflect both the scale and systemic significance of banks. Institutions deemed "globally systemic" (G-SIBs) now face supplementary capital reserves—mandatory holdings exceeding baseline risk-adjusted requirements—to mitigate economy-wide fallout from potential failures [5, 16]. Within this framework, G-SIBs assume a pivotal role as custodians of international financial stability. Their systemic importance stems from their capacity to maintain operational resilience—ensuring uninterrupted execution of financial intermediation even during economic volatility or crisis scenarios. By absorbing shocks triggered by macroeconomic disruptions, these institutions mitigate spillover effects, thereby reducing systemic risk contagion across interconnected markets [15].

With the designation of the G-SIBs, the Basel Committee on Banking Supervision (BCBS) carries out an evaluation process on these banks according to pre-established criteria, generating a score in each established criterion and then the simple average between these criteria is made to designate the rankings of the banks best evaluated by the BCBS. With this, the following problem arises: Can the current evaluation of G-SIBs be improved by different mathematical approaches?

The multicriteria decision aid (MCDA), field of Operations Research (OR) uses mathematical modelling to support the decision taking. Mathematical and computational modelling studies enable the implementation of complex algorithms as a form of treatment in data analysis [25, 33, 34] and decision-making in an objective, trivial and transparent way. According to [21, 31], MCDA methods have been developed for different problems [3, 10, 12, 22, 32]. As well, many applications of multicriteria methods were applied in various areas, like investment [24, 26], machine learning classification [10], in additive manufacturing [14], demand forecasting for steel companies [35], and others.

The objective of this study is to propose a new MCDA model for the evaluation of G-SIBs using the PSI-WISP method to evaluate them, using the software developed [2] for the method to evaluate the alternatives. The results obtained intend to shed light on a need for complexity in the evaluation by these actors, making use of a robust mathematical tool that has already been scientifically proven in several areas of activity.

The work is organized 5 sections, in Sect. 1 the introduction of the subjects is made, in Sect. 2 a literature review on the G-SIBs, the Current evaluation method of the G-SIBs, and the hybrid MCDA PSI-WISP method is carried out. Section 3 addresses the methodology applied, Sect. 4 addresses the method application, Sect. 5 the results are discussed, and Sect. 6 presents the conclusions and future studies.

2 Description of the Problem

The construction of the evaluation process goes through several stages, and for the definition of a real problem, it is of paramount importance to structure the problem to be addressed, which strengthens the relevance of the research and the proposed hybrid method.

To organize and structure the reasoning, Fig. 1 presents a mind map seeking the relationships between MCDA method and G-SIB Assessment, seeking to understand the main activities and where they are located and can provide a richer structuring of the application.

In this process of analysis of the evaluation branches of G-SIBs, 5 main approaches can be observed. The construction of mental minds approaches the MCDA evaluation process, a general evaluation process, and the G-SIBs evaluation process. In the final interpretation of this process emerges the idea that using MCDA to evaluate these countries can generate more gains and trust. The purple balloon referees to decision-making processes, and the green balloon explains more about the G-SIBs evaluation, finally, he Blue and de Salmon Balloon emphasizes evaluating the expected results with this methodology.

2.1 Current Evaluation Method of G-SIBs

Empirical analyses underpinning the Basel III reforms, utilizing aggregated financial data, demonstrated that enhanced regulatory measures targeting liquidity management and capital advantages correlate with a reduced incidence of system-wide banking collapses [6].

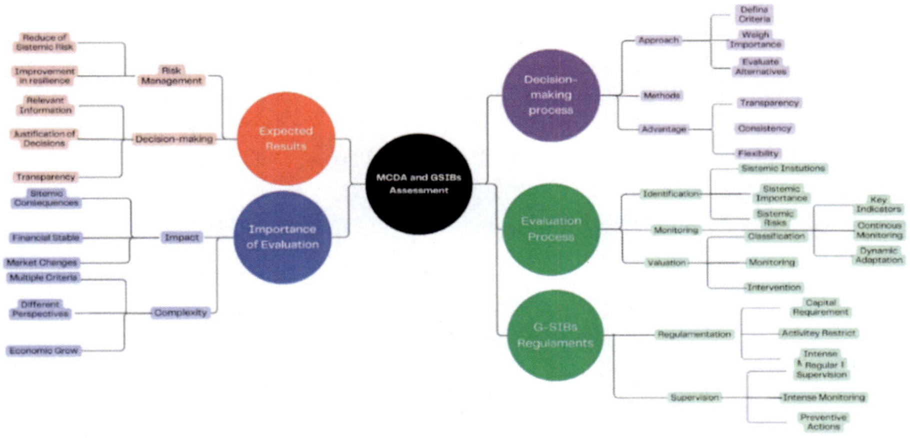

Fig. 1. MCDA mind map and G-SIB assessment

The detrimental spillover effects of institutions labeled "too big to fail" due to their scale, interdependencies, operational complexity, irreplaceability, or transnational reach—are well-documented. When financial entities prioritize private gains without internalizing broader societal costs, their individually rational decisions may collectively exacerbate systemic inefficiencies. Furthermore, implicit government guarantees engender moral hazard, distorting market incentives by encouraging excessive risk-taking, undermining disciplinary mechanisms, and fostering uneven competitive landscapes. These dynamics not only heighten the probability of future financial instability but also compound fiscal burdens, as taxpayers ultimately shoulder both direct bailout costs and the indirect repercussions of distorted market behavior [5].

Systemically critical global banks inherently pose significant resolution challenges, owing to their intricate operational and organizational architectures. Their multinational footprint often leads to jurisdictional discrepancies in bankruptcy frameworks and multifaceted fiscal implications, as regulatory regimes and tax policies vary across host countries [41].

Furthermore, the functional significance of liquidity reserves versus capital adequacy mechanisms diverges systemically between domestic and global banks. Unlike their regional counterparts, multinational institutions capitalize on asynchronous fluctuations in macroeconomic cycles and monetary policies across diverse markets [17, 19]. This geographic diversification enables them to leverage internal funding networks, strategically redistributing financial resources among subsidiaries to optimize stability and mitigate localized shocks [41].

MCDA methods provide a robust framework for dealing with the complexity and multidimensionality involved in assessing G-SIBs. They allow for the simultaneous consideration of multiple critical factors such as size, interconnectivity, complexity, substitutability, and the extent of international operations. The use of MCDA facilitates the integration of quantitative and qualitative data, providing a holistic view of the health and risks associated with each bank.

Another critical aspect is MCDA's ability to adapt to changes in the financial and regulatory environment. The criteria and weights used in the valuation can be updated regularly to reflect new information and market conditions, ensuring that valuations remain relevant and accurate over time.

The current method of evaluation of G-SIBs, carried out by the BCBS, considers criteria of Trans-jurisdictional Activity, Size, Interconnectivity, Substitutability and Complexity. The method and its weightings are described in Table 1.

Table 1. Current evaluation of G-SIB [5]

Category	Individual indicator	Indicator weighting (%)
Cross-jurisdictional activity (20%)	Cross-jurisdictional claims	10
	Cross-jurisdictional liabilities	10
Size (20%)	Total exposures as defined for use in the Basel II leverage ratio	20
Interconnectedness (20%)	Intra-financial system assets	6.67
	Intra-financial system liabilities	6.67
	Securities outstanding	6,67
Substitutability financial institution infrastructure (20%)	Assets under custody	6.67
	Payments activity	6.67
	Underwritten transactions in debt and equity markets	3.33
	Trading volume	3.33
Complexity (20%)	Notional amount of OTC derivatives	6.67
	Level 3 assets	6.67
	Trading and available-for-sale securities	6.67

Within the BCBS evaluation, an analysis is made by base points, in each criterion a normalization is made by dividing the value of the alternative in determined by the total sum of the criterion. As the result of this division is a percentage number in relation to the total size of the population, it is multiplied by 10,000 to bring the data to base points, to which it is exposed to a maximum score of 1,000 (which would be 100% in the evaluation of the given criterion. Finally, after obtaining the score for each criterion, the simple average is performed in each criterion for each alternative, in what the committee identifies as overall.

3 Theoretical Background

In this session, the literature review will be addressed regarding the MCDA methods that will be addressed in this work, including the methodological construction of the tools and their mathematical modelling, for application in the sessions ahead.

3.1 Preference Selection Index Method

This study employs the Preference Selection Index (PSI) methodology to derive criterion weights. Unlike traditional multicriteria approaches, the PSI framework eliminates the need for explicit prioritization of attributes. Instead, it calculates aggregate preference scores for each criterion through statistical analysis of their inherent variability [28]. These scores enable the computation of a composite PSI value for every alternative, with the highest-scoring option emerging as the optimal choice [28].

The first step is the definition of the decision matrix, and after that, the normalisation processes arise, which are given by Eqs. (1) and (2):

$$r_{ij} = \frac{x_{ij}}{\text{max}x_{ij}} \tag{1}$$

for Profit criteria and.

$$r_{ij} = \frac{minx_{ij}}{x_{ij}} \tag{2}$$

for Cost criterion.

To continue the method, the quadratic variation of preferences is calculated, given by Eq. (3):

$$PV_j = \sum_{i=1}^{N} [R_{ij} - \overline{R}_j]^2 \tag{3}$$

where is the mean of the normalized attribute j and $\overline{R}_j = \frac{1}{N}\sum_{i=1}^{N} R_{ij}$

In the next step, Eq. (4), the calculation of ϕ_j:

$$\phi_j = 1 - PV_j \tag{4}$$

In sequence, the General Preference Calculation is made, given by Eq. (5):

$$\Psi_j = \frac{\phi_j}{\sum_{j=1}^{M} \phi_j} \tag{5}$$

Finally, the Preference Index is calculated, given by Eq. (6):

$$I_i = \sum_{j=1}^{M} (R_{ij} * \Psi_j) \tag{6}$$

3.2 Simple Weighted Sum Product Method

The Simple Weighted Sum Product (WISP) method was created by Stanujkic et al. [38]. Figure 2 shows the connections of the seminal article, linking its main references for creation and addressing the works that followed new applications of the method. The articles that's has not connection with the seminal articles in Fig. 2, was used like reference but don't participate directly of the seminal paper.

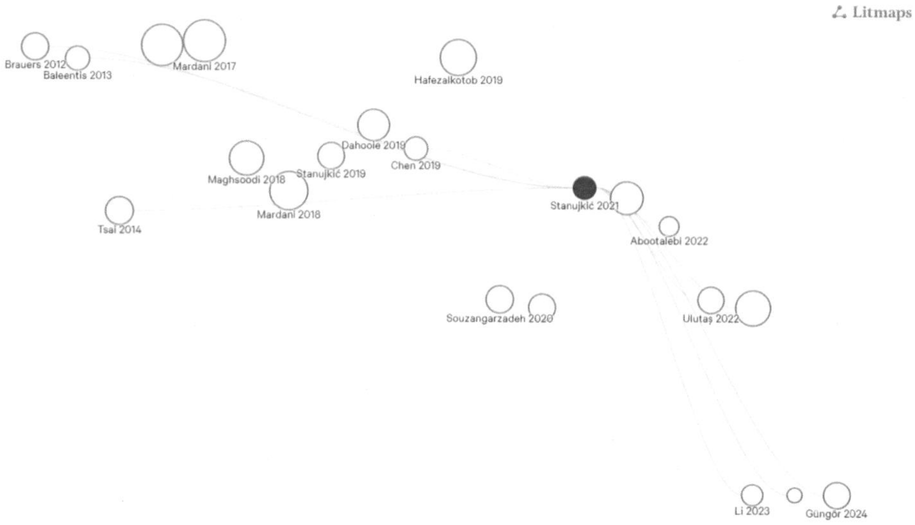

Fig. 2. WISP method litmaps

The Method, created in 2021, has a robust literature, dating from [7], and which emerges as a basis for new publications in the years 2022, until currently in the year 2024.

The WISP methodology synthesizes four evaluative interactions between beneficious and cost criteria to compute the composite utility score of each alternative [38].

The construction of the model follows the following steps, shown in Table 2.

The second step already brings a particularity of the method, the standardization process. This process is essential to take the dimensionality of the data and bring it to a base that allows different criteria to be compared. The normalisation process is given by Eq. (7):

$$r_{ij} = \frac{x_{ij}}{max_i x_{ij}} \tag{7}$$

In step 3, the weighted sum and weighted product of the standard classifications of the beneficial and non-beneficial criteria for each alternative are calculated, as in Eqs. (8), (9), (10) and (11):

$$ws_i^{max} = \sum_{j \in \Omega_{max}} r_{ij} w_j \tag{8}$$

Table 2. WISP method step-by-step [38]

Step	Description
1	Construction of the initial decision matrix
2	Normalization of the initial decision matrix
3	Calculate the weighted sum and weighted product of the normalizaed ratings of beneficial and non-beneficial criteria for each alternative
4	Calculate the values of utility measures
5	Recalculate the values of four utility measures
6	Determine the overall utility of each alternative
7	Rank the alternatives in descending order and select the most suitable one

$$ws_i^{min} = \sum_{j\in\Omega_{min}} r_{ij}w_j \tag{9}$$

$$wp_i^{max} = \prod_{j\in\Omega_{max}} r_{ij}w_j \tag{10}$$

$$wp_i^{min} = \prod_{j\in\Omega_{min}} r_{ij}w_j \tag{11}$$

In continuity with the method, step 4 calculates the value of the measured utilities, given by Eqs. (12), (13), (14) and (15):

$$u_i^{sd} = ws_i^{max} - ws_i^{min} \tag{12}$$

$$u_i^{pd} = wp_i^{max} - wp_i^{min} \tag{13}$$

$$u_i^{sr} = \frac{ws_i^{max}}{ws_i^{min}} \tag{14}$$

$$u_i^{pr} = \frac{wp_i^{max}}{wp_i^{min}} \tag{15}$$

where and denote the differences between the weighted sum and the weighted product of the normalized classifications of alternative *i*, respectively. Similarly, where and denote the ratios between the weighted sum and the weighted product of the normalized classifications of alternative $u_i^{sd} u_i^{pd} u_i^{sr} u_i^{pr}$ i, respectively.

When the decision-making problem does not involve non-beneficial criteria, Eqs. (16) and (17) should be expressed as follows:

$$u_i^{sr} = ws_i^{max} \tag{16}$$

$$u_i^{pr} = wp_i^{max} \tag{17}$$

Similarly, in the case of solving decision-making problems that do not involve non-beneficial criteria, the equations should take the following form:

$$u_i^{sr} = \frac{1}{ws_i^{min}} \tag{18}$$

$$u_i^{pr} = \frac{1}{wp_i^{min}} \tag{19}$$

In step 5, the values of the 4 utility measures are recalculated by Eqs. (20), (21),(22) and (23):

$$\overline{u}_i^{sd} = \frac{1 + u_i^{sd}}{1 + max_i u_i^{sd}} \tag{20}$$

$$\overline{u}_i^{pd} = \frac{1 + u_i^{pd}}{1 + max_i u_i^{pd}} \tag{21}$$

$$\overline{u}_i^{sr} = \frac{1 + u_i^{sr}}{1 + max_i u_i^{sr}} \tag{22}$$

$$\overline{u}_i^{pr} = \frac{1 + u_i^{pr}}{1 + max_i u_i^{pr}} \tag{23}$$

where $\overline{u}_i^{sd}$, $\overline{u}_i^{pd}$, and denote the recalculated values of, $\overline{u}_i^{sr}\overline{u}_i^{pr}$ $u_i^{sd}u_i^{pd}$, u_i^{sr}, u_i^{pr}

The values of u_i^{sd} and u_i^{sr} can be positive, negative, or zero. In such cases, they should be mapped to the range (0,1) using Eqs. (14) and (15) before determining the overall usefulness of each alternative u_i in Eq. (24).

$$u_i = \frac{1}{4}(\overline{u}_i^{sd} + \overline{u}_i^{pd} + \overline{u}_i^{sr} + \overline{u}_i^{pr}) \tag{24}$$

In the last step of the method, the alternatives are ranked in descending order and the most appropriate value is selected. The alternative with the most utility is preferred.

4 Methodology

Figure 3 shows in more detail the entire methodological construction that justifies the application of the hybrid PSI-WISP method in the evaluation of G-SIBs.

The methodological process of the proposal for the evaluation of the G-SIBs included the introduction and study of the relevance of the theme, a detailed description of the problem in question, and a literature review of both the evaluation process of the G-SIBs and the multicriteria method PSI-WISP.

Data were collected from the BCBS database with the evaluation data of the G-SIBs for the end of 2022, which are the most recent data available by the committee.

After applying the method, it will be possible to analyse the results and compare the two evaluation processes and the contributions that the MCDA method had to the application.

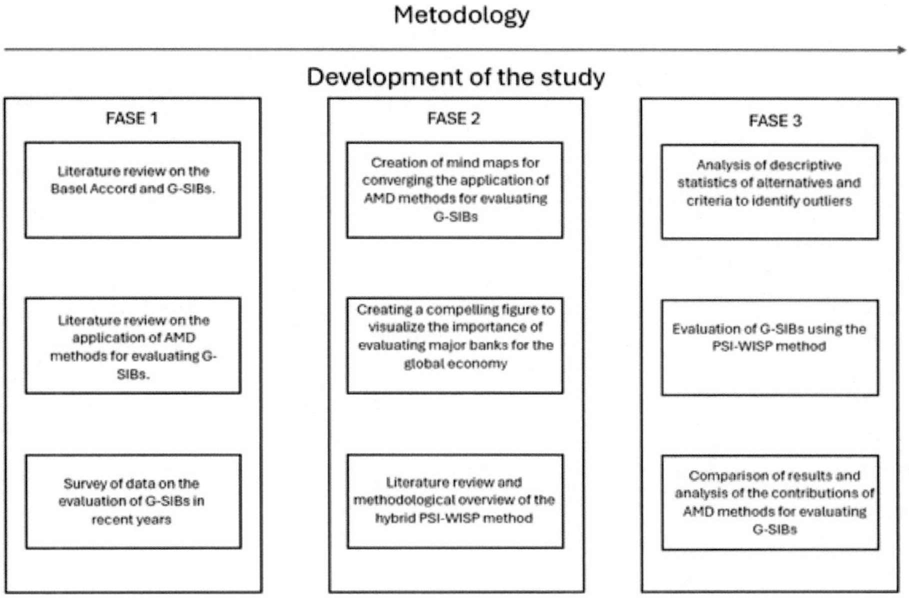

Fig. 3. Methodology

After all the literature review carried out on the evaluation method of the G-SIBs and the hybrid method PSI-WISP, the application of the method is done directly, taking advantage of the BCBS evaluation process, adding the multicriteria tool in the final step of evaluation of alternatives and criteria. Therefore, the decision matrix, given in Table 3, presents the data after the final treatment of the BCBS, where the multicriteria method will be applied.

The decision matrix brings with it in the first column the ranking made by the BCBS, making the simple average of the criteria established by the committee.

Table 3. Decision matrix

Bank	Size	Interconnectedness	Substitutability	Complexity	Cross-jurisdictional
JP Morgan (US)	404	427	500	529	410
Citigroup (US)	269	309	500	366	512
HSBC (GB)	280	240	282	357	680
Bank of America (US)	329	255	496	405	249
BNP Paribas (FR)	256	273	224	356	568
ICBC (CN)	557	406	217	210	137
Bank of China (CN)	414	367	188	207	262
Barclays (GB)	161	184	238	427	284
MUFG (JP)	287	218	182	244	327
China Construction (CN)	484	269	235	195	63
Deutsche Bank (DE)	122	125	253	384	357
Goldman Sachs (US)	171	209	253	368	206
Agricultural Bank (CN)	469	313	217	131	42
Crédit Agricole (FR)	236	255	135	299	219
Morgan Stanley (US)	130	176	240	329	166
Société Generale (FR)	146	227	132	284	230
UBS (CH)	96	220	207	214	237
Santander (ES)	174	147	42	102	483
SMFG (JP)	202	228	55	159	246

(*continued*)

Table 3. (*continued*)

Bank	Size	Interconnectedness	Substitutability	Complexity	Cross-jurisdictional
RBC (CA)	142	151	160	157	279
Wells Fargo (US)	206	167	249	174	56
Mizuho (JP)	189	136	127	174	206
Toronto Dominion (CA)	141	104	85	118	317
BNY Mellon (US)	41	121	500	29	58
ING Bank (NL)	104	124	53	84	376
State Street (US)	30	106	500	22	62
Standard Chartered (GB)	85	121	75	125	293
BoComm (CN)	192	186	98	143	53
BPCE (FR)	148	141	52	221	100
Credit Suisse (CH)	65	110	178	144	122
Unicredit (IT)	93	146	44	87	209
CITIC (CN)	136	151	106	124	24
Industrial Bank (CN)	147	193	98	90	11
Bank of Montreal (CA)	87	103	86	82	163
China Merchants (CN)	153	119	108	93	27
Bank of Nova Scotia (CA)	102	106	44	58	160
Shaghai Pudong (CN)	132	171	90	50	16

(*continued*)

Table 3. (*continued*)

Bank	Size	Interconnectedness	Substitutability	Complexity	Cross-jurisdictional
Intesa (IT)	101	131	28	105	83
DZ Bank (DE)	54	190	29	130	41
Nordea (FI)	55	100	53	89	137

5 Results and Discussion

As explained in Sect. 4, the application of the hybrid PSI-WISP method will be made from the evaluation of the criteria made by the BCBS, thus, instead of using the weighted average of the criteria, the MCDA method seen in this work will be used.

In the first moment, the PSI method will be used to generate the weights, which will later be used in the WISP method to evaluate the alternatives. The application of the method is done using the PSI and MPSI-MARA For Decision Making software [2].

In the first step of the method, the weights are defined using the PSI method, with application directly by the software. The weights generated by the method are given by Table 4.

Table 4. Weights generated by the PSI method

Criteria	Weights generated by PSI method
Size	0.1718
Interconnectedness	0.0837
Substitutability	0.3505
Complexity	0.3505
Cross-Jurisdictional	0.1953

After generating the weights, the application of the WISP method begins, following all the steps designated in the theoretical foundation. All operations will be carried out through the MEREC-WISP software [13].

Table 5 shows the standardized matrix of the method, developed using the Eq. (1) for each criterion and alternative:

After normalization, the step 3 brings the weighted sum and weighted product of the standard classifications, with Eq. (8), (9), (10) and (11). The aggregation process is carried out according to utility measures, and using Eqs. (12), (13), (14), (15), (22) and (23). The ranking can be observed in Table 6.

In the final evaluation process, in the last step of the method, the Eq. (24) is used to calculate the average of the values used for the recalculation of the utility measures. With this, the application of the method ends, and the results obtained will be discussed in the next session.

Table 5. Normalized matrix

	C1	C2	C3	C4	C5		C1	C2	C3	C4	C5
A1	0.725	1.000	1.000	1.000	0.603	A21	0.370	0.391	0.498	0.329	0.082
A2	0.483	0.724	1.000	0.692	0.753	A22	0.339	0.319	0.254	0.329	0.303
A3	0.503	0.562	0.564	0.675	1.000	A23	0.253	0.244	0.170	0.223	0.466
A4	0.591	0.597	0.992	0.766	0.366	A24	0.074	0.283	1.000	0.055	0.085
A5	0.460	0.639	0.448	0.673	0.835	A25	0.187	0.290	0.106	0.159	0.553
A6	1.000	0.951	0.434	0.397	0.201	A26	0.054	0.248	1.000	0.042	0.091
A7	0.743	0.859	0.376	0.391	0.385	A27	0.153	0.283	0.150	0.236	0.431
A8	0.289	0.431	0.476	0.807	0.418	A28	0.345	0.436	0.196	0.270	0.078
A9	0.515	0.511	0.364	0.461	0.481	A29	0.266	0.330	0.104	0.418	0.147
A10	0.869	0.630	0.470	0.369	0.093	A30	0.117	0.258	0.356	0.272	0.179
A11	0.219	0.293	0.506	0.726	0.525	A31	0.167	0.342	0.088	0.164	0.307
A12	0.307	0.489	0.506	0.696	0.303	A32	0.244	0.354	0.212	0.234	0.035
A13	0.842	0.733	0.434	0.248	0.062	A33	0.264	0.452	0.196	0.170	0.016
A14	0.424	0.597	0.270	0.565	0.322	A34	0.156	0.241	0.172	0.155	0.240
A15	0.233	0.412	0.480	0.622	0.244	A35	0.275	0.279	0.216	0.176	0.040
A16	0.262	0.532	0.264	0.537	0.338	A36	0.183	0.248	0.088	0.110	0.235
A17	0.172	0.515	0.414	0.405	0.349	A37	0.237	0.400	0.180	0.095	0.024
A18	0.312	0.344	0.084	0.193	0.710	A38	0.181	0.307	0.056	0.198	0.122
A19	0.363	0.534	0.110	0.301	0.362	A39	0.097	0.445	0.058	0.246	0.060
A20	0.255	0.354	0.320	0.297	0.410	A40	0.099	0.234	0.106	0.168	0.201

As proposed for this study, this section compares the current evaluation process of the G-SIBs, elaborated by the BCBS, face to face with the ranking generated by the hybrid multicriteria PSI-WISP method. Table 8 shows the difference in the evaluation of the two different methodologies.

There were changes in the ranking when using the MCDA method, the first and second places remained at the top, and although these are not considerable changes in the evaluation, you can see a change in the top ten global banks. From this point of view, the use of a multicriteria methodology for the evaluation of G-SIBs was valid. A mathematical robustness can be observed in the mathematical foundation of the methods approached, which can contribute to a greater accreditation of the evaluation process. In this study, a concordance was made between the evaluation made by the BCBS and the increment of the PSI-WISP multicriteria model.

Table 6. Ranking generated by the PSI-WISP method

A	Alternative	PSI-WISP	BCBS rank	A	Alternative	PSI-WISP	BCBS rank
A1	JP Morgan (US)	1	1	A20	RBC (CA)	21	20
A2	Citigroup (US)	2	2	A22	Mizuho (JP)	22	22
A4	Bank of America (US)	3	4	A18	Santander (ES)	23	18
A3	HSBC (GB)	4	3	A19	SMFG (JP)	24	19
A5	BNP Paribas (FR)	5	5	A23	Toronto Dominion (CA)	25	23
A6	ICBC (CN)	6	6	A30	Credit Suisse (CH)	26	30
A8	Barclays (GB)	7	8	A25	ING Bank (NL)	27	25
A11	Deutsche Bank (DE)	8	11	A27	Standard Chartered (GB)	27	27
A7	Bank of China (CN)	9	7	A28	BoComm (CN)	29	28
A12	Goldman Sachs (US)	**10**	12	A29	BPCE (FR)	30	29
A10	China Construction (CN)	11	10	A32	CITIC (CN)	31	32
A9	MUFG (JP)	12	9	A35	China Merchants (CN)	32	35
A13	Agricultural Bank (CN)	13	13	A33	Industrial Bank (CN)	33	33
A24	BNY Mellon (US)	14	24	A34	Bank of Montreal (CA)	34	34
A15	Morgan Stanley (US)	15	15	A31	Unicredit (IT)	35	31
A26	State Street (US)	16	26	A37	Shaghai Pudong (CN)	36	37

(*continued*)

Table 6. (*continued*)

A	Alternative	PSI-WISP	BCBS rank	A	Alternative	PSI-WISP	BCBS rank
A14	Crédit Agricole (FR)	17	14	A36	Bank of Nova Scotia (CA)	37	36
A17	UBS (CH)	18	17	A40	Nordea (FI)	38	40
A16	Société Generale (FR)	19	16	A38	Intesa (IT)	39	38
A21	Wells Fargo (US)	20	21	A39	DZ Bank (DE)	40	39

6 Conclusion and Future Work

The study and constant evaluation of G-SIBs is of paramount importance for global financial stability, the Basel III agreement brought the methodology to measure and monitor the risk of these main banks. The great contribution of this work was the use of methodologies already tested and proven by the academy, with great mathematical robustness, to assist in the evaluation of G-SIBs.

MCDA methods are already used in various sectors of society to support decision-making; many applications can be observed in the military field, in health, in the environmental sector and many others that offer various benefits to society. In this line, MCDA methods have considerable exploratory potential within the financial market and can contribute to the regulation of the global financial market.

This research contributes both theoretically and practically. Theoretically, it advances the application of multicriteria decision aid (MCDA) to financial regulation, demonstrating its robustness in systemic risk evaluation. Practically, it provides regulators and policymakers with an alternative tool that enhances transparency and consistency in G-SIB rankings.

Despite its contributions, this study has limitations, such as reliance on BCBS's 2022 dataset and potential methodological biases. Future research should explore PSI-WISP's application using primary data sources and test its performance during financial crises to refine global evaluation metrics and improve risk assessment methodologies.

Acknowledgements. The authors acknowledge Fundação para a Ciência e a Tecnologia (FCT) for its financial support—UIDB/50022/2020 (LAETA Base Funding) and Coordenação de Aperfeiçoamento de Pessoal de Nível Superior (CAPES—Brazil) for the sandwich doctoral scholarship.

References

1. Abootalebi, S., Hadi-Vencheh, A., Jamshidi, A.: Ranking the alternatives with a modified TOPSIS method in multiple attribute decision making problems. IEEE Trans. Eng. Manag. **69**(5), 1800–1805 (2022). https://doi.org/10.1109/TEM.2019.2933593

2. Araujo, T.M.P.C., Gomes, C.F.S., Santos, M.: PSI and MPSI-MARA For Decision Ma-king (v1). Universidade Federal Fluminense, Niterói, Rio de Janeiro (2023)
3. Baidya, J., Garg, H., Saha, A., et al.: Selection of third party reverses logistic providers: an approach of BCF-CRITIC-MULTIMOORA using Archimedean power aggregation operators. Complex Intell. Syst. **7**, 2503–2530 (2021). https://doi.org/10.1007/s40747-021-00413-x
4. Baležentis, T., Zeng, X.: Group multi-criteria decision making based upon interval-valued fuzzy numbers: an e tension of the MULTIMOORA method. Expert Syst. Appl. **40**(2), 543–550 (2013). ISSN 0957–4174. https://doi.org/10.1016/j.eswa.2012.07.066
5. Basel Committee on Banking Supervision.: The G-SIB assessment methodology-scorecalculation (2014). http://www.bis.org/bcbs/publ/d296.pdf
6. Basel Committee of Banking Supervision.: An Assessment of the Long-Term Eco-nomic Impact of Stronger Capital and Liquidity Requirements. Bank of International Settlements (2010)
7. Brauers, W.K.M., Zavadskas, E.K.: Robustness of MULTIMOORA: a method for multi-objective optimization. Informatica **23**(1), 1–25 (2012). https://doi.org/10.15388/Informatica.2012.346
8. Casciello, R., Maffei, M., Ziebart, D.A.: Regulatory and contextual factors influencing earnings and capital management decisions: evidence from the European banking sector. Rev. Quant. Financ. Acc. **63**, 87–146 (2024). https://doi.org/10.1007/s11156-024-01253-9
9. Chen, S.X., Wang, J.Q., Wang, T.l.: Cloud-based ERP system selection based on extended probabilistic linguistic MULTIMOORA method and Choquet in-tegral operator. Comp. Appl. Math. **38**, 88 (2019). https://doi.org/10.1007/s40314-019-0839-z
10. Costa, I.P.A., Basílio, M.P., Maêda, S.M.N., Gomes, C.F.S., Santos, M.: Algorithm selection for machine learning classification: an application of the MELCHIOR multicriteria method. Front. Artif. Intell. Appl. **341**, 154–161 (2021). https://doi.org/10.3233/FAIA210243
11. Dahooie, J.H., Zavadskas, E.K., Firoozfar, H.R., Vanaki, A.S. Mohammadi, N., Brauers, W.K.M.: An improved fuzzy MULTIMOORA approach for multi-criteria decision making based on objective weighting method (CCSD) and its application to technological forecasting method selection. Eng. Appl. Artif. Intell. **79**, 114–128 (2019). ISSN 0952-1976.https://doi.org/10.1016/j.engappai.2018.12.008
12. De Almeida, I.D.P., Hermogenes, L.R.D.S., De Araújo Costa, I.P., ...Costa, D.D.O., Gomes, I.J.A.: Assisting in the choice to fill a vacancy to compose the PROANTAR team: Applying VFT and the CRITIC-GRA-3N methodology. Proced. Comput. Sci. **214**, 478–486 (2022). ISSN 1877–0509. https://doi.org/10.1016/j.procs.2022.11.202
13. De Paula, L.S., Da Silva, W.A., Silva, B., Santos, M.: MEREC-WISP: A Multiple Criteria Decision Making Support Website
14. Drumond, P., Basílio, M.P., De Araújo Costa, I.P., Gomes, C.F.S., Dos Santos, M.: Multicriteria analysis in additive manufacturing: an ELECTRE-MOr based approach. Front. Artif. Intell. Appl. **341**, 126–132 (2021). https://doi.org/10.3233/FAIA210240
15. Dzhagityan, E., Orekhov, M.: Global systemically important banks: do they still pose risks for financial stability? Int. Organ. Res. J. **17**(3) (2022). https://doi.org/10.17323/1996-7845-2022-03-03
16. Favara, G., Ivanov, I., Rezende, M.: GSIB surcharges and bank lending: evidence from US corporate loan data. J. Financ. Econ. **142**, 1426–1443 (2021). https://doi.org/10.1016/j.jfineco.2021.06.026Basel. Committee on Banking Supervision.: The G-SIB assessment methodology-score calculation (2014). http://www.bis.org/bcbs/publ/d296.pdf
17. Garcia-Herrero, A., Vazquez, F.: International diversification gains and home bias in banking. J. Bank. Finance **37**, 2560–2571 (2013)
18. Gordy, M.B.: Arisk-factor model foundation for ratings-based bank capital rules. J. Financ. Intermediation **12**, 199–232 (2003). https://doi.org/10.1016/S1042-9573(03)00040-8

19. Griffith-Jones, S., Segoviano, M., Spratt, S.: Basel II and Developing Countries: Diversification and Portfolio Effects. Working Paper, The London School of Economics (2002)
20. Güngör, E.: Prioritizing management strategies for laurel harvesting to enhance forest-based bioeconomy: a hybrid framework. Forests **15**, 1165 (2024). https://doi.org/10.3390/f15071165
21. Hafezalkotob, A., Hafezalkotob, A., Liao, H., Herrera, F.: An overview of MULTIMOORA for multi-criteria decision-making: theory, developments, applications, and challenges. Inf. Fusion **51**, 145–177 (2019). ISSN 1566–2535. https://doi.org/10.1016/j.inffus.2018.12.002
22. Herodoutou, H., Chen, Y., Lu, J.: A survey on automatic parameter tuning for big data processing systems. ACM Comput. Surv. **53**(2l), 1–37, 31 (2021)
23. Li, Z.-S., Chang, K.-H.: A novel pythagorean fuzzy set-based risk-ranking method for handling human cognitive information in risk-assessment prob-lems. Systems **11**, 402 (2023). https://doi.org/10.3390/systems11080402
24. Lucas, F.F., Santos, M., Gomes, C.F.S., Costa, A.P.A., Braga, G.O., Costa, L.M.A., Silva, L.F., Costa, V.P.A.: Valuation of real estate investment trusts using the PSI-CoCoSo multicriteria method. Proced. Comput. Sci. **242**, 881–887 (2024). ISSN 1877–0509. https://doi.org/10.1016/j.procs.2024.08.264
25. Lucas, S.F., Neves, J.C.S., Neves, G.C.: Robust solution identification for uncertainty management in MOLP – An interactive approach. Pesquisa Operacional **42**, e247885 (2022). https://doi.org/10.1590/0101-7438.2022.042.00247885
26. Maêda, S.M.N., Basílio, M.P., Costa, I.P.A., Moreira, M.A.L., Santos, M., Gomes, C.F.S., De Al-meida, I.D.P., Costa, A.P.A.: Investments in times of pandemics: an approach by the SAPEVO-M-NC method. Front. Artif. Intell. Appl. **341**, 162–168 (2021). https://doi.org/10.3233/FAIA210244
27. Maghsoodi, A.I., Kavian, A., Khalilzadeh, M., Brauers, W.K.M.: CLUS-MCDA: a novel framework based on cluster analysis and multiple criteria decision theory in a supplier selection problem. Comput. Ind. Eng. **118**, 409–422 (2018). ISSN 0360–8352. https://doi.org/10.1016/j.cie.2018.03.011
28. Maniya, K., Bhatt, M.G.: A selection of material using a novel type decision-making method: preference selection index method. Mater. Des. **31**, 1785–1789 (2010). https://doi.org/10.1016/j.matdes.2009.11.020
29. Mardani, A., Nilashi, M., Zakuan, N., Loganathan, N., Soheilirad, S., Saman, M.Z.M., Ibrahim, O.: A systematic review and meta-analysis of SWARA and WASPAS methods: theory and applications with recent fuzzy developments. Appl. Soft Comput. **57**, 265–292 (2017). ISSN 1568–4946. https://doi.org/10.1016/j.asoc.2017.03.045
30. Mardani, A., et al.: Determining the utility in management by using multi-criteria decision support tools: a review. Econ. Res. -Ekonomska Istraživanja **31**(1), 1666–1716 (2018). https://doi.org/10.1080/1331677X.2018.1488600
31. Mohammed Abdelkader, E., Zayed, T., El Fathali, H., Alfalah, G., Al-Sakkaf, A., Moselhi, O.: An integrated multi-criteria decision making model for the assessment of public private partnerships in transportation projects. Mathematics **11**, 3559 (2023). https://doi.org/10.3390/math11163559
32. Narayanamoorthy, S., Anuja, A., Kang, D., Kureethara, J.V., Kalaiselvan, S., Manirathinam, T.: A distinctive symmetric analyzation of improving air quality using multi-criteria decision making method under uncertainty conditions. Symmetry **2020**, 12 (1858). https://doi.org/10.3390/sym12111858
33. Neves, J.C.S., Lucas, S.F., Neves, G.C., Weissmann, J.: Bi-level model: new approach to dynamic vehicle allocation in supply chain. Pesquisa Operacional **43**, e266695 (2023). https://doi.org/10.1590/0101-7438.2023.043.00266695
34. Pereira, D.A.M., et al.: Multicriteria and statistical approach to support the outranking analysis of the OECD countries. IEEE Access **10**, 69714–69726 (2022). https://doi.org/10.1109/ACCESS.2022.3187001

35. Silva, L.F., Costa, I.P.A., Oliveira, T.E.S., Rangel, G.C., Lucas, F.F., Costa, L.M.A., Corriça, J.V.P., Costa, A.P.A., Santos, M.: Demand forecasting for steel company spare items with time series templates. Proced. Comput. Sci. **242**, 57–64 (2024). ISSN 1877–0509. https://doi.org/10.1016/j.procs.2024.08.230
36. Souzangarzadeh, H., Jahan, A., Rezvani, M.J., et al.: Multi-objective optimiza-tion of cylindrical segmented tubes as energy absorbers under oblique crushes: D-optimal design and integration of MULTIMOORA with combinative weighting. Struct. Multidisc. Optim. **62**, 249–268 (2020). https://doi.org/10.1007/s00158-020-02486-7
37. Stanujkic, D., Karabasevic, D., Zavadskas, E.K., Smarandache, F., Brauers, W.K.: A bipolar fuzzy Extension of the MULTIMOORA Method. Informatica **30**(1), 135–152 (2019). https://doi.org/10.15388/Informatica.2019.201
38. Stanujkic, D., Popovic, G., Karabasevic, D., Meidute-Kavaliauskiene, I.,Ulutaş, A.: An integrated simple weighted sum product method—WISP. In: IEEE Transactions on Engineeing Management, vol. 70, no. 5, pp. 1933–1944 (2021b). https://doi.org/10.1109/TEM.2021.3075783
39. Tsai, W.H., Lee, P.L., Shen, Y.S., et al.: A combined evaluation model for encour-aging entrepreneurship policies. Ann. Oper. Res. **221**, 449–468 (2014). https://doi.org/10.1007/s10479-011-1029-6
40. Ulutaş, A., Topal, A., Pamučar, D., Stević, Ž, Karabašević, D., Popović, G.: A new integrated multi-criteria decision-making model for sustainable supplier selection based on a novel grey WISP and grey BWM methods. Sustainability **14**, 16921 (2022). https://doi.org/10.3390/su142416921
41. Vazquez, F., Federico, P.: Bank funding structures and risk: evidence from the global financial crisis. J. Bank. Finance **61**, 1–14 (2015). https://doi.org/10.1016/j.jbankfin.2015.08.023

Energy and Environment

Efficiency and Energy Consumption in Suspended Load Movement Control

Diogo Baptista[1,2,3(✉)] and Carlos Neves[1,2]

[1] ESTG, Polytechnic University of Leiria, Leiria, Portugal
diogo.baptista@ipleiria.pt
[2] INESC Coimbra, Coimbra, Portugal
[3] CIMA-UE, Évora, Portugal

Abstract. The use of lifting devices is common in many areas of economic activity, from industrial logistics to construction and port operation. In this study, a laboratory test rig will be used to analyze energy consumption associated with controlling the movement of suspended loads. The objective is to assess the energy efficiency of different control methods employed in industrial applications. By studying various control techniques, including proportional-integral-derivative (PID) control, pole placement and optimal control, their respective energy consumption patterns were compared. This comparative analysis will enable informed decisions regarding the most sustainable and efficient control strategy for managing suspended loads in industrial settings. The findings of this study hold significant implications for optimizing energy usage and promoting sustainability in industrial operations.

Keywords: Crane control · Suspended load motion · Energy consumption

1 Introduction

The transportation of suspended loads is a common task in many industrial sectors, such as logistics, construction, and port operations. [1]. The unactuated nature of some of the degrees of freedom on one such mechanic system, implies a difficult manual control which requires highly trained operators [2], rendering this problem a good target for automatic control with several controller flavours: e.g. [3] uses a supervisory scheme for control, [4] chose PID-Like controllers, [5] present a unit magnitude input control and in [6] a very comprehensive survey can be found. Efficient control of the movement of these loads is essential to ensure the safety and productivity of these processes. However, different control approaches can significantly influence the energy consumption associated with these operations.

The aim of this paper is to investigate and compare various types of control in the transportation of suspended loads, with an emphasis on optimizing energy consumption. The intention is to determine which control methods are most energy-efficient and, consequently, more sustainable for practical applications.

© The Author(s), under exclusive license to Springer Nature Switzerland AG 2027
M. Calvete Gaspar et al. (Eds.): Regional Helix 2024, LNME, pp. 173–187, 2027.
https://doi.org/10.1007/978-3-032-36365-7_12

This study builds upon previous work in this area which highlight the importance of efficient control approaches to minimize energy consumption in DC motors, such as [7] where a comparative study of control techniques is done or [8], where the LQR is applied in an industrial application. By applying different control methods and comparing the results obtained, a contribution will be made to the development of more sustainable and energy-efficient solutions for the transportation of suspended loads.

The work describes the experimental design, measurement methods, and data analysis techniques used, as well as the results obtained in simulation, and the conclusions drawn from these results. The aim is to provide valuable insights for the development of more efficient and sustainable suspended load transportation systems in the future.

2 Modelling

2.1 Development of the Model for the System

Independently of the specific controllers used, modelling is always a starting point for control systems applications. In this case, modelling of suspended load motion in literature is rich, and encompasses simple solutions, as a simple, rigid, fixed length pendulum on a cart, with a single actuator (cart displacement) through to complex 3D dynamics including the flexibility of the cable into the modelling (see [6] and [9] for comprehensive surveys).

The present study adopts an intermediate solution, with two degrees of freedom (cart movement and pendulum length). Here, under the assumption of small angular displacements, the cable, although flexible, is considered as rigid for modelling effects.

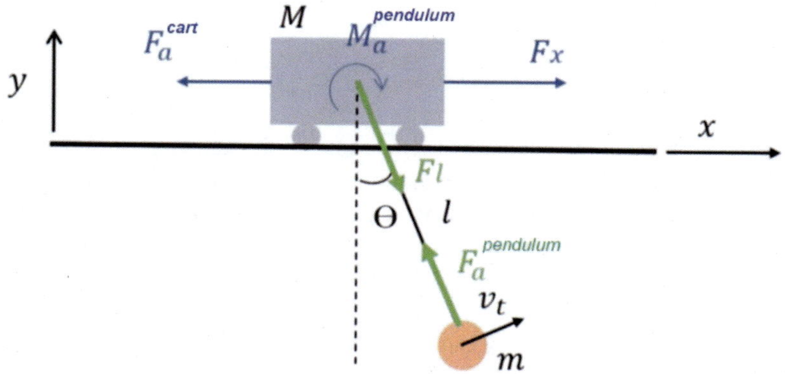

Fig. 1. Simplified model of the system

The system consists of a cart with linear displacement, which moves when a force is applied, and a pendulum (or cable) connected to the cart, which can move vertically when a force is applied (see Fig. 1). The suspended load is located at the end of the cable.

The model depends on the following parameters:

- M is the mass of the cart, in kilograms (kg);
- m is the mass of the suspended load, in kilograms (kg);
- θ is the angle between the vertical position and the instantaneous position of the pendulum, in radians (rad);
- l is the length of the pendulum, in meters (m);
- F_x is the external force applied to the cart, in Newtons (N);
- F_a^{cart} is the frictional force of the cart, in Newtons (N);
- F_l is the external force applied to the pendulum, in Newtons (N);
- F_a^{pend} is the frictional force of the pendulum, in Newtons (N);
- M_a^{pend} is the frictional moment of the pendulum, originating from the suspension point of the load, in Newton meters $(N.m)$;
- v_t is the tangential velocity at the end of the cable, in meters per second (m/s).

To simplify the model, the mass of the suspension rod was not considered, as it is much smaller than the load, and thus its moment of inertia is set to zero, and it was treated as a rigid body. It was assumed that the mass of both the cart and the suspended load are concentrated at their respective centers of mass. Additionally, the elongation (elastic) effect between the cart and the load, due to the tensile force on the rod, was ignored.

The equation of motion along the generalized coordinate of the cart's linear position, x, can be obtained using the $Euler - Lagrange$ equation:

$$\frac{d}{dt}\left(\frac{\partial L}{\partial q\dot{x}}\right) - \frac{\partial L}{\partial qx} = Qext \tag{1}$$

The generalized coordinates of the system, q, correspond to the three degrees of freedom (x, θ, l) of the system under study, and the generalized external forces correspond to non-conservative external forces. For each generalized coordinate,

$$Q_{ext} = \begin{bmatrix} F_{cart} - F_a^{cart} \\ -M_a^{pend} \\ F_{pend} - F_a^{pend} \end{bmatrix} = \begin{bmatrix} F_x - D_x\dot{x} \\ D_\theta\dot{\theta} \\ F_l - D_l\dot{l} \end{bmatrix}, \tag{2}$$

where:

- D_x, D_θ and D_l are the viscous damping coefficients (N·s/m or N·s/rad) associated with the motion in the $x, \theta andl$ directions, respectively.
- $\dot{x}$, $\dot{\theta}$ and $\dot{l}$ are the time derivatives (velocities) of the generalized coordinates.

Therefore, by determining each term, we obtain the equation of motion with respect to $x, \theta andl$, given respectively by:

$$\ddot{x}(M + m) + m\left(\ddot{l}sin\theta + 2\dot{l}\dot{\theta}cos\theta + l\ddot{\theta}cos\theta - l\dot{\theta}2sin\theta\right) = F_C - D_x\dot{x} \tag{3}$$

$$m\ddot{x}lcos\theta + 2ml\dot{l}\theta + ml2\ddot{\theta} + mglsin\theta = -D_\theta\dot{\theta} \tag{4}$$

$$m\ddot{x}sin\theta + m\ddot{l} - ml\dot{\theta}^2 - mgcos\theta = F_l - D_l\dot{l} \tag{5}$$

According to the characteristics and advantages of state space representation, the mathematical model was chosen to be represented in state space with 6 state variables, $x, \dot{x}, \theta, \dot{\theta}, l, and\ \dot{l}$, which form the state vector

$$[x_1\, x_2\, x_3\, x_4\, x_5\, x_6]^T = \left[x\dot{x}\theta\dot{\theta}l\dot{l}\right]^T$$

Since the system will operate around an equilibrium point, it is possible to approximate the nonlinear system to a linear system, considering the equilibrium point of the system as point $[0000l_0 0]^T$, where l_0 corresponds to the length of the pendulum at the equilibrium point, it is possible to obtain the state equation of the linearized model

$$\dot{x} = Ax + Bu \tag{6}$$

where

$$A = \begin{bmatrix} 0 & 1 & 0 & 0 & 0\,0 \\ 0 & \frac{-D_x}{M} & -\frac{mg}{M} & \frac{D_\theta}{Ml_0} & 0\,0 \\ 0 & 0 & 1 & 0 & 0\,0 \\ 0 & \frac{D_x}{Ml_0} & -\frac{g}{l_0} & -\frac{D_\theta(M+m)}{mMl_0^2} & 0\,0 \\ 0\,0 & & 0\,0 & & 0 \quad 1 \\ 0\,0 & & 0\,0 & & 0 \quad \frac{-D_l}{m} \end{bmatrix}$$

and

$$B = \begin{bmatrix} 0 & 0 \\ \frac{1}{m} & 0 \\ 0 & 0 \\ -\frac{1}{ml_0} & 0 \\ 0 & 0 \\ 0 & \frac{1}{m} \end{bmatrix}$$

In the given system, the direct transmission matrix D is zero because the input variable is a force that has no direct relation to the output variables. Thus, the output equation of the model is

$$y = Cx \tag{7}$$

Since the existing instrumentation only allows for the measurement of the variables $x, \theta\ and\ l$, the matrix C was defined as

$$C = \begin{bmatrix} 1\,0\,0 & 0\,0\,0 \\ 0\,0\,1 & 0\,0\,0 \\ 0\,0\,0 & 0\,1\,0 \end{bmatrix}$$

However, since the command input to the motor drivers is given in terms of velocity, it was necessary to reformulate the system model with velocity input, that is, the velocities applied to the cart and the cable.

Writing the linear model obtained in the form of a system of differential equations,

$$\begin{cases} \dot{x} = \dot{x} \\ \ddot{x} = -\frac{D_x}{M}\dot{x} - \frac{mg}{M}\theta + \frac{D_\theta}{Ml_0}\dot{\theta} + \frac{1}{M}F_c \\ \dot{\theta} = \dot{\theta} \\ \ddot{\theta} = \frac{D_x}{Ml_0}\dot{x} - \frac{g}{l_0}\theta - \frac{D_\theta(M+m)}{mMl_0^2}\dot{\theta} - \frac{1}{Ml_0}F_c \\ \dot{l} = \dot{l} \\ \ddot{l} = -\frac{D_l}{m}\dot{l} + \frac{1}{m}F_l \end{cases} \tag{8}$$

Analyzing the system in (Eq. 8), it is possible to rewrite the second equation to determine the force exerted by the motor as a function of the linear velocity and the pendulum angle, resulting in the following,

$$\frac{1}{M}F_c = \ddot{x} + \frac{D_x}{M}\dot{x} + \frac{mg}{M}\theta - \frac{D_\theta}{Ml_0}\dot{\theta} \tag{9}$$

by substituting into the fourth equation of the system in (Eq. 8), one gets

$$\ddot{\theta} = \left(-\frac{D_\theta(M+m)}{mMl_0^2} + \frac{D_\theta}{Ml_0^2}\right)\dot{\theta} + \left(-\frac{g}{l_0} - \frac{mg}{Ml_0}\right)\theta - \frac{1}{l_0}\dot{v_x} \tag{10}$$

Following Vaccaro [10], by integrating the equation (Eq. 10) twice and setting the first state variable as $X_1 = \theta$ and defining the second state variable X_2 as

$$X_2 = \left(-\frac{g}{l_0} - \frac{mg}{Ml_0}\right)\int \theta$$

one gets the first state equation of the model

$$\dot{X_1} = \left(-\frac{D_\theta(M+m)}{mMl_0^2} + \frac{D_\theta}{Ml_0^2}\right)X_1 + X_2 - \frac{1}{l_0}\dot{v_x} \tag{11}$$

By deriving X_2, the second state equation can be obtained. The third state variable X_3 corresponds to the derivative of the position, and the fourth state variable, X_4, is the derivative of the length of the cable.

Therefore, the inputs of the new model are the velocities v_x and v_l and its outputs are defined as the load oscillation angle X_1, the cart position X_3, and the cable length X_4. Thus, it is possible to write a new linear model of the system in state space as follows:

$$\dot{X} = AX + BV$$

that is,

$$\begin{bmatrix} \dot{X_1} \\ \dot{X_2} \\ \dot{X_3} \\ \dot{X_4} \end{bmatrix} = \begin{bmatrix} \left(-\frac{D_\theta(M+m)}{mMl_0^2} + \frac{D_\theta}{Ml_0^2}\right) & 1 & 0 & 0 \\ \left(-\frac{g}{l_0} - \frac{mg}{Ml_0}\right) & 0 & 0 & 0 \\ 0 & 0 & 0 & 0 \\ 0 & 0 & 0 & 0 \end{bmatrix} \begin{bmatrix} X_1 \\ X_2 \\ X_3 \\ X_4 \end{bmatrix} + \begin{bmatrix} -\frac{1}{l_0} & 0 \\ 0 & 0 \\ 1 & 0 \\ 0 & 1 \end{bmatrix} \begin{bmatrix} v_x \\ v_l \end{bmatrix} \tag{12}$$

and the output equation, it is given by:

$$\begin{bmatrix} x \\ \theta \\ l \end{bmatrix} = \begin{bmatrix} 0 & 0 & 1 & 0 \\ 1 & 0 & 0 & 0 \\ 0 & 0 & 0 & 1 \end{bmatrix} \begin{bmatrix} X_1 \\ X_2 \\ X_3 \\ X_4 \end{bmatrix} \tag{13}$$

2.2 Development of a Force Observer for the Load Movement System

Starting from the linear model developed given in (Eq. 8), the transfer function of a model that allows estimating the values of the forces involved in the operation of the system can be found. By estimating these values, one can calculate the instantaneous mechanical power involved and the total energy transferred over time, knowing that $p = F.V$, where p is the instantaneous mechanical power (W, watts), F is the applied force (N, newtons) and V is the velocity of the object (m/s), that is,

$$p(t) = F_c v_x(\mathrm{t}) + F_l v_l(t) \tag{14}$$

and the total energy E (J, joules) over a time interval $[t_1, t_2]$ is given by

$$E = \int_{t_1}^{t_2} p(t)dt \tag{15}$$

Taking the linear model developed during the work of Vicente Caneiro[11] and extended by Daniela Pedrosa in [12], it is possible to express the force applied to the cart from the second equation of the system in (Eq. 8):

$$\begin{aligned} F_c &= M\ddot{x} + D_x\dot{x} + \mathrm{mg}\ \theta - \frac{D_\theta}{l_0}\dot{\theta} \\ &= M\ v_x + D_x v_x +\ \mathrm{mg}\ \theta - \frac{D_\theta}{l_0}\dot{\theta} \end{aligned} \tag{16}$$

Considering the fourth equation of the system given in (Eq. 8) and applying the Laplace Transform (L.T.), results in:

$$s^2\Theta(s) = \frac{D_\theta}{Ml_0^2}\left(1 - \frac{M+m}{m}\right)s\Theta(s) - \frac{g}{l_o}\left(1 + \frac{m}{M}\right)\Theta(s) - \frac{1}{l_0}sV_x(S) \tag{17}$$

that is

$$\Theta(s) = -\frac{s}{l_0 s^2 + \frac{D_\theta}{Ml_0^2}\left(\frac{M+m}{m} - 1\right)s + \frac{g}{l_0}\left(1 + \frac{m}{M}\right)}V_x(s) \tag{18}$$

Similarly applying Laplace Transform (L.T.) to equation (Eq. 16), one obtains:

$$F_c(s) = MsV_x(s) + D_x V_x(s) +\ \mathrm{mg}\Theta(\mathrm{s}) - \frac{D_\theta}{l_0}s\Theta(s)$$

$$= (Ms + D_x)V_x(s) + \left(mg - \frac{D_\theta}{l_0}s\right)\Theta(s) \tag{19}$$

Now substituting in (Eq. 19) the result from (Eq. 18), yields:

$$F_c(s) = \left[Ms + D_x + \frac{\left(\frac{D_\theta}{l_0}s^2 - mgs\right)}{l_0 s^2 + \left(\frac{D_\theta(M+m)}{mMl_0} - \frac{D_\theta}{Ml_0}\right)s + \left(g + \frac{mg}{M}\right)}\right]V_x(s) \tag{20}$$

Returning to the initial system of differential equations in (Eq. 8), taking the last equation and expressing the force,

$$F_l = m\ddot{l} + D_l\dot{l} = m\dot{v}_l + D_l v_l \tag{21}$$

applying the Laplace Transform (L.T.) again, the result is:

$$F_l(s) = (ms + D_l)V_l(s) \tag{22}$$

3 Controller Development

The control system for the movement of the suspended load, as illustrated in Fig. 2, employs a controller that regulates the three main variables: the load oscillation angle, the cart position, and the cable length. These variables are grouped into the state vector $X = \left[\theta\ x\ l\right]^T$, while the reference values for each variable are represented by the reference vector $X_{ref} = \left[\theta_{ref}\ x_{ref}\ l_{ref}\right]^T$.

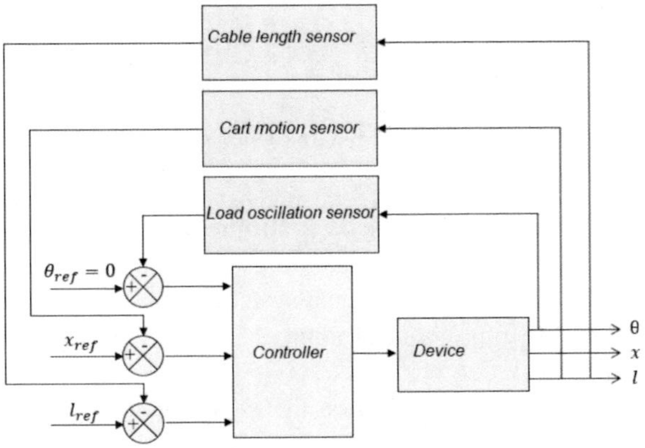

Fig. 2. Control system diagram

The control system for the movement of the cart and the length of the cable uses an error signal between the reference value and the measurement taken by sensors. For the load oscillation angle controller, the reference θ_{ref} is set to 0°, while for the other two inputs, x_{ref} and l_{ref} which correspond to the movement of the cart and the cable, a step input signal is used.

Simulations of the system were conducted with different controllers, including PID controller, the State Feedback (SF) controller, and the Model Predictive Control (MPC), using MATLAB and Simulink, allowing for the analysis of the system's response to various control parameters.

Regarding the State Feedback (RE) controller, the pole placement method was employed to obtain the values that best fit the model's performance. Additionally, it was necessary to modify the model to enable the system to receive a reference different from zero by incorporating the difference $X_{ref} - X$, allowing this to be regulated to zero. The equations that represent the system in state space are

$$\begin{aligned} \dot{X} &= (A - BK)X + BKX_{ref} \\ y &= CX \end{aligned} \quad (23)$$

and the inputs to the system are given by

$$V = K\left(X_{ref} - X\right) \quad (24)$$

where K is the gain matrix determined by the pole placement method. In Fig. 3, a block diagram for this state feedback control is presented.

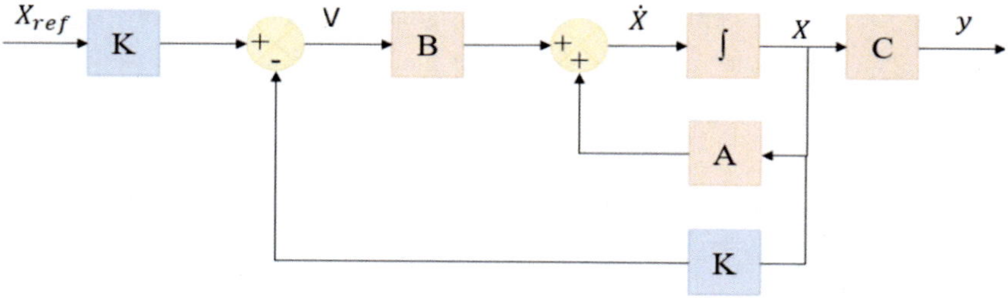

Fig. 3. State feedback control with pole placement, given a reference to the system.

On the other hand, MPC is an advanced control technique that uses a system model to predict future behavior of its outputs and adjust the control inputs in an optimized way. MPC predicts the future behavior of the system outputs over a prediction horizon of N time steps, based on future control inputs. (In Fig. 4 a simple intuition into the MPC strategy is presented.)

The goal is to minimize a cost function that penalizes the difference between the predicted output and the desired reference, as well as the control effort. To allow for flexibility in how much weight is given to the tracking error and control effort, the cost function includes weights. At each time step, MPC solves an optimization problem to determine the optimal sequence of control inputs that minimizes the cost function over the prediction horizon. Only the first control input is applied to the system, and the process is repeated at the next time step.

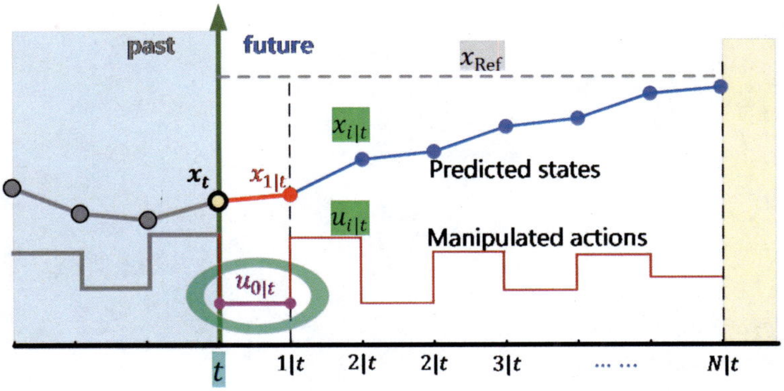

Fig. 4. MPC controller strategy[1]

For linear systems, the MPC formulation is simplified, which leads to a quadratic programming (QP) problem that can be solved efficiently. In the case of regulation, the cost function of MPC is

$$J = \min \frac{1}{2} X_{t+N}^{T} S X_{t+N} + \frac{1}{2} \sum_{k=0}^{N-1} \left(X_{t+k}^{T} Q X_{t+k} + V_{t+k}^{T} R V_{t+k} \right) \tag{25}$$

In a tracking problem, in the cost function J in (Eq. 25, the state variable is replaced by the error variable, thus obtaining

$$J = \min \frac{1}{2} E_{t+N}^{T} S E_{t+N} + \frac{1}{2} \sum_{k=0}^{N-1} \left(E_{t+k}^{T} Q E_{t+k} + V_{t+k}^{T} R V_{t+k} \right) \tag{26}$$

where

$$E_k = X_{ref} - y_k$$

and S, Q are the weighting matrices applied to the tracking error and R is the weighting matrix on the control effort.

4 Experimental Setup and Implementation

The tests were developed around a laboratorial experiment, consisting of an 18 m length rail with a load suspended from a cart. The cart and a reel mounted on it are actuated by electric motors, allowing for horizontal cart motion [11] and load hanging length variation [12]. Part of this laboratory equipment can be seen in Fig. 5.

The controllers were implemented on a microcontroller development board (Arduino Mega) to establish the usability limits of this platform. These are discrete versions of

[1] Tsesea, CC BY-SA 4.0 <https://creativecommons.org/licenses/by-sa/4.0>, via Wikimedia Commons.

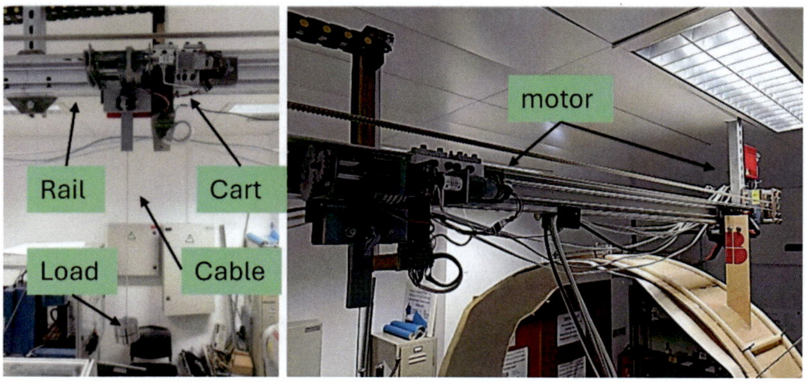

Fig. 5. Laboratorial experiment

the developments of Sect. 3, with a sampling time of 10ms, which was fast enough for the control and affordable in terms of the computational power of the used platform.

The simulations were conducted considering two distinct scenarios, each with specific objectives for the movement of the cart and the coupled pendulum.

In the first scenario, the cart was required to move to a reference position of 2.25 m, while the load followed a predefined trajectory, as shown in Fig. 6.

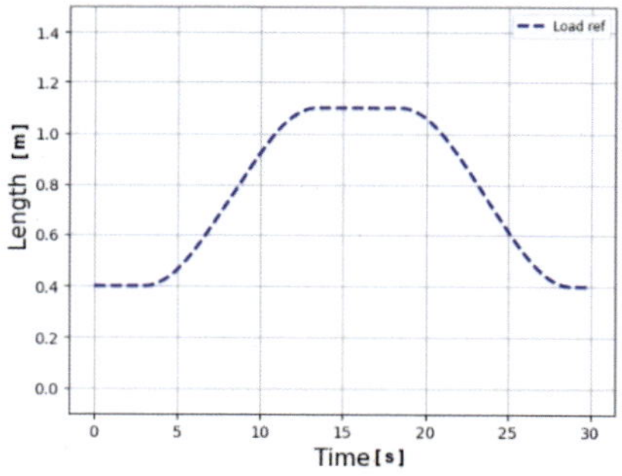

Fig. 6. Load reference for the 1st scenario

For this case, the 2.25 m position reference was imposed through a step input. The control parameters were adjusted to ensure that the cart reached its target position within a time frame of 20 to 30 s, while maintaining system stability and minimizing load oscillation.

In the second scenario, both the cart position and load trajectory were driven by time-varying references. The specified trajectory for this scenario is illustrated in Fig. 7.

This setup aimed to simulate a more complex movement, where synchronization between the cart's motion and controlled load oscillation was critical to achieve the desired performance outcomes.

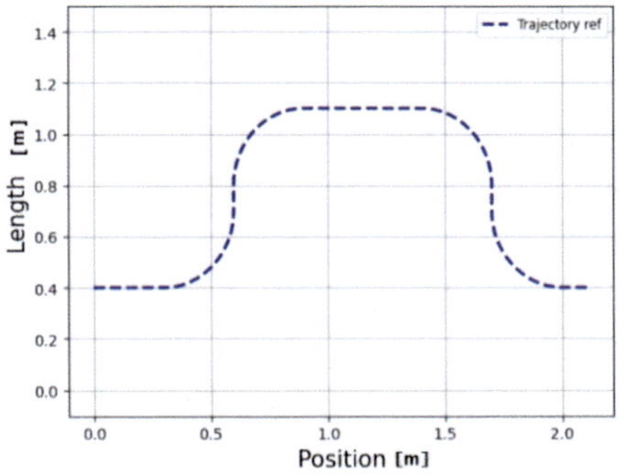

Fig. 7. Trajectory reference for 2nd scenario

5 Results

The results from the simulations for both scenarios are presented in the following figures, from Figs. 8, 9, 10, 11 and 12, allowing for a clear comparison of system performance under varying control conditions.

In the first scenario, three distinct situations, labeled A, B, and C, are presented in Figs. 8, 9, and 10, respectively. The purpose of having these three variations is to provide a more comprehensive assessment of the performance of the Model Predictive Control (MPC). By progressively adjusting the MPC in each situation, one can observe how the system responds in terms of accuracy in position and energy efficiency, comparing these results with other control methods.

In this comparison, the PID and State-Space (SS) controllers remained unchanged across all three situations, ensuring a consistent benchmark for evaluation. The only control strategy that was modified was the MPC, allowing the effect of its tuning on system performance to be isolated and observed.

In case A, the MPC controller is configured in such a way that it takes longer to reach the reference position when compared to the PID and SS controllers. Case B presents an intermediate configuration, where the MPC achieves the reference in a time frame between the other two controllers. Finally, in case C, the MPC is tuned for a faster response, outperforming both the PID and SS controllers in terms of speed in reaching the target position.

In the second scenario, two situations, labeled A and B, are analyzed. Given that the objective was to follow a predefined trajectory, the MPC was specifically tuned to closely replicate the behavior of the other two controllers.

In situation A, the MPC adopts a more conservative approach, ensuring that the cart's position remains strictly within the predefined bounds, it avoids exceeding the 2.25 m limit. This configuration prioritizes safety and constraint adherence over trajectory matching (Fig. 11).

In contrast, situation B presents a more aggressive tuning of the MPC. Here, the controller allows a slight overshoot beyond the 2.25 m mark to achieve a trajectory more closely aligned with the behavior of the PID and SS controllers. This configuration aims to reduce tracking error while still maintaining system stability and efficiency (Fig. 12).

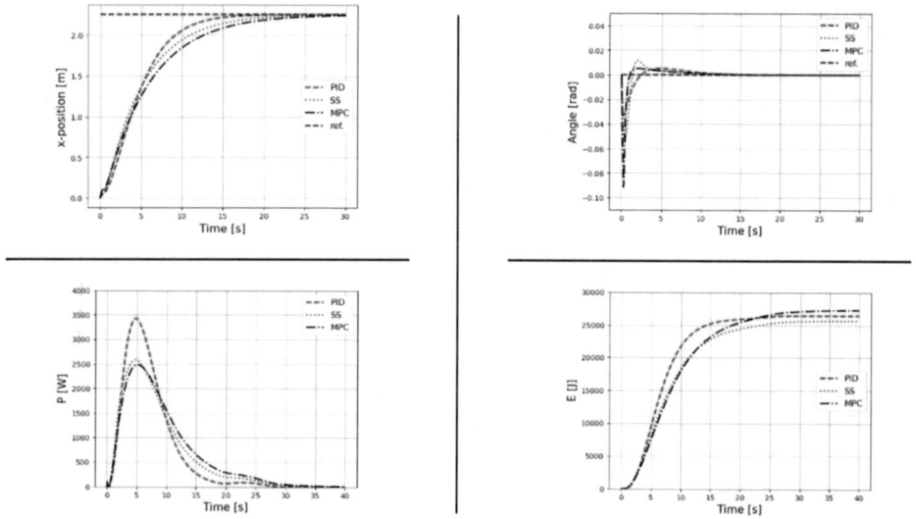

Fig. 8. Performance comparison between PID, SS, and MPC strategies in case A of first scenario. The top-left plot shows the cart's position tracking, and the top-right plot displays the load oscillation angle. The bottom-left plot presents the instantaneous mechanical power required, and the bottom-right plot shows the accumulated energy consumption.

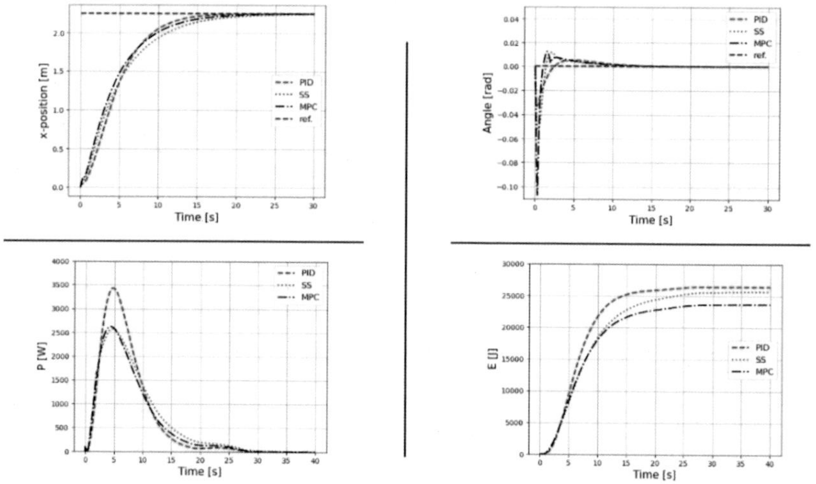

Fig. 9. Performance comparison between PID, SS and MPC strategies in case B of first scenario. The top-left plot shows the cart's position tracking, and the top-right plot displays the load oscillation angle. The bottom-left plot presents the instantaneous mechanical power required, and the bottom-right plot shows the accumulated energy consumption.

Overall, these two situations highlight the flexibility of the MPC in balancing trajectory accuracy and constraint management, with both configurations maintaining the controller's characteristic advantage in energy efficiency.

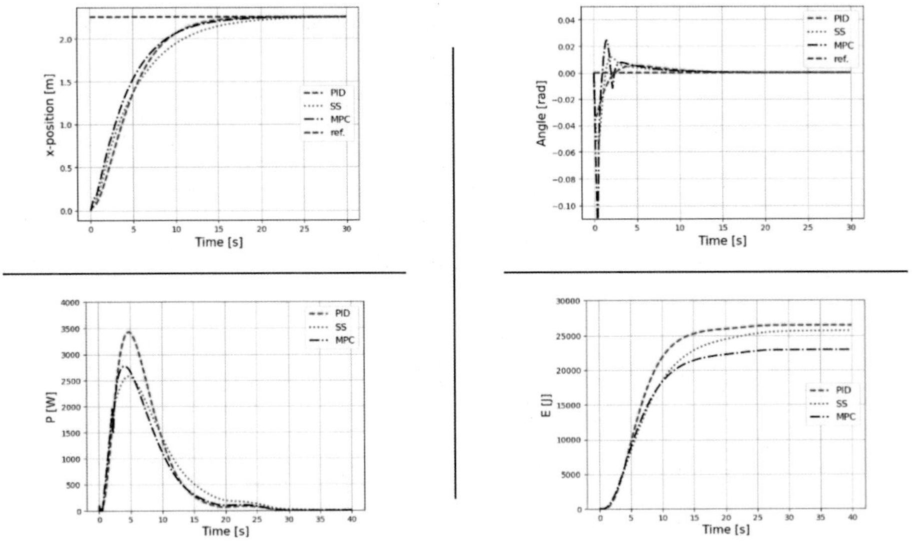

Fig. 10. Performance comparison between PID, SS and MPC strategies in case C of first scenario. The top-left plot shows the cart's position tracking, and the top-right plot displays the load oscillation angle. The bottom-left plot presents the instantaneous mechanical power required, and the bottom-right plot shows the accumulated energy consumption.

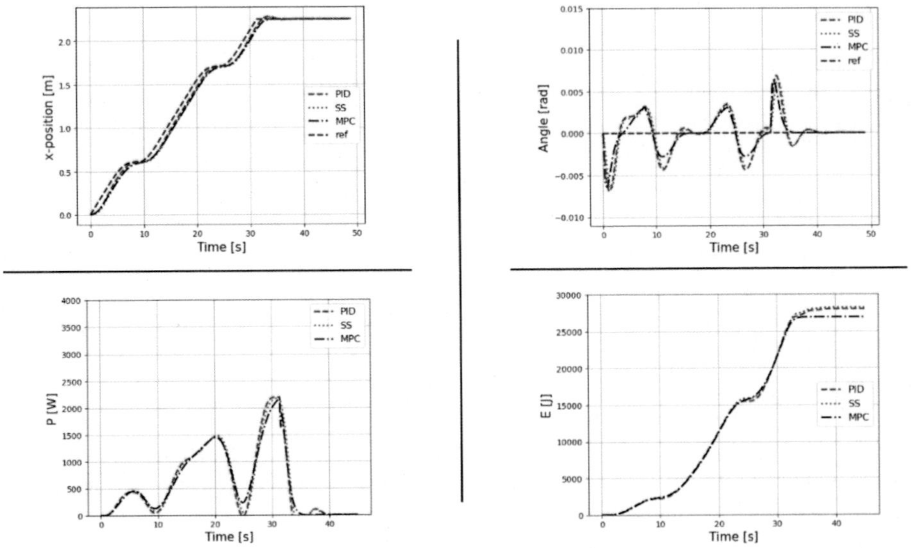

Fig. 11. Performance comparison between PID, SS and MPC strategies in case A of second scenario. The top-left plot shows the cart's position tracking, and the top-right plot displays the load oscillation angle. The bottom-left plot presents the instantaneous mechanical power required, and the bottom-right plot shows the accumulated energy consumption.

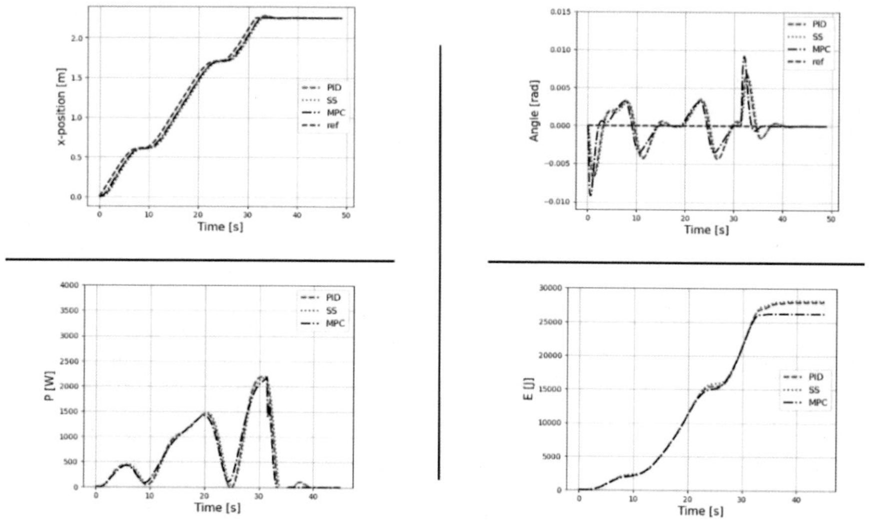

Fig. 12. Performance comparison between PID, SS and MPC strategies in case B of second scenario. The top-left plot shows the cart's position tracking, and the top-right plot displays the load oscillation angle. The bottom-left plot presents the instantaneous mechanical power required, and the bottom-right plot shows the accumulated energy consumption.

6 Conclusions and Further Work

This study presented an analysis of suspended load movement control, focusing on energy efficiency. Various control methods, such as PID, State Feedback control, and Model Predictive Control (MPC), were compared using the simulation of a laboratory platform. The results demonstrated that the choice of control method significantly influences energy consumption, with predictive control showing superior performance in terms efficiency in more complex load movement scenarios at the cost of a very small increase on the maximum oscillation angle.

It is important to highlight that the tests conducted considered the use of a microcontroller platform which, despite its limitations in computational capacity, can implement efficient solutions for suspended load control, allowing for optimized performance even in resource-constrained environments.

In this way, this work contributes to the development of more sustainable and optimized systems, thus providing a solid foundation for future research and industrial applications. The use of advanced control techniques not only improves motion accuracy but also reduces energy consumption, which is crucial for sustainability in industrial operations.

The results presented are encouraging and foster the pursuit of further work in applying MPC control in more complex situation and experimenting with more detailed cost functions specifically tailored for energy consumption minimization. Moreover, the model used here follows closely the operation of the real system as much as the cable length is near the linearization point (half the maximum cable length), slightly degrading as it offsets from that point. Even if, for the experiment used, this is not very important and the close loop control shows it can cope with this modelling discrepancy, the use of a piecewise linear model could improve this point. Also, in terms of microcontroller implementation, more work will be done to better establish the relationship between the mathematical formulation of the controller and the computation load of its implementation. Finally, the good results obtained in terms of the control results encourage the implementation of a full 3D test rig.

References

1. Ramli, L., Mohamed, Z., Abdullahi, A.M., Jaafar, H.I., Lazim, e I.M.: Control strategies for crane systems: a comprehensive review. Mech. Syst. Signal Process. **95**, 1–23, out (2017)
2. Lee, H.-H.: A new design approach for the anti-swing trajectory control of overhead cranes with high-speed hoisting. Int. J. Control **77**, 931–940 (2004)
3. Man, Y., Liu, Y.: Positioning and antiswing control of overhead crane systems: a supervisory scheme. J. Frankl. Inst. **360**(18), 14329–14343 (2023). https://doi.org/10.1016/j.jfranklin.2023.10.038
4. Zhang, S., He, X., Zhu, H., Li, X., Liu, e.X.: PID-like coupling control of underactuated overhead cranes with input constraints. Mech. Syst. Signal Process. **178**, 109274 (2022)
5. Maghsoudi, M.J., Ramli, L., Sudin, S., Mohamed, Z., Husain, A.R., Wahid, e H.: Improved unity magnitude input shaping scheme for sway control of an underactuated 3D overhead crane with hoisting. Mech. Syst. Signal Process. **123**, 466–482 (2019)
6. Rasool Mojallizadeh, M., Brogliato, B., Prieur, e C.: Modeling and control of overhead cranes: a survey. INRIA Grenoble; UGA (Université Grenoble Alpes); GIPSA Lab; INP Grenoble; LJK/Grenoble University – INRIA. https://hal.inria.fr/hal-03817363
7. Maghfiroh, H., Ataka, A., Wahyunggoro, O., Cahyadi, A.I.: Optimal energy control of DC motor speed control: comparative study. In: 2013 International Conference on Computer, Control, Informatics and its Applications (IC3INA), pp. 89–93. Jakarta, Indonesia (2013). https://doi.org/10.1109/IC3INA.2013.6819154
8. Maghfiroh, H., Gunawan, M., Anwar, M.: Optimal energy control of DC-drive conveyor using LQR method. AIP Conf. Proc. **2217**(1), 030145 (2020). https://doi.org/10.1063/5.0000619
9. Abdel-Rahman, E.M., Nayfeh, A.H., Masoud, Z.N.: Dynamics and control of cranes: a review. J. Vib. Control **9**(7), 863–908 (2003). https://doi.org/10.1177/1077546303009007007
10. Vaccaro, R.J.: Digital control. McGraw-Hill, A State-Space Approach (1995)
11. Caneiro, V.: Comando Avançado do Movimento de Cargas Suspensas, Instituto Politécnico de Leiria, Leiria (2021). https://iconline.ipleiria.pt/bitstream/10400.8/6120/1/VicenteCaneiro_Relatorio_Projeto_Mestrado_2172023.pdf
12. Pedrosa, D.: Movimentação de Cargas Suspensas, Inst. Politécnico de Leiria, Leiria (2024). https://iconline.ipleiria.pt/bitstreams/4d318349-5cb6-46c1-bf3f-e6421b37410d/download

Optimization of the Thermal Efficiency of a Domestic Pellet Boiler

Daniel Gouveia[1], Tânia Ferreira[1], and Carlos Pereira[1,2](✉)

[1] ESTGV, Polytechnic Institute of Viseu, Campus Politécnico de Repeses, Viseu, Portugal
ccatorze@estgv.ipv.pt

[2] Centre for Research in Digital Services (CISeD), Polytechnic Institute of Viseu, Viseu, Portugal

Abstract. Certain types of domestic biomass combustion equipment is characterized by low efficiency and high pollutant emissions due to issues related to sizing and operating conditions. In this context, the present work aimed to optimize the thermal efficiency of a pellet boiler by varying the boiler's power and excess air. Thus, the boiler's performance was evaluated with different air excess levels at two distinct thermal loads: 3 and 5. Thermal efficiency was determined using the direct method and assessed during the steady-state phase. The results showed that, in general, increasing the boiler's power increased its thermal efficiency and decreased carbon monoxide (CO) emissions. However, at the lower thermal load, reducing the excess air led to increased thermal efficiency and a significant decrease in CO emissions. Therefore, it was concluded that the boiler was not fully optimized by the manufacturer regarding excess air adjustments at all thermal loads, highlighting the need for further adjustments to achieve optimal performance under varying operating conditions.

Keywords: Energy · Combustion · Pellets · Boiler optimization

1 Introduction

The balance between energy supply and demand remains of fundamental importance to society, being a constant that endures and will naturally continue to be crucial in the future [1]. Until the mid-19th century, coal and wood were humanity's primary energy sources. However, the Industrial Revolution provided a technological and economic context favorable to the emergence and growth of the petrochemical industry [2].

Driven by the constant increase in population and industrialization, global energy consumption has been rising quickly. From the second half of the 20th century to the present day, consumption has increased by 400% [3, 4]. Fossil fuels have emerged as the primary energy source, standing out for their high energy content and low cost [5].

© The Author(s), under exclusive license to Springer Nature Switzerland AG 2027
M. Calvete Gaspar et al. (Eds.): Regional Helix 2024, LNME, pp. 188–200, 2027.
https://doi.org/10.1007/978-3-032-36365-7_13

Today, many countries heavily depend on fossil fuels, which form the basis of their energy production systems. However, the excessive use of these fuels carries severe and irreversible consequences for the planet, manifesting in air pollution and global warming [6]. Around 75% of global greenhouse gas emissions are estimated to come from burning fossil fuels for energy production [4]. Furthermore, it is widely agreed that fossil fuel energy sources are finite and subject to fluctuations arising from the interaction between supply and demand, geopolitical events, and government policies that directly influence the stability of their prices [7].

To adopt more sustainable practices, it is essential to increase the use of renewable energy sources. In this context, biomass emerges as one of the most promising sources, standing out as the only renewable energy source capable of being converted into solid, liquid, and gaseous fuels and subsequently into thermal and electrical energy [8]. Moreover, biomass has wide availability in both urban and rural areas and can be stored and used as needed, a characteristic not shared by other renewable energy sources [9].

Some of the primary biomass sources include forest products, agricultural by-products, and waste from wood processing industries. The energy recovery of these materials through combustion can be done in different ways: by direct burning without processing the raw material; through chip-burning systems, where the raw material is pre-shredded; or by the production and burning of pellets [11]. Pellets are a solid fuel derived from biomass, highly versatile, and designed to compete with fossil fuels [11, 12]. Their quality varies according to the raw material used, durability, and moisture content [13]. Among solid fuels derived from biomass, pellets exhibit the most efficient combustion due to their lower moisture content and high bulk density, resulting in lower pollutant emissions and reduced ash formation [14, 15].

Some biomass combustion equipment, particularly domestic systems, is characterized by low efficiency and high pollutant emissions [16]. This occurs due to the prioritization of design over performance, with constructions that do not incorporate existing knowledge and technologies, demonstrating an apparent lack of research, monitoring, and control by manufacturers. Additionally, the efficiency of these systems may be impacted by issues related to installation, inadequate maintenance, and adverse operational conditions, resulting in performance that falls short of initial expectations [17].

Considering global efforts to promote sustainability and reduce dependence on fossil fuels, improving the efficiency of biomass combustion systems is crucial, not only for reducing pollutant emissions but also for contributing to a more efficient and sustainable energy sector.

In this context, the main objective of this work was to optimize the thermal efficiency of a domestic pellet boiler experimentally. Specifically, the study focused on the influence of thermal load and excess air on the boiler's performance and pollutant emissions.

2 Materials and Methods

2.1 Pellets Characterization

Initially, the pellets used were characterized by evaluating their physical and chemical properties. The tests followed the ISO 17225-2 standard [18], which establishes specifications for wood pellets and serves as the basis for the ENplus standard classes.

The moisture content of the pellets was determined using the mass loss principle following the EN 14774-2:2010 standard [19]. Three samples with different amounts of pellets were placed in a Venticell laboratory oven, model LSIS-B2V, at 105 °C ± 2 °C, and their mass was periodically measured. The moisture content, on a wet basis (wb), was determined using the following equation:

$$MC_{wb}(\%) = \frac{m_w - m_d}{m_w} \times 100 \tag{1}$$

where m_w is the initial mass of the sample before being placed in the oven, and m_d is the final mass after the drying process.

Three sample groups were selected to determine the pellet's dimensions, and 20 pellets were randomly chosen for each group. Using a digital caliper with a resolution of 0.1 mm, the length and average diameter of each selected pellet were measured. The mass of each pellet was determined using a Sartorius digital scale, model BP 310P, with a resolution of 0.005 g. The bulk density of the pellets was assessed according to the dimensional method proposed by Rabier et al. [20], which consists of dividing the mass of each pellet by its corresponding volume, as a pellet can be approximated to a cylinder.

Following the EN 15210-1 standard [21], the fines content was determined by sieving, and the mechanical durability of the pellets was evaluated using a tumbling device, as specified by ASAE S269.4 [22].

The elementary and immediate analysis of the pine pellets was carried out by LE.CBE - Testing Laboratory of the Center for Biomass for Energy. In Table 1, it is possible to observe the main properties of the pellets and their calorific value.

Table 1. Physical and chemical properties of the pine pellets.

	Pine pellets
Moisture content$^{1)}$ (%)	4.92
Length (mm)	18.80
Diameter (mm)	6.09
Fines (%)	0.02
Mechanical durability (%)	99.40
Ashes$^{2)}$ (%)	0.5
Total carbon$^{2)}$ (%)	50.5
Hydrogen$^{2)}$ (%)	6.2

(continued)

Table 1. (*continued*)

	Pine pellets
Nitrogen2) (%)	0.12
Oxygen2) (%)	42.6
Sulfur2) (%)	0.013
Lower heating value3) (MJ/kg)	18.94

1) Wet basis. 2) On a dry mass basis. 3) Dry basis

2.2 Experimental Setup

The experimental setup used in this work is based on that previously used in the work conducted by Ferreira et al. [23], Pinto et al. [24], and Tomás et al. [25]. The main component is a domestic pellet boiler whose manufacturer announces a nominal power of 21.4 kW distributed between water heating (15.0 kW) and space heating (6.4 kW) and an overall efficiency of 89%.

In the experimental installation used in previous works, the boiler operated in an open circuit; mains water entered directly into the boiler, always cold, and the hot water that left the boiler was discharged into the sewer without its energy being used. However, in an actual installation, water enters the boiler and then passes through the heat exchanger above the combustion chamber. During this process, the water absorbs the heat generated by the combustion of the pellets, increasing its temperature. Subsequently, it is directed into a closed circuit, distributing the heat and returning to the boiler at a lower temperature.

To carry out this work, it was necessary to modify the installation to simulate the actual operating conditions of the boiler, as the continuous entry of cold water reduced the temperature in the combustion chamber and negatively affected its performance. In the current configuration, the hot water leaving the boiler is mixed with cold water from the network through a thermostatic valve to regulate the return temperature to the boiler. However, the pressure inside the piping was low, hindering the water flow. A pressure-regulating valve and a circulation pump were incorporated to address this issue. A new flow meter was also installed to measure the mains water entering the thermostatic valve. Three type K thermocouples were also incorporated to measure the water inlet (T_1) and outlet (T_2) temperatures in the boiler and the exhaust gas temperature (T_3). To acquire data from these sensors, a system was developed that used an ESP32 controller that sent the data via Wi-Fi to an SQL database on a computer. The evolution of the data was monitored in real-time in the form of tables and graphs. To measure the concentration of exhaust gases, a probe was placed in the chimney, connected to a gas analyzer, Testo 350. Figure 1 shows a scheme of the experimental setup used.

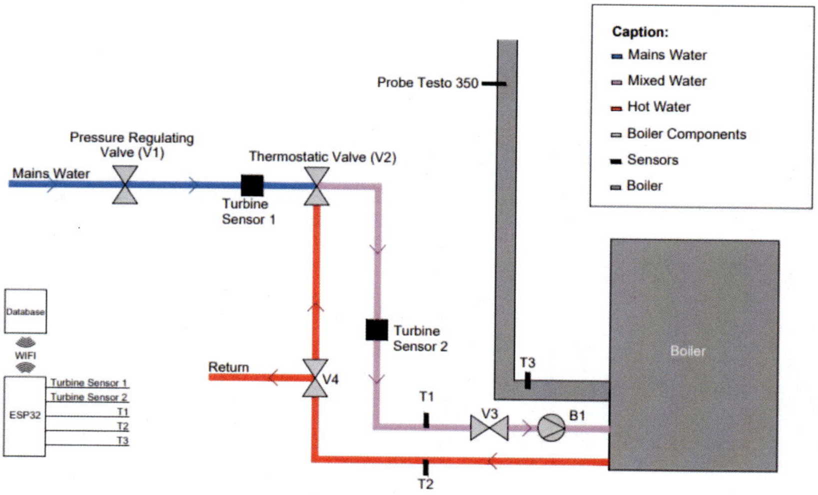

Fig. 1. Experimental setup.

2.3 Experimental Procedure

The boiler has five thermal loads predefined by the manufacturer; however, in this study, only thermal loads 3 and 5 were tested with different excess air values. The thermal load was selected on the boiler control panel, and it defines the speed of the worm screw that feeds the pellets, while excess air was controlled by adjusting the power of the exhaust fan. The exhaust fan powers defined by the manufacturer for thermal loads 3 and 5 are 38 and 48%, respectively. Each test condition was repeated three times to check the repeatability of the results. Before starting each test, the ash, combustion chamber, and heat exchanger were cleaned. The previously weighed pellets were then placed in the boiler hopper, and the desired thermal load and excess air were adjusted. In all tests, water circulation was maintained at a pressure of 1.2 bar, and its flow rate was 9.3 l/min (±0.12 l/min). After ignition, it took around 30 min for the system to reach a steady state. Under these conditions, the water entry temperatures into the boiler were 50 and 60ºC (±1.9 ºC) at thermal loads 3 and 5, respectively. After 30 min of steady-state data acquisition, the boiler was turned off, the unconsumed pellets were weighed, and the procedure for a new test was restarted.

2.4 Calculation Methodology

The boiler's thermal efficiency, $\eta_{thermal}$, was determined using the direct method under steady-state conditions, according to Eq. 2.

$$\eta_{thermal} = \frac{\dot{Q}_{output}}{\dot{Q}_{input}} \tag{2}$$

where $\dot{Q}_{output}$ represents the thermal power transferred to the water in the boiler circuit (kW) and $\dot{Q}_{input}$ represents the thermal power released during the combustion of the pellets (kW). The thermal power released during the combustion of the pellets, $\dot{Q}_{input}$ (kW), was calculated using Eq. 3:

$$\dot{Q}_{input} = \dot{m}_{pellets} \cdot PCI_{\text{pellets}} \tag{3}$$

where $\dot{m}_{pellets}$ represent the mass flow rate of the pellets (kg/s) and $PCI_{pellets}$ the lower heating value of the pellets (kJ/kg). For calculation purposes, the heating value of the pellets was used on a dry basis. As for the thermal power transferred to the water, $\dot{Q}_{output}$, it was calculated from the energy balance that results in the Eq. 4:

$$\dot{Q}_{output} = \dot{m}_{H_2O} \cdot \overline{c}_{H_2O}.(T_2 - T_1) \tag{4}$$

where $\dot{\boldsymbol{m}}_{\boldsymbol{H_2O}}$ is the average mass flow rate of water in the boiler circuit (kg/s), $\overline{c}_{H_2O}$ the specific heat capacity of water at the average temperature (kJ/kg K) and T_2 and T_1 represent the average temperature of the water at the outlet and inlet of the boiler, respectively (°C).

The excess air was obtained based on the composition of combustion products and the uncertainties associated with the measurements taken in this work were calculated according to Coleman and Steele [26] and are shown in Table 2.

Table 2. Uncertainties associated with the measurements of the variables under study.

Measured variable	Load 3	Load 5
O_2 molar concentration (% O_2 v/v)	0.2	0.2
CO molar concentration (% CO v/v)	0.00717	0.00120
NO molar concentration (% NO v/v)	0.0005	0.0005
CO_2 molar concentration (% CO_2 v/v)	0.3890	0.3620
Fuel gases temperature (%)	0.80	0.78
Water temperature at the boiler inlet (%)	0.98	0.92
Water temperature at the boiler outlet (%)	0.93	0.86
Water mass flow rate (%)	0.0316	0.0316
Pellets mass flow rate (%)	0.00715	0.01320

3 Results and Discussion

All tests were conducted under similar operating conditions regarding the average mass flow rate of water, internal water pressure in the boiler, and the position of the thermostatic valve.

To simplify the reading, the tests were designated by the operation loads, followed by the exhaust fan power at which the test was conducted, and a numerical index representing the test number under the specified conditions. For example, P5E401 represents the first test performed with the boiler at thermal load 5 and 40% of the fan power.

In the initial phase of the tests, during the system's heating, it was possible to observe the variation of the combustion gas temperatures and the inlet and outlet water temperature. Figure 2 illustrates the evolution of the exhaust gas temperature during this phase at the two thermal loads for the various air excesses under study.

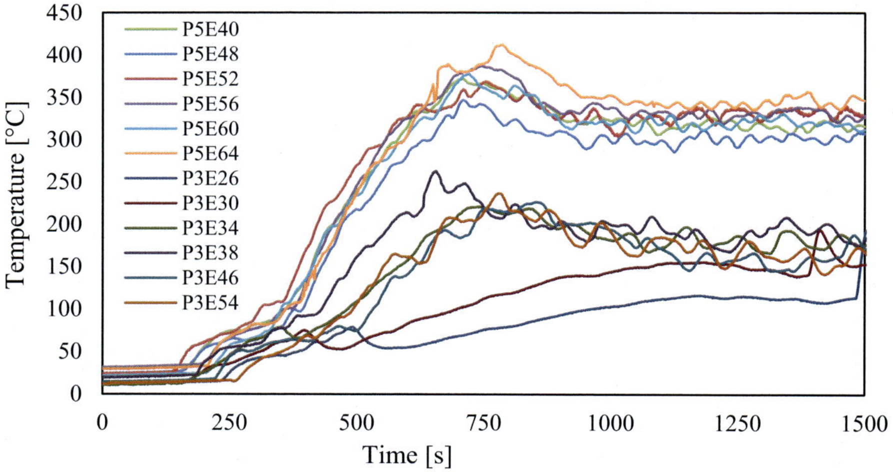

Fig. 2. Exhaust gas temperature during the start-up phase.

As expected, it is possible to observe that the maximum temperature reached is higher in the tests conducted at load 5.

Regarding the inlet and outlet water temperatures (Win and Wout), Fig. 3 illustrates their evolution during tests carried out at both thermal loads. As expected, higher temperatures were reached at thermal load 5.

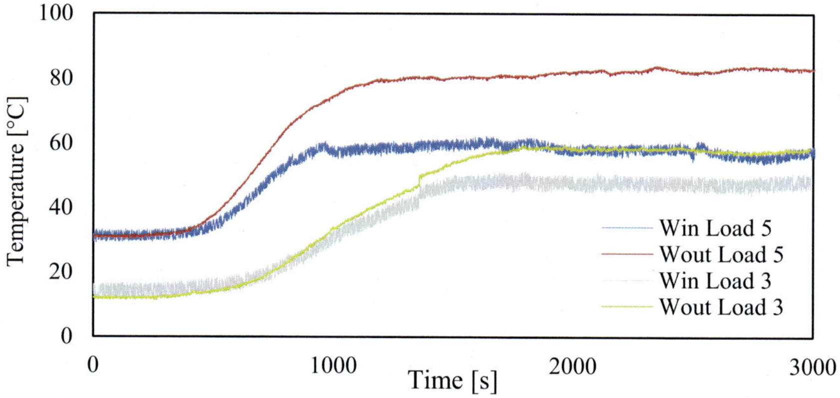

Fig. 3. Evolution of inlet and outlet water temperatures during tests at thermal loads 3 and 5.

3.1 Influence of Excess Air and Thermal Load on Boiler Thermal Efficiency

In the installation used, excess air is controlled by the exhaust fan power measured as a percentage of the maximum fan power. Figure 4 illustrates the relationship between the exhaust fan power and excess air for thermal load 3.

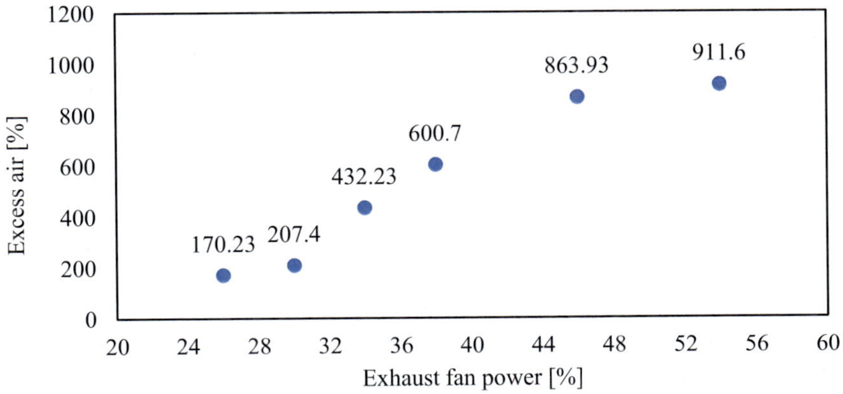

Fig. 4. Excess air as a function of the exhaust fan power for thermal load 3.

Figure 5 illustrates the influence of excess air on the boiler's thermal efficiency for thermal loads 3 and 5. The values represent the average results obtained from the tests conducted for each excess air level studied. As can be seen, for load 3, the reduction of excess air led to a significant increase in the boiler's thermal efficiency. At this thermal load, the highest efficiency, 76.35%, was obtained with the fan power set at 30%. For load 5, the variation in excess air did not significantly impact the boiler performance. However, there is a maximum value of 76.64% obtained for an exhaust fan power of 56% and a tendency for efficiency to decrease for higher values of excess air.

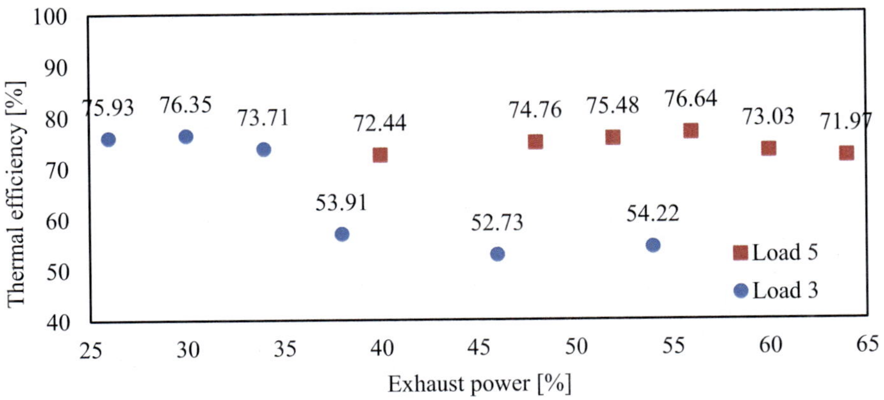

Fig. 5. Thermal efficiency as a function of exhaust fan power for the thermal load 3 and 5.

Analyzing the manufacturer's specifications with the results obtained, it can be seen at thermal load 5, the exhaust fan power set at the factory (48%) allows to obtain a thermal efficiency close to 75%; however, with a more adequate excess air, it can reach values close to 77%. As for thermal load 3, for the exhaust fan power set at the factory (38%), a thermal efficiency close to 57% was obtained, while with an adequate excess of air, it is possible to reach values above 76%.

The European standard EN 14785 [27] sets minimum thermal efficiency requirements for domestic pellet boilers, specifying efficiency limits of 70% for minimum and 75% for maximum load. From the analysis of the previous figures, it can be observed that only the tests P5E52, P5E56, P3E26, and P3E30 comply with the standards set by the mentioned regulation.

3.2 Influence of Excess Air on Emissions

Carbon monoxide (CO) and nitrogen oxides (NO_x) are two of the most common pollutants associated with the combustion of solid fuels. According to Loo and Koppejan [28], a high concentration of CO in the combustion products generally indicates incomplete combustion. The main factor influencing CO formation is the excess air present during combustion. A high excess air value causes a decrease in combustion temperature, consequently leading to the formation of carbon monoxide [29]. However, other factors may also be related, such as reduced residence time of the reactants in the combustion chamber, low temperature, or low turbulence in the process [30].

Figures 6 and 7 represent the average carbon monoxide emissions for the various fan power settings at each operating thermal load. Comparing the previously mentioned figures, it is evident that the boiler's operating power has a significant influence on carbon monoxide emissions. The higher the boiler power, the lower the CO emissions. However, it should be noted that the variation in excess air at load 5 did not result in a significant change in CO emissions. This may be related to the temperature in the combustion chamber, as excess air does not significantly influence the exhaust gas temperature. On the other hand, at load 3, the impact of excess air on carbon monoxide emissions is more evident. The reduction in excess air led to a relevant decrease in CO emissions.

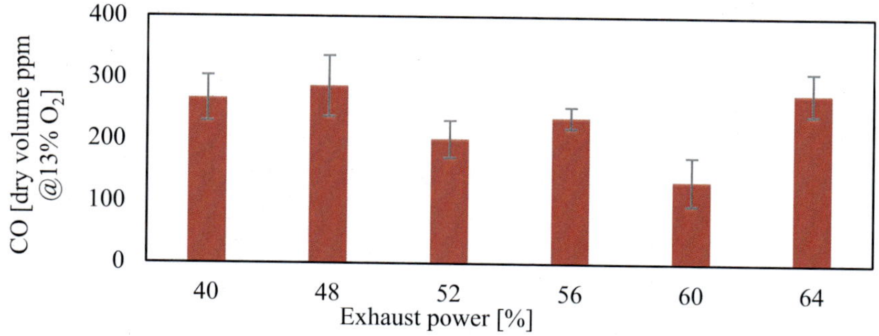

Fig. 6. Carbon monoxide emissions for different fan power at load 5.

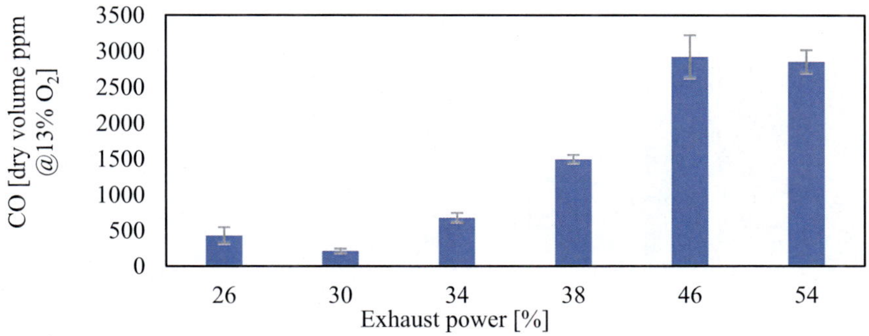

Fig. 7. Carbon monoxide emissions for different fan power at load 3.

In both thermal loads, an appropriate adjustment of the excess air allowed for a significant reduction in CO emissions compared to the values observed in the tests conducted under the excess air conditions recommended by the manufacturer. Specifically, in thermal load 5, this adjustment resulted in a reduction of CO emissions by approximately half. In thermal load 3, however, the reduction was even more pronounced.

The EN 14785 standard also limits carbon monoxide emissions at minimum and maximum thermal loads, specifying 600 ppm at 13% O_2 and 400 ppm at 13% O_2, respectively. In the case of maximum thermal load, load 5, all tests complied with the standard. For load 3, it was assumed that the allowable emissions would be 500 ppm at 13% O_2, based on the interpolation of the values specified in the standard. Thus, only tests P3E26 and P3E30 would meet the standard's requirements.

Regarding emissions of nitrogen oxides, Figs. 8 and 9 present the average values obtained for the various excess air levels at each thermal load.

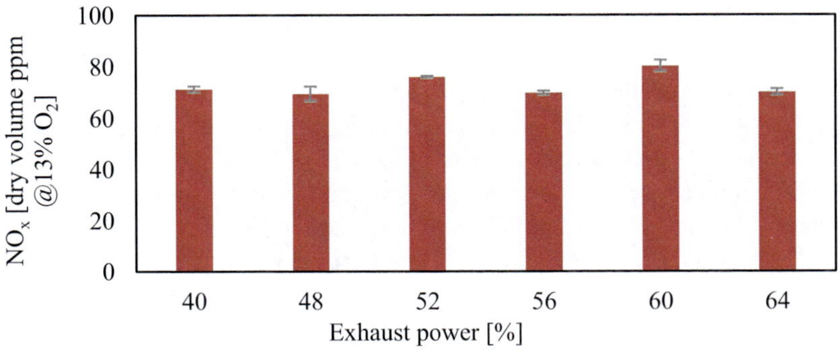

Fig. 8. Nitrogen oxides emissions for different fan power at load 5.

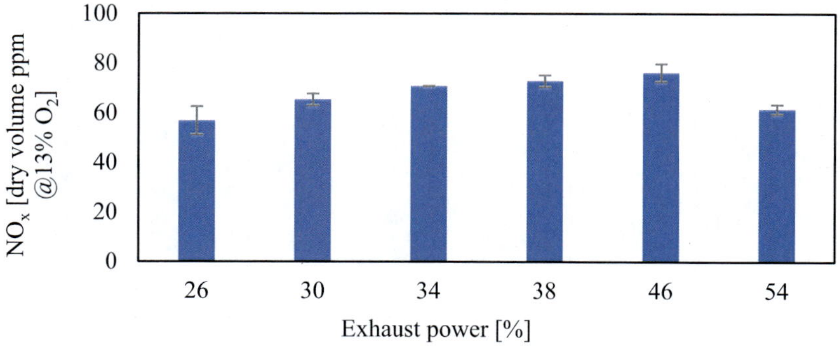

Fig. 9. Nitrogen oxides emissions for different fan power at load 3.

For both thermal loads, the variation in excess air does not significantly influence nitrogen oxide emissions. The similarity in results is explained by the fact that, in the combustion process, the formation of nitrogen oxide (NO) through the thermal mechanism requires combustion temperatures in the range of 1300 °C [31, 32]. If the temperatures do not reach this value, the formation will occur due to the fuel mechanism. Since the fuel used has a low nitrogen content, 0.12% (w/w), the formation of NO through this mechanism is insignificant.

4 Conclusions

In this study, the influence of excess air on the thermal efficiency of a domestic pellet boiler and the emissions resulting from the combustion of pellets at two different thermal loads were analyzed.

In the previously carried out works, the results indicated that increasing the thermal load led to an increase in the boiler's thermal efficiency. However, when adjusting the excess air appropriately, it was observed that the thermal efficiency tends to be very similar regardless of the thermal load used, suggesting that the boiler was not optimized initially. Only tests P5E52, P5E56, P3E26, and P3E30 complied with the EN 14785 standard, achieving a thermal efficiency of at least 75%.

Regarding pollutant emissions, at load 3, it was observed that the excess air used have a significant effect on carbon monoxide emissions. CO emissions were very high in tests P3E38, P3E46, and P3E54, indicating poor combustion, possibly caused by insufficient temperature in the reaction zone of CO with air, potentially due to the combustion chamber temperature and the flow dynamics. The reduction of excess air at the lower thermal load improved the combustion conditions, leading to a significant decrease in carbon monoxide emissions. All tests conducted at the higher thermal load complied with the EN 14785 standard regarding CO emissions.

It was observed that thermal load and excess air did not influence NO_x emissions, as NO formation was primarily due to the fuel mechanism.

To optimize the combustion efficiency of the boiler under study, it is recommended to set the fan power to 56% and 30% at thermal loads 5 and 3, respectively. This adjustment will not only maximize the boiler's efficiency but also result in a reduction of pollutant emissions.

These results highlight the need for careful tuning of this type of boiler at all thermal loads. Manufacturers must pay special attention to intermediate and low thermal loads, as these are what users often select because they think this is a way to save fuel.

Acknowledgements. This work was financed by National Funds through FCT – Foundation for Science and Technology, I.P., under the project Ref.ª UIDB/05583/2020. We additionally thank the Digital Services Research Center (CISeD) and the Polytechnic Institute of Viseu for the support provided.

References

1. Nieuwlaar, E., Blok, K.: Introduction to Energy Analysis, 3rd ed. Routledge (2020)
2. Mesa, J., Rocha, J., Olivares, E., Barboza, L., Brossard, L., Brossard Junior, L.: Pirólise Rápida em Leito Fluidizado: Uma Opção Para Transformar Biomassa em Energia Limpa. Rev. Anal. 32–36 (2003)
3. Smil, V.: Energy Transitions: Global and National Perspectives. Bloomsbury Publishing USA (2016)
4. Ritchie, H., Rosado, P., Roser, M.: Energy Mix. Our World in Data (2020)
5. Ediger, V.: An integrated review and analysis of multi-energy transition from fossil fuels to renewables. Energy Procedia **156**, 2–6 (2019)
6. Martins, F., Felgueiras, C., Smitková, M.: Fossil fuel energy consumption in European countries. Energy Procedia **153**, 107–111 (2018)
7. Yoon, S.: On the interdependence between biofuel, fossil fuel and agricultural food prices: evidence from quantile tests. Renew. Energy **199**, 536–545 (2022)
8. Faaij, A.: Modern biomass conversion technologies. Mitig. Adapt. Strat. Glob. Change **11**(2), 343–375 (2006)
9. Zafar, S.: The Importance of Biomass Energy in Energy Mix (2022). https://www.bioenergyconsult.com/a-glance-at-biomass-energy/. Last accessed 09 January 2024
10. Saidur, R., Abdelaziz, E., Demirbas, A., Hossain, M.S., Mekhilef, S.: A review on biomass as a fuel for boilers. Renew. Sustain. Energy Rev. **15**(5), 2262–2289 (2011)
11. Jenkins, D.: Wood Pellet Heating Systems. Routledge (2010)
12. Obernberger, I., Thek, G.: The Pellet Handbook. Earthscan (2010)
13. Hansen, M., Jein, A., Hayes, S., Bateman, P.: English HandBook for Wood Pellet Combustion. National Energy Foundation (2009)
14. Oliveira, C.: Biomassa Bioenergia Briquete Woodpellets (2016)
15. Ferreira, T., Pereira, C., Paiva, J.M., Pinho, C.: Simple methodology to quantify the fragmentation on batches of char pellets during fluidized bed combustion. Energy Fuels **31**(5), 5073–5078 (2017)
16. European Environment Agency. Small combustion (2016)
17. Greenpro. Bioenergia—manual sobre tecnologias, projecto e instalação. Altener (2004)
18. ISO 17225-2:202. Solid biofuels—fuel specifications and classes Part 2: Graded wood pellets (2021)
19. EN 14774-2:2010—Solid biofuels—Methods for the determination of moisture content—oven dry method—part 2: total moisture—simplified method (2015)
20. Rabier, F., et al.: Particle density determination of pellets and briquettes. Biomass Bioenerg. **30**, 954–963 (2006)
21. EN 15210-1:2010 - Solid biofuels—Methods for the determination of mechanical durability of pellets and briquettes—Part 1: Pellets (2015)

22. ASAE Standards. S269.4, Cubes, pellets and crumbles—definitions and methods for determining density, durability and moisture content, St. Joseph, MI, USA (1998)
23. Ferreira, T., Paiva, J., Pinho, C.: Performance assessment of invasive Acacia dealbata as a fuel for a domestic pellet boiler. Chem. Eng. Trans. **42**, 73–78 (2014)
24. Pinto, A.: Avaliação da Biomassa de Paulownia Enquanto Combustível na Forma Peletizada. Master's Thesis, Instituto Politécnico de Viseu. Escola Superior de Tecnologia e Gestão de Viseu (2015)
25. Tomás, A.: Estudo da Combustão de Peletes de Mimosa numa Caldeira Doméstica de 20 kW. Master's Thesis, Instituto Politécnico de Viseu. Escola Superior de Tecnologia e Gestão (2019)
26. Coleman, H., Steele, W.: Experimentation and Uncertainty Analysis for Engineers, 2nd ed. John Wiley and Sons (1999)
27. EN 14785: 2006—Residential space heating appliances fired by wood pellets requirements and test methods (2022)
28. Van Loo, S., Koppejan, J.: The Handbook of Biomass Combustion and Co-firing, 2nd ed. Earthscan, United Kingdom (2008)
29. Johansson, L., Leckner, B., Gustavsson, L., Cooper, D., Tullin, C., Potter, A.: Emission characteristics of modern and old-type residential boilers fired with wood logs and wood pellets. Atmos. Environ. **38**(25), 4183–4195 (2004)
30. Karchniwy, E., Haugen, N., Klimanek, A., Langørgen, Ø., Sładek, S.: The effect of turbulence on mass transfer in solid fuel combustion: RANS model. Combust. Flame **227**, 65–78 (2021)
31. Meiller, M., Jakuttis, M., Binder, S., Hornung, A.: Fuel Staging in biomass heating systems—a primary measure to reduce NOx emissions. In: European Biomass Conference and Exhibition Proceedings, 21st EUBCE-Copenhagen 2013, pp. 550–554 (2013)
32. Rabaçal, M., Fernandes, U., Costa, M.: Combustion and emission characteristics of a domestic boiler fired with pellets of pine, industrial wood wastes and peach stones. Renew. Energy **51**, 220–226 (2013)

Valorisation of Limestone Quarry Waste for a Local Sustainable Precipitated Calcium Carbonate Production

N. S. Oliveira[1,2](✉), S. Monteiro[1,2], M. L. Heleno[1], M. L. Alves[1,3], A. Veiga[1,4], and A. Silva[1,5]

[1] School of Technology and Management, Polytechnic of Leiria, Leiria, Portugal
nelson.oliveira@ipleiria.pt

[2] LSRE-LCM—Laboratory of Separation and Reaction Engineering, Polytechnic of Leiria; ALiCE—Associate Laboratory in Chemical Engineering, Faculty of Engineering, University of Porto, Porto, Portugal

[3] INESC Coimbra—Institute for Systems Engineering and Computers at Coimbra, Polytechnic of Leiria, Leiria, Portugal

[4] CGEO—Geosciences Center, University of Coimbra, Coimbra, Portugal

[5] CEFAGE-UE, Center for Advanced Studies in Management and Economics, Évora; MARE, Marine and Environmental Sciences Centre, Polytechnic of Leiria; CIIC, Computer Science and Communication Research Centre, Polytechnic of Leiria, Leiria, Portugal

Abstract. The limestone extraction industry generates significant waste, evident from large heaps of discarded material. Reducing and preventing this waste should be a priority, aligning with circular economy strategies to enhance sustainability and environmental preservation. It is crucial to find economically attractive alternatives that reclassify this waste as by-products or secondary raw materials and to convert into a value-added product, such as precipitated calcium carbonate (PCC). The demand for calcium carbonate is projected to grow by 3.08% until 2029, which will also increase limestone waste, particularly in Serra de Aire e Candeeiros, Portugal. Here, the industry, focused on ornamental rock, produces about 48% waste from limestone extraction. Certain applications require specific properties that natural calcium carbonate (GCC) cannot offer, unlike PCC, whose production allows control over properties such as purity, granulometry, and whiteness. However, the environmental sustainability of PCC depends on factors like transport from quarry to production site and energy consumption during conversion. This study is a preliminary study, at laboratory level, of the production of PCC from limestone waste in Serra de Aire e Candeeiros. The preliminary results show the capacity to reduce undesirable elements from limestone secondary raw materials, achieving about 98.1 up to 100% ($\pm$0.13%) of calcium carbonate (calcite) in PCC and provided relevant information for improving the capacity of the transformation process. This is an essential step towards creating the conditions for local production of PCC, and in this way the environmental impact can be minimised, providing a sustainable solution to manage and utilize limestone secondary raw material effectively.

Keywords: Limestone waste · Circular economy strategy · Precipitated calcium carbonate

© The Author(s), under exclusive license to Springer Nature Switzerland AG 2027
M. Calvete Gaspar et al. (Eds.): Regional Helix 2024, LNME, pp. 201–212, 2027.
https://doi.org/10.1007/978-3-032-36365-7_14

1 Introduction

Incorporating ornamental stone waste into the industry or other sectors is an environmentally sustainable approach to waste management, fostering circular economy strategies and industrial symbiosis. The transformation of waste from the ornamental stone industry into secondary raw materials for other industries contributes to sustainability and environmental preservation, which is in line with Sustainable Development Goal 12 (SDG 12), namely achieving sustainable management and efficient use of natural resources, and their environmentally sound management throughout their life cycle, in order to minimize their negative impacts on human health and the environment (United Nations 2024).

Waste from the extractive industry can be classified into mining waste and non-mining waste from supporting operations. Mining waste is directly generated from quarrying activities such as extraction, treatment, or transformation, but it primarily originates at the blasting front due to block chips and non-viable material from carving falls. In the ornamental stone sector, up to 3 m^3 of waste is produced for every m^3 of marketable rock (Assimagra 2015). According to ANIET (2018), waste in the marble industry can represent between 80 and 90% of the total extracted material.

Currently, the extractive industries are using strategies to reduce this waste, aiming for maximum profitability, without forgetting that this natural resource is finite, and that it is present in different geological forms. For instance, in the "Pé da Pedreira" quarrying nucleus, covering 391 hectares with 32 companies, the estimated ornamental limestone reserves are about 59,533,400 m^3, with usable reserves around 50% or 29,766,700 m^3 (APA 2018). Notably, limestone constitutes approximately 10% of all rocks on Earth's surface (Parker 1967).

Generally, residues from limestone extraction can be classified into soils and rocks (such as vegetal soil and materials from karstification), unfeasible stone blocks (large quantities of stones without ornamental value), and sludge. Effective management of extraction residues should focus on prevention, reduction, and recovery, with disposal in heaps as a last resort, Only in the cases where recovery is neither technically nor financially viable. Heap locations should consider factors like natural heritage, landscape impacts, and resource management. Despite high waste volumes, many heaps are oversized and lack proper separation by typology (land or blocks), posing security and land use concerns (Assimagra 2015).

Valorising materials in heaps locally can significantly reduce the economic and environmental impacts associated with transporting waste. By processing and repurposing waste on-site or within a close geographic area, industries can cut down on transportation costs, reduce carbon emissions, and minimize wear and tear on infrastructure. This localized approach not only promotes environmental sustainability but also enhances the economic viability of waste management practices. For instance, converting quarry waste into secondary raw materials for local construction projects or industrial applications can create a sustainable loop that benefits both the quarry operators and the surrounding community. The use of localized waste valorisation techniques can also foster stronger community relations and contribute to regional development. Local businesses can benefit from a steady supply of raw materials, leading to the creation of jobs and boosting the local economy. Moreover, the environmental benefits of reduced transportation emissions and lower energy consumption align with broader regional and national sustainability goals. In regions where transportation infrastructure is limited or underdeveloped, local processing and utilization of waste can be particularly advantageous, ensuring that the environmental footprint of quarrying activities is kept to a minimum.

Calcium carbonate, naturally obtained from limestone, chalk, marble, and other sedimentary rocks, can be processed into ground calcium carbonate and precipitated calcium carbonate (GCC and PCC, respectively). Both materials are in high demand in industries like paper, packaging, construction, transportation, and industrial applications, with expected compound annual growth rate (CAGR) of 3.08% during the forecast period 2024–2029 (Mordor Intelligence 2024). Calcium carbonate has a long commercial history, dating back to its use in prehistoric cave paintings. The industrial revolution expanded its applications, leading to the development of PCC to complement natural calcium carbonate. PCC has been crucial for industries such as adhesives and sealants, paper, plastics and rubber, food, paints, and pharmaceuticals.

This research focuses on producing of PCC through a comparative analysis to identify advantageous conditions for producing and commercializing PCC in a quarry context. This approach seeks to add value to limestone waste from extraction operations by transforming it into by-products. However, while not all limestone quarries in Serra de Aire e Candeeiros are suitable for PCC production, five extraction poles have been identified as having potential raw materials, involving several companies dedicated to quarrying in these areas. This localized production could optimize the environmental sustainability of PCC and provide a practical solution for managing limestone waste effectively.

The Table 1 shows the areas of specific intervention (AI), the area intervened by quarrying (AQ), and the volume of existing extraction waste in heaps (VW) of 475 heaps in the five areas of specific intervention (Assimagra 2015). These data are strong indicators of the potential and relevance of recovering limestone extraction waste. By leveraging these heaps, the industry can minimize the environmental impact of waste transport, foster economic growth in local communities, and adhere to sustainable development principles.

Table 1. Characterization of the heaps existing in the areas of specific intervention (Assimagra 2015). AI: area of specific intervention; AQ: area intervened by quarrying, VW: volume of extraction waste in heaps.

Extraction poles	AI (ha)	AQ (ha)	VW (m^3)
Moleanos	147	17.03	95 700
Portela das Salgueiras	63	45.5	138 700
Pé da Pedreira	1374	510.4	3 402 000
Cabeça Veada	29	20.3	252 700
Codaçal	98	71.8	898 940

These data underscore the significant potential for utilizing limestone extraction waste, thereby promoting sustainable practices and reducing environmental impact.

2 Methodology

Using different limestone quarries located in the Serra de Aires and Candeeiros, 11 samples of limestone were collected from different parts of the extraction area. The samples were taken in an area where secondary raw materials accumulate, i.e. raw materials that are not suitable for use in the ornamental stone processing industry. This unfeasibility is essentially due to the karstification and small size of the rock fragments (Monteiro et al. 2023).

During pre-treatment, the samples were geologically characterized, reduced to fragments of less than 20 mm in size and dried in an oven at 105 °C until the mass stabilized. In the next phase, the samples were calcined at 900 °C until a constant mass was obtained. The samples were hydrated using an aqueous solution of ammonium chloride in a mass ratio sample NH_4Cl of 1:2. The solution obtained after hydration was filtered, ensuring that insoluble impurities were removed, and then carbonated with CO_2 until neutral pH was reached. The solid obtained by precipitation in the carbonation phase was filtered and dried (PCC) in order to meet the conditions for its chemical characterization.

For each 11 samples, the raw material without treatment (MP), calcinated and PCC were pulverized for chemical characterization by X-ray fluorescence (XRF) in the HERCULES laboratory. The characterization by XRF was performed using a Bruker S2 PUMA with internal standard (15 certified standards), where the samples were analyzed in the form of beads to minimize interference from their respective matrices. The beads were prepared with 12.000 g of flux and 1.2000 g of sample calcined at 1050 °C. The flux used was a 1:1 mixture of lithium metaborate and lithium tetraborate. The mixture was homogenized and melted in a fusion device (Claisse® LeNeo®). For the determination of Loss on Ignition (LOI), 5 g of each sample were weighed and placed in a muffle furnace at a temperature of 1065 °C for 30 min.

Based on the results of LOI of the previously dried samples, the percentage of calcium carbonate in MP and PCC was estimated using Eq. 1:

$$\%CaCO_3 = \left(\frac{LOI}{M(CO_2)} - \frac{\%MgO}{M(MgO)}\right)M(CaCO_3) \tag{1}$$

The molecular weights of calcium carbonate, carbon dioxide, and magnesium oxide are represented by $M(CaCO_3)$, $M(CO_2)$, and $M(MgO)$, respectively. The equation corrects the presence of magnesium carbonate, and it assumes that the mass loss is due to the emissions of carbon dioxide from calcination of calcium carbonate.

3 Raw Material Characterization

The samples used for this study come from the Maciço Calcário Estremenho, a geomorphological structure where lithologies from the middle to upper Jurassic predominate. Not all limestone quarries in this region are suitable as raw material sources for PCC production. In general, the higher the calcium carbonate content in the limestone, the more efficient the production process, resulting in lower production costs and increased profitability (Monteiro et al. 2023).

The samples collected correspond to limestones from quarries in operation, in three extractions poles, with the commercial names of "Branco Real", "Semi-rijo" and "Moca creme". The samples correspond to limestones with different states of alteration and texture. Table 2 shows the identification and macroscopic description of the samples, their origin and the acronym to be used in the analysis of data and results.

Figure 1 shows some examples of illustrative limestone samples. The selection of sampling quarries was based on the high percentage of calcium carbonate present in these rock varieties and low percentage of impurities. The limestone samples were selected from the quarry waste (karstification zones, small fragments and powder resulting from the cutting of blocks).

Fig. 1. Limestone rock samples used for the production of PCC.

4 Results

The samples of limestone (MP) and precipitated calcium carbonate (PCC) were calcinated and characterized by X-ray fluorescence (XRF), where the oxides of aluminum (Al_2O_3), silicium (SiO_2), phosphorous (P_2O_5), sodium (Na_2O), magnesium (MgO), potassium (K_2O), calcium (CaO), titanium (TiO_2), manganese (MnO), iron (Fe_2O_3) and the elements of sulfur (S), strontium (Sr), barium (Ba) and zirconium (Zr) were determined.

The Fig. 2 presents the elementary chemical composition (% (w/w)) by XRF of each oxide after calcination of the limestone rock samples. The content of iron, silicium and aluminum seems to be correlated, and the sample MP1 and MP8 presents the highest content on iron, silicium and aluminum, a possible correlation to the karstification characteristic of this samples. The magnesium can be found in the natural stone as magnesium carbonate (magnesite), and it decompose in carbon dioxide during calcination. All samples present similar content of magnesium and phosphorous. The content of sodium, potassium, titanium, manganese is very low or not measurable for all samples. The Fig. 3 presents the elementary chemical composition (% (w/w)) by XRF of PCC produced in this work. The content of silicium and iron reduced significantly, with silicium reducing about 80 up to 98%, exception to PCC5 with a reduction of 40%. Iron was not measurable in all samples with exception of sample PCC8 due to higher concentration in MP8 (karstification). Magnesium reduced about 10 up to 39%. Although aluminum content has reduced, the aluminum become similar to all PCC. The content of sodium increases in PCC1, PCC2, PCC4, PCC6, PCC7 and PCC11, possible due to a contamination of the materials used in the procedure, although the results are very close to the XRF quantification limits. In the elements of S, Sr, Ba and Zr there has also been a reduction in their composition, with a median reduction of 30%.

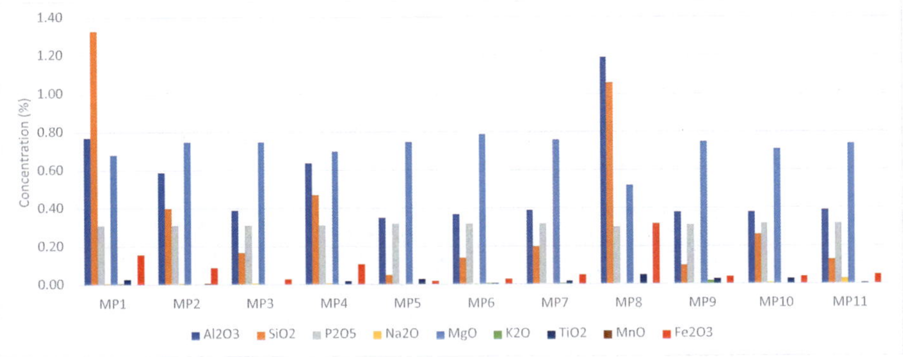

Fig. 2. Elementary chemical composition by XRF of limestone rock samples (MP).

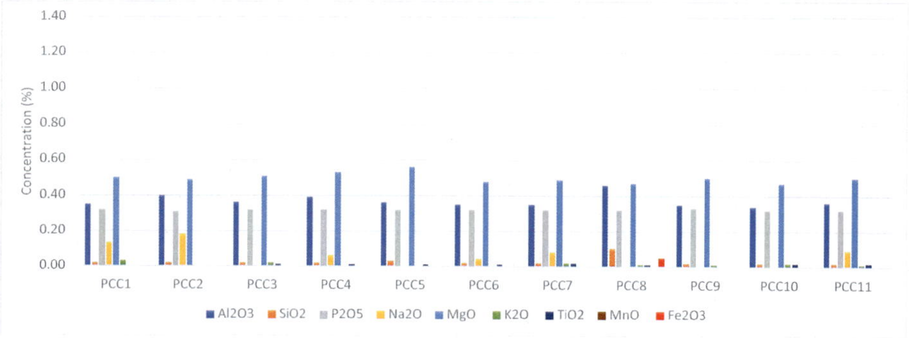

Fig. 3. Elementary chemical composition by XRF of PCC.

The results show an overall decrease of all elements. Among the various samples, the production of PCC8 was not effective, since some iron and silicon were measured in the final product. This fact raises the need to improve some steps of the experimental procedure before moving on to a larger scale.

Table 2. Identification and macroscopic description of the collected samples by extraction pole.

Extraction Pole	Macroscopic description of limestone rock samples	Acronym
Salgueira, Arrimal	*Branco real* with karstification - whitish cream color, fine grain, with iron oxides and an orange surface (karstification)	MP1
	Branco real semi-rijo - whitish cream color, with bioclast and iron oxides. Fine grain	MP2
	Branco real with texture - white color with oxidation surfaces, with grayish bioclasts and with geoids of calcite crystals. Fine to medium grain	MP3
	Branco real from a karst wall - white and orange color, sheets about 1 cm thick between karst surfaces	MP4
	Powder resulting from cutting blocks, light cream color and medium to coarse grain	MP5
Codaçal	Semi-rijo RR - light cream color, fine and uniform grain, with grayish bioclasts	MP6
	Semi-rijo CM - light orange cream color, with grayish bioclasts and iron oxides. Fine grain	MP7
Pé da Pedreira	*Moca creme* from an karstification surface - calcite from recrystallization, orange cream color	MP8
	Moca creme (4M) - cream color, with grayish bioclasts and iron oxides. Fine to medium grain	MP9
	Moca creme (Perlina) - light cream color, with bioclasts and iron oxides. Fine to medium grain	MP10

(continued)

Table 2. (*continued*)

Extraction Pole	Macroscopic description of limestone rock samples	Acronym
	Moca Creme (3G) - light cream color, with grayish bioclasts and iron oxides. Fine to medium grain	MP11

One of the most relevant elements is the presence of calcium, since it is directly related to the production of PCC, and their quality. The higher the calcium oxide content in the chemical analysis, the greater the probability of obtaining quality PCC. The Fig. 4 compares the results of calcium content by XRF. The content of calcium increased in some PCC samples (PCC3, PCC4, PCC5, PCC8-PCC11), slightly decreased in PCC1, PCC6 and PCC7. In the PCC2, the content of calcium decreased significantly, although the loss on ignition (LOI) increased.

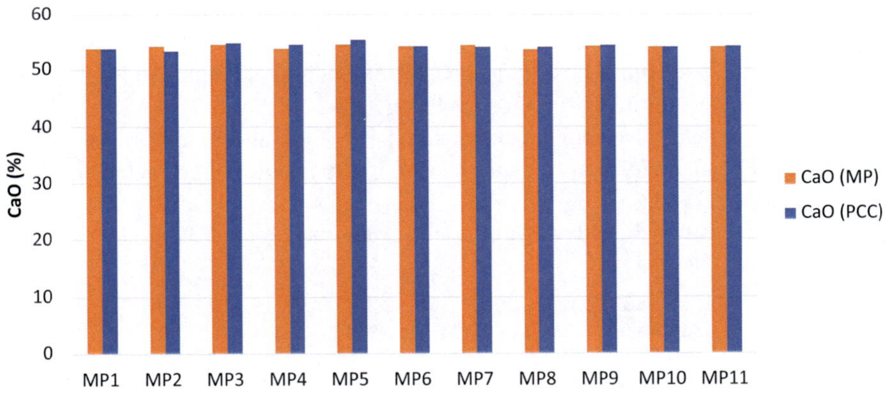

Fig. 4. Calcium oxide composition by XRF of limestone rock samples comparable with PCC.

The Fig. 5 presents the results of LOI from the calcination of the limestone rock and PCC, showing an increase of LOI. This is related to the emission of carbon dioxide from the thermal decomposition of carbonate, and it is directly correlated to the content of calcium carbonate, after correction of the estimated magnesium carbonate.

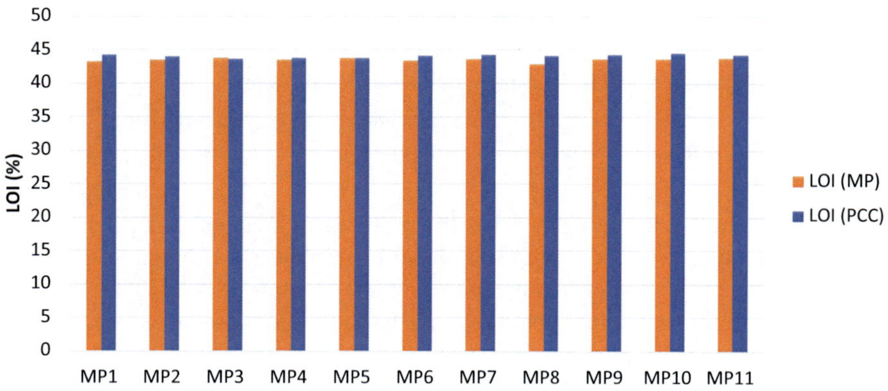

Fig. 5. Loss on ignition (LOI) from the calcination of limestone rock samples (MP) comparable with PCC.

The percentage of calcium carbonate on MP and PCC samples shown on Table 3 increase from MP to PCC, from 0.5 up to 3%, from percentages from 96.3 up to 97.9% in MP to percentages from 98.1% up to 100.2% of calcium carbonate in PCC, with an uncertainty of 0.26% for a 95% confidence level.

The results by XRF also show a relationship between Sr and S, so it can be deduced that they could be in the form of celestine ($SrSO_4$), responsible for a delicate blue color.

Table 3. Percentage of calcium carbonate on the MP and PCC.

Acronym	% $CaCO_3$ (MP)	% $CaCO_3$ (PCC)
MP1	96.6	99.4
MP2	97.3	99.0
MP3	97.6	98.1
MP4	97.2	98.2
MP5	97.7	98.2
MP6	96.7	99.3
MP7	97.3	99.5
MP8	96.3	99.2
MP9	97.5	99.5
MP10	97.6	100.2
MP11	97.9	99.4

The overall results show a significant reduction in some relevant elements, such as iron and silicium, especially in samples with high content of karstification material and regular crystalline production of calcite, reducing the presence of magnesite. However, it presents a significant variance in the presence of calcium oxide content, and by consequence, calcium carbonate (98.1 up to 100%). The experimental procedure demonstrated high capacity to reduce relevant elements and capacity to produce near 100% calcium carbonate.

Further work, the experimental procedure should be improved in order to obtain less variation in the PCC content between the various samples, making it possible to obtain content of over 99%. To study the consistency of particles size of PCC by Scanning Electron Microscopy (SEM) and to present alternatives for the use or exploitation of ammonium chloride in the calcium oxide hydration process, promoting a more sustainable production process. Should be also considered an alternative scenario for calcination without the consumption of energy, since the source of energy can be problematic in the quarries. One of the alternatives to be explored is the chemical decomposition of limestone with hydrochloric acid and carbonatation with ammonium hydroxide and carbon dioxide, taking the advantages of the equilibrium between ammonium hydroxide, ammonium chloride and hydrochloric acid (Grimes et al. 2020).

5 Conclusions

The study demonstrates the potential to effectively convert limestone waste from the Serra de Aire e Candeeiros into PCC, attributing an add-value by removing most of the contaminants. The preliminary results indicate that the conversion process can achieve a calcium carbonate purity of approximately 98.1–100% (±0.13%), highlighting the efficiency of the transformation from limestone waste to PCC. Significant reductions in undesirable elements such as iron, silicon and magnesium were observed. This purification is crucial for producing high-quality PCC, which is preferred in various industrial applications for its controlled properties.

Localized production of PCC in Serra de Aire e Candeeiros can minimize environmental impact by reducing transportation needs and energy consumption associated with the production process. This aligns with circular economy principles by promoting the reuse of industrial waste and reducing overall carbon footprint. The growing demand for calcium carbonate, projected to increase by 3.08% annually until 2029, presents an economic opportunity for utilizing limestone waste as a resource for PCC production. This demand justifies further investment and development in this area.

The study suggests that further optimization of the experimental procedure is necessary to achieve more consistent calcium carbonate content in PCC. Additionally, exploring alternatives for the use of ammonium chloride in the hydration process could enhance the sustainability and cost-effectiveness of the production method. The findings contribute valuable insights into the feasibility and benefits of converting limestone waste into high-value PCC, supporting both environmental sustainability and economic growth in the limestone extraction industry.

Acknowledgements. The authors express appreciation to the Stone by Portugal - Plano de Recuperação e Resiliência "Agendas para A Inovação Empresarial", Portaria n.º 43-A/2022 January 19th, for the support on their work. The authors thank Muanassa Jorge for the preliminary tests and HERCULES Laboratory represented by José Mirão for the contribution in the chemical and mineralogical characterization.

This work was financially supported by LA/P/0045/2020 (ALiCE), UIDB/50020/2020, and UIDP/50020/2020 (LSRE-LCM), funded by national funds through FCT/MCTES (PIDDAC), UIDB/00308/2020 (INESCC) - with the https://doi.org/10.54499/UIDB/00308/2020 and UIDB/00073/2020 (CGEO).

References

ANIET.: Guia de potenciais destinos para as lamas de processamento da pedra incluindo simbioses industriais. Associação Nacional da Industria extrativa e Transformadora. Associação Nacional da Industria extrativa e Transformadora (2018)

APA.: Declaração de Impacte Ambiental de Projeto Integrado do núcleo de Exploração de Pedreiras de Pé da Pedreira. Agência Portuguesa do Ambiente

Assimagra.: Plano de gestão de resíduos de extração - Projeto de sustentabilidade da industria extrativa - exploração sustentável de recursos no Maciço Calcário Extremenho. Assimagra (2015)

Grimes, C.J., Hardcastle, T., Manga, M.S., Mahmud, T., York, D.W.: Calcium carbonate particle formation through precipitation in a stagnant bubble and a bubble column reactor. Cryst. Growth Des. **20**, 5572–5582 (2020)

Monteiro, S., Jorge, M., Heleno, M., Oliveira, N., Alves, M., Veiga, A., Silva, A.: Sustainability Study of Limestone Quarry Waste Into Value-added Products: PCC AND GCC. In: 7th Global Stone Congress. Batalha, Portugal (2023)

Mordor Intelligence (2024) Calcium Carbonate Market Size—Industry Report on Share, Growth Trends & Forecasts Analysis (2024–2029). Retrieved July 2024, from Mordor Intelligence: https://www.mordorintelligence.com/industry-reports/calcium-carbonate-market

Parker, R.L.: Composition of the Earth's crust. In: Fleischer, M. (ed.) Data of geochemistry. Geological Survey Professional Paper 440-D, USA (1967)

United Nations (2024) SDG 12, Ensure sustainable consumption and production patterns. Retrieved June 2024, from Sustainable Development: https://sdgs.un.org/goals/goal12

Model Proposal for Separation of Plastics by Dissolution/Precipitation Process

Lizete Heleno[1,2(✉)], Joel Correia Vasco[1,3], Nelson Simões Oliveira[1,4,5], Marcelo Gaspar[1], and Ana Silveira[2,6]

[1] School of Technology and Management, Polytechnic Institute of Leiria, Leiria, Portugal
lizete.heleno@ipleiria.pt
[2] MARE—Marine and Environmental Sciences Centre, ARNET—Aquatic Research Network Associate Laboratory, NOVA School of Science and Technology, NOVA University Lisbon, Caparica, Portugal
[3] Institute for Polymers and Composites, Universidade do Minho, Braga, Portugal
[4] LSRE-LCM—Laboratory of Separation and Reaction Engineering, Polytechnic Institute of Leiria, Leiria, Portugal
[5] Faculty of Engineering, ALiCE—Associate Laboratory in Chemical Engineering, University of Porto, Porto, Portugal
[6] Department of Environmental Sciences and Engineering, NOVA School of Science and Technology, NOVA University of Lisbon, Caparica, Portugal

Abstract. Plastic consumption continues to rise globally, leading to increased waste generation. According to the OECD, in 2022, the worldwide recycling rate for plastics was only 9%, mainly due to the complexity of plastic waste, which often consists of mixed polymers and contamination from other materials This work focuses on developing a plastic waste recycling process using solvent-based methods, through dissolution/precipitation, which require solvents with effective solvation properties, low environmental impact and no harmful effects on health. To recycle plastic waste mixtures containing polyethylene (PE), polypropylene (PP), polyvinyl chloride (PVC), polyethylene terephthalate (PET), and polystyrene (PS), a total of 105 solvents have been investigated. Among these, 25 demonstrated adequate solvation capacity for the polymers studied. However, only two solvents were classified as non-hazardous according to the Globally Harmonized System (GHS) for chemical hazard classification. The remaining solvents were classified into hazard levels from 1 to 4 based on the GHS criteria. The results of this work allowed us to present a proposal for separating a mixture of PVC, PET, PE, PS and PP polymers, with 1 of the solvents classified as non-hazardous, 2 solvents of hazard level 1 and two solvents of hazard level 2.

Keywords: Green solvents · Plastic recycling · Selective dissolution-precipitation

1 Introduction

In 2019, the estimated use of plastics worldwide was 460 Mt, of which 429 Mt were fossil origin (primary plastics), 2 Mt were bioplastics, and 29 Mt were plastics from recycling (secondary plastics). It was also found that the estimated production of plastic

© The Author(s), under exclusive license to Springer Nature Switzerland AG 2027
M. Calvete Gaspar et al. (Eds.): Regional Helix 2024, LNME, pp. 213–229, 2027.
https://doi.org/10.1007/978-3-032-36365-7_15

waste was 353 Mt, of which only 9% was recycled. Although around 15% of plastic waste was collected for recycling, only 60% of this amount was effectively recycled. As a result, the overall final destinations of plastic waste including waste generated during recycling processes were landfill (49%), mismanaged or uncollected litter (22%), and incineration (19%) (OECD 2022). This waste sent to recycling was then converted into materials able to be incorporated in the plastics industry, with the remaining fraction classified as recycling waste and losses. This means that only 32 Mt of the 353 Mt of plastic waste was recycled in 2019 (OECD 2022). In the EU27+3, more than 29 Mt of post-consumer plastic waste was collected in 2020. Of this total, 34.6% were recycled, 42.0% were incinerated with energy recovery, and 23.4% were sent to landfill (OECD 2022). Disposal of waste in landfills and the environment can produce physical problems for wildlife resulting from ingestion or entanglement in plastic, the leaching of chemicals from plastic products, and the potential for plastics to transfer chemicals into ecosystems (Okunola et al. 2019).

Plastic waste is categorized into two primary types: post-industrial (or post-production) and post-consumer. Post-industrial plastic waste can be more easily recycled, as it has a lower level of contamination when compared to post-consumer waste (Chawla et al. 2022). Consequently, post-consumer plastic waste must undergo pre-treatment, washing, and separation processes, before recycling. Contamination is characterized by the existence of additives such as fillers, antioxidants, plasticizers, pigments, flame retardants, and even organic contamination. Moreover, post-consumer plastic comprises many types of plastics, including polyethylene (PE), polypropylene (PP), and polyethylene terephthalate (PET) (Roosen et al. 2020). The recycling, recovery, and management of plastic waste is a critical topic of discussion in the present context (Singh et al. 2017; OECD 2022).

Plastic recycling processes potentiate an economy of circular plastics, being a viable method for dealing with large-scale plastic waste with environmental benefits, namely, net decrease in greenhouse gas emissions, net energy conservation, and reduced consumption of natural resources relative to alternative waste management solutions (Jones et al. 2021; Jung et al. 2023). Nonetheless, the recycling sector encounters many challenges concerning increased implementation costs and the availability of technology to manage variations in the quality of recycled materials. The low cost of virgin plastics may also limit the utilization of recycled plastics (Jones et al. 2021; Jubinville et al. 2020).

The categories of plastic recycling linked to the plastic lifecycle are classified according to ISO 15270 and ASTM D7209-06 in mechanical recycling (primary and secondary), chemical recycling (tertiary), and energy recovery (quaternary) (Maris et al. 2018).

Primary recycling includes primary re-extrusion, and it is related to plastic waste, post-industrial, clean and uncontaminated, exhibiting equivalent properties to the original material. When such a process takes place within a closed recycling loop, the disposable waste can be effectively reintegrated into the manufacturing cycle. Secondary recycling encompasses collecting, separation and sorting, as well as a washing procedure to eliminate organic or other types of contaminants, and the grinding of the material. To ensure greater efficiency in mechanical recycling, it is not convenient to use heavily contaminated plastics. Tertiary recycling comprises chemical recycling, which transforms plastic materials into smaller molecules, typically in the form of liquids or gases, utilized as feedstock to produce chemicals and fuels. Quaternary recycling, or incineration, is a method which significantly reduces waste volume and allows energy recovery from plastic waste by combustion. This usually takes place when the waste is significantly contaminated and cannot be recycled using alternative methods (Evode et al. 2021; Maris et al. 2018).

One of the limitations of primary and secondary recycling is the very complexity of plastics, which can contain various types of plastics, fillers and additives, as well as the difficulties in separating some types of plastics that could be a subsequent source of contamination at the end of the primary and secondary recycling process. This research focuses on the treatment of the most common plastic waste by dissolution/precipitation with suitable solvents (secondary recycling process), allowing the various types of plastic to be separated and purified for optimised pelletisation by type of plastic, namely PS, PET, PP, PE and PVC. In dissolution processes using green solvents, the process parameters (e.g. dissolution temperature) and the plastics to be separated (type and diversity) must be considered (Klotz et al. 2024).

In the first phase of the study, it was necessary to define criteria for classification of adequate green solvents and in the second phase it was defined the plastic separation process taking into consideration the solubility of polymers in the selected solvents.

Choosing green solvents is fundamental to promoting sustainability in chemistry and industry. Due to the specific characteristics of solvents, selection must consider a comprehensive set of criteria that consider the intended objective, the environmental impact, and the occupational health of workers involved in the process. The increasing concern for human health and environmental protection has resulted in the creation of several guidelines and manuals, such as those from GlaxoSmithKline, Pfizer, Sanofi, and AstraZeneca, which serve as references for selecting less harmful solvents (Winterton 2021; Ikegwu et al. 2024). Gu and Jérôme (2013) propose the subsequent criteria for evaluating solvent acceptability: i) Accessible in the required quantity with a reliable long-term supply; ii) Technical performance (including solvency) no inferior to the corresponding conventional solvent; iii) Stability during usage and storage; iv) Low or non-flammable; v) Competitively priced; vi) Capable of being recycled; vii) Purity suitable for intended use; viii) Resource and energy-efficient production, that should preferably assessed over the life cycle; ix) Derived from renewable intermediates and feedstocks; x) Established acceptable toxicity and ecotoxicity profiles adequate for regulatory compliance; xi) Completely biodegradable to harmless products; and xii) Complies with transportation standards and regulations.

2 Methodology

Green solvents are characterized by a lower environmental impact and a lower risk to human health throughout their life cycle. In the research done, several methods were identified with this purpose. A combined proposal of the Environment, Health, and Safety (EHS) assessment method and life cycle assessment (LCA) is noted, with the aim of identifying inherent risks of solvents and quantifying the use of resource and respective emissions related to solvent production (Capello et al. 2007; Byrne et al. 2016). The usage of the Globally Harmonized System of Classification and Marking of Risk Pictograms of Chemical Products (GHS) is referenced to assess if the solvent poses a risk to human health (Wang et al. 2019). To assess the level of danger of solvents based on the GHS criteria (United Nations 2021), the information available on the ECHA website regarding properties of concern and hazard classification: physical, environmental, and health was consulted. Physical hazards are determined by the inherent characteristics of the chemical, categorized as follows: explosive, flammable, oxidizing, pressurized gases, and corrosive to metals. Environmental hazards include aquatic toxicity and ozone depletion categories. Health hazards include corrosive, toxic, harmful and irritating categories (United Nations 2021). A point was allocated to each item based on this information, determining the hazard level of solvents, with a maximum score of 4 and a minimum of 0. A score of 0 is considered for non-hazardous solvents.

For the identification of the most appropriate solvents taking into consideration their solubility in polymers will be considered the Hansen method, which determines the three types of intermolecular interactions involved in solubility, which are (δ_P), dispersive (δ_D) and hydrogen bonding (δ_H) (Gao et al. 2021). This methodology was enhanced by Hildebrand's technique, which is applicable solely to non-polar molecules without hydrogen bonds. (Sánchez-Camargo et al. 2019). To identify whether a solvent can dissolve a determinate solute, the value of the relative energy difference (RED) will be lower than 1, and bad solvents have an RED value higher than 1. RED is calculated as the ratio R_a/R_o, where R_o is the experimental sphere radius for the polymer and R_a is the distance in Hansen three-dimensional space, between two molecules, a solute i and a solvent j, according to Eq. 1.

$$Ra = \sqrt{4(\delta_{Di} - \delta_{Dj})^2 + (\delta_{Pi} - \delta_{Pj})^2 + (\delta_{Hi} - \delta_{Hj})^2} \quad (1)$$

In this work, the Microsoft Excel document developed by Abbott (2013) has been modified for the purpose of selecting an appropriate solvent for the separation of various polymers, specifically PE, PP, PVC, PET, and PS.

Figure 1 summarizes the solvent evaluation methodology, combining hazard assessment based on GHS criteria with polymer solubility analysis using the Hansen method and the Relative Energy Difference (RED) to classify solvent suitability.

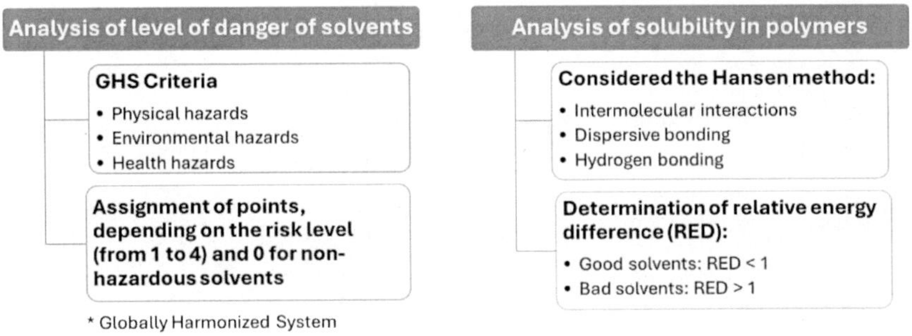

Fig. 1. Schematic overview of the solvent selection methodology based on hazard assessment (GHS criteria) and polymer solubility evaluation using Hansen solubility parameters.

3 Results

From the original list developed by Abbott (2013), were added more solvents based on their origin (biobased and/or renewable) and their Hansen solubility parameters were taken from the literature and Abbott webpage, totaling 105 solvents studied for PS, PE, PP, PET and PVC. Table 1 presents the list of solvents with a RED (based on bibliographic data) minor than 1 for at least one of the indicated polymers (values in bold and marked in green). The other colours indicate solubility values greater than 1. This method determines the polymer selectivity and defines which solvent(s) are desired.

Table 1. Relative energy difference (RED) and dangerousness classification.

Solvent	Polymer					GHS criteria*	
	PS	PE	PP	PET	PVC	D/NH**	Hazards level ***
Benzene	1.15	**0.80**	**0.54**	2.35	2.40	D	4
Benzyl Benzoate	**0.96**	1.98	2.08	1.25	1.17	D	2
Butyl Benzoate	**0.71**	1.54	1.97	0.74	0.82	NH	0
Chloroform	**0.86**	1.03	1.63	1.20	1.48	D	4
Cyclohexane	1.56	**0.68**	**0.60**	2.74	2.76	D	3
Cyclohexanone	1.17	2.04	2.46	0.78	0.45	D	2
Di-isoButyl Ketone	1.30	**0.91**	1.70	1.71	1.87	D	2
Dimethyl Cyclohexane	1.71	**0.62**	**0.99**	2.71	2.79	D	3
1,3-Dioxolane	1.70	2.26	2.85	**0.39**	1.06	D	2
Ethyl Benzene	1.10	**0.57**	**0.39**	2.34	2.36	D	2
Isophorone	1.27	1.88	2.41	**0.96**	**0.85**	D	4
d-Limonene	1.00	**0.48**	1.23	1.72	1.92	D	4
Methyl Cyclohexane	1.75	**0.67**	1.03	2.74	2.82	D	3
Methyl Oleate	1.23	**0.93**	1.73	1.58	1.76	D	1

(*continued*)

Table 1. (*continued*)

Solvent	Polymer					GHS criteria*	
	PS	PE	PP	PET	PVC	D/NH**	Hazards level ***
N-Methyl-2-Pyrrolidone (NMP)	2.24	3.13	3.56	1.27	**0.99**	D	4
Dichloromethane	1.47	1.95	2.59	**0.63**	**0.94**	D	4
Tetrahydrofuran (THF)	1.56	1.79	2.53	0.81	1.31	D	4
Toluene	0.84	0.60	0.61	2.09	2.10	D	4
Xylene	0.99	0.36	0.84	2.00	2.12	D	2
Cyrene™ (Dihydrolevoglucosenone)	2.25	3.25	3.60	1.34	**0.97**	D	1
Dimethyl isosorbide (DMI)	1.40	2.00	2.59	**0.32**	**0.78**	NH	0
2-Methyltetrahydrofuran	**0.88**	1.11	1.74	1.25	1.33	D	2
Bromotrichloromethane	1.18	**0.99**	**0.25**	2.56	2.47	D	1
p-cymene	1.56	**0.72**	**0.67**	2.74	2.74	D	3
Eucalyptol	**0.91**	**0.97**	1.57	1.51	1.54	D	4

* https://echa.europa.eu/pt/information-on-chemicals
** D—Dangerous; NH—No hazards have been classified
*** Awarding of 1 point based on the following criteria: hazardous properties, physical hazards, environmental hazards, and health hazards (According to Annex)

Results show that it will be possible to apply 25 solvents to separate polymers using dissolution, followed by precipitation. Nevertheless, there are merely 2 solvents among the 25 exhibit non-hazardous properties in accordance with GHS criteria. In the solvents analysis on what refers to their danger level based on the GHS criteria presented, among the remaining 23 solvents, 3 are classified as hazard level 1, 7 as hazard level 2, 4 as hazard level 3, and 9 as danger level 4. It has been confirmed that among the 23 solvents, all present health hazards, 13 provide physical hazards, and 9 present environmental hazards.

Considering the solvents that present a lower level of danger and the RED value lower than 1 for each polymer, Table 2 presents the proposed solvents selected for the separation of the mixture of PS, PE, PP, PET and PVC. The origin (renewable or not) of the selected solvents was also considered.

Table 2. Selected solvents proposal to separate polymer mixture.

Solvent	Dissolved polymer	RED	Hazards level	Renewable origin
Cyrene™ (Dihydrolevoglucosenone)	PVC	0.97	1	Yes

(*continued*)

Table 2. (*continued*)

Solvent	Dissolved polymer	RED	Hazards level	Renewable origin
Dimethyl isosorbide (DMI)	PET	0.32	0	Yes
Methyl Oleate	PE	0.93	1	Yes
2-Methyltetrahydrofuran	PS	0.88	2	Yes
Ethyl Benzene	PP	0.39	2	In development*

* Qin et al. (2024)

In the first interaction, the possibility of selecting non-hazardous solvents was verified, noting that Butyl Benzoate presents similar RED values (<1) for PS, PET and PVC polymers, therefore its use would not be viable. On the other hand, Dimethyl isosorbide (DMI) would be a good option for separating PET and PVC polymers, with a hazard level of 0. Next, it was analyzed which solvents could be suitable for separating PET and/or PVC, and was considered the best option, separating PVC with Cyrene, and then separating PET with Dimethyl isosorbide (DMI). For the second interaction, solvents with hazard level 1 were identified, verifying that of the three solvents, Methyl Oleate has a RED value < 1 for PE. It is also verified that Bromotrichloromethane presents a RED value < 1 for PE and PP polymers, with this solvent not being considered, since its origin is not renewable. The last interaction will be to select solvents with hazard level 2 and which have a RED value < 1 for PS and PP polymers. Therefore, 2-Methyltetrahydrofuran was considered for the separation of PS, as it is the solvent that presents the lowest RED value and also because the PE has already been dissolved and separated at this stage. For PP, the solvent Ethyl Benzene was selected, as of the solvents analyzed it is also the one with the lowest RED value.

It is worth noting that the proposal for separating polymers by dissolution using specific solvents still lacks analysis of the potential degradation of dissolved polymers at the solvent dissolution temperatures. And also evaluate the solvent recovery process, in order to guarantee the sustainability of the solution presented.

Therefore, it will be necessary to continue to investigate other types of solvents and use other methodologies to assess environmental and health risks in order to ensure their use in an industrial context.

4 Conclusions

With the increase in the production of plastic waste, there is the development of recycling processes for this waste, to enhance the circularity of plastic. This study identified a significant set of solvents, requiring the assessment of their applicability, considering their solubility in polymers and their low environmental impact and absence of harmful effects on health. As a result, around 105 solvents were identified, only 25 have solubility properties in polymers and from this list of 25 solvents, only 2 showed non-hazardous characteristics according to the GHS criteria, for separating PE, PP, PVC, PET and

PS. The remaining solvents were classified into hazard levels from 1 to 4 based on the GHS criteria. The results of this work allowed us to present a proposal for separating a mixture of PVC, PET, PE, PS and PP polymers, with 1 of the solvents classified as non-hazardous, 2 solvents of hazard level 1 and two solvents of hazard level 2.

Future work will involve continuing research into other green solvents, as well as finding methodologies for selecting solvents, in compliance with the criteria established by Gu and Jérôme (2013). It is also worth mentioning the need to evaluate the potential degradation of dissolved polymers at the solvent dissolution temperatures, as well as to evaluate the degradation potential of dissolved polymers and solvent recovery strategies.

Acknowledgements. The authors acknowledge the financial support of the Mobilizing Project "Better Plastics – Plastics in a circular economy" – ref. 46091 – Compete 2020.

Annex

GHS criteria (https://echa.europa.eu/pt/information-on-chemicals)

Solvent	Properties of concern	Hazards classification				
		Description	P	E	H	Hazards level
Benzene	Carcinogenic and Mutagenic	This substance may be fatal if swallowed and enters airways, may cause genetic defects, may cause cancer, causes damage to organs through prolonged or repeated exposure, is a highly flammable liquid and vapour, causes serious eye irritation and causes skin irritation	X	X	X	4

(*continued*)

(*continued*)

Solvent	Properties of concern	Hazards classification				
		Description	P	E	H	Hazards level
Benzyl Benzoate	–	This substance is toxic to aquatic life with long lasting effects and is harmful if swallowed	–	X	X	2
Butyl Benzoate	–	–	–	–	–	0
Chloroform	Suspected to be Carcinogenic and Suspected to be Toxic to Reproduction	This substance is toxic if inhaled, causes damage to organs through prolonged or repeated exposure, is harmful if swallowed, causes serious eye irritation, is suspected of causing cancer, is suspected of damaging the unborn child and causes skin irritation	X	X	X	4

(*continued*)

(*continued*)

Solvent	Properties of concern	Hazards classification				
		Description	P	E	H	Hazards level
Cyclohexane	–	This substance may be fatal if swallowed and enters airways, is very toxic to aquatic life, is very toxic to aquatic life with long lasting effects, is a highly flammable liquid and vapour, causes skin irritation and may cause drowsiness or dizziness	X	X	X	3
Cyclohexanone	–	This substance is a flammable liquid and vapour and is harmful if inhaled	X	–	X	2
Di-isoButyl Ketone	–	This substance is a flammable liquid and vapour and may cause respiratory irritation	X	–	X	2

(*continued*)

(*continued*)

Solvent	Properties of concern	Hazards classification				
		Description	P	E	H	Hazards level
Dimethyl Cyclohexane	–	This substance may be fatal if swallowed and enters airways, is toxic to aquatic life with long lasting effects, is a highly flammable liquid and vapour, causes skin irritation and may cause drowsiness or dizziness	X	X	X	3
1,3-Dioxolane	–	This substance is a highly flammable liquid and vapour and may damage fertility or the unborn child and causes serious eye damage	X	–	X	2
Ethyl Benzene	–	This substance may be fatal if swallowed and enters airways, is a highly flammable liquid and vapour, is harmful if inhaled and may cause damage to organs through prolonged or repeated exposure	X	–	X	2

(*continued*)

(*continued*)

Solvent	Properties of concern	Hazards classification				
		Description	P	E	H	Hazards level
Isophorone	Suspected to be Carcinogenic and Under assessment as Endocrine Disrupting	This substance is harmful if swallowed, is harmful in contact with skin, causes serious eye irritation, is suspected of causing cancer and may cause respiratory irritation	–	–	X	4
d-Limonene	Skin sensitising	This substance may be fatal if swallowed and enters airways, is very toxic to aquatic life, is a flammable liquid and vapour, is harmful to aquatic life with long lasting effects, causes skin irritation and may cause an allergic skin reaction	X	X	X	4

(*continued*)

(*continued*)

Solvent	Properties of concern	Hazards classification				
		Description	P	E	H	Hazards level
Methyl Cyclohexane	–	This substance may be fatal if swallowed and enters airways, is toxic to aquatic life with long lasting effects, is a highly flammable liquid and vapour, causes skin irritation and may cause drowsiness or dizziness	X	X	X	3
Methyl Oleate	–	This substance causes serious eye irritation	–	–	X	1
N-Methyl-2-Pyrrolidone (NMP)	Toxic to Reproduction	This substance may damage the unborn child, causes serious eye irritation, causes skin irritation and may cause respiratory irritation	–	–	X	4
Dichloromethane	Suspected to be Carcinogenic and Under assessment as Endocrine Disrupting	This substance causes serious eye irritation, causes skin irritation and may cause drowsiness or dizziness	–	–	X	4

(*continued*)

(*continued*)

Solvent	Properties of concern	Hazards classification				
		Description	P	E	H	Hazards level
Tetrahydrofuran (THF)	Suspected to be Carcinogenic	This substance is a highly flammable liquid and vapour, causes serious eye irritation, is suspected of causing cancer and may cause respiratory irritation	X	–	X	4
Toluene	Suspected to be Toxic to Reproduction	This substance may be fatal if swallowed and enters airways, is a highly flammable liquid and vapour, is suspected of damaging the unborn child, may cause damage to organs through prolonged or repeated exposure, causes skin irritation and may cause drowsiness or dizziness	X	–	X	4

(*continued*)

(*continued*)

Solvent	Properties of concern	Hazards classification				
		Description	P	E	H	Hazards level
Xylene	–	This substance is a flammable liquid and vapour, is harmful in contact with skin, is harmful if inhaled and causes skin irritation	X	–	X	2
Cyrene™ (Dihydrolevoglucosenone)	–	This substance causes serious eye irritation	–	–	X	1
Dimethyl isosorbide (DMI)	–	–	–	–		0
2-Methyltetrahydrofuran	–	This substance is a highly flammable liquid and vapour, is harmful if swallowed, causes serious eye damage and causes skin irritation		–	X	2
Bromotrichloromethane	–	This substance is harmful if swallowed, causes serious eye irritation and causes skin irritation	–	–	X	1

(*continued*)

(*continued*)

Solvent	Properties of concern	Hazards classification				
		Description	P	E	H	Hazards level
p-cymene	–	This substance may be fatal if swallowed and enters airways, is toxic if inhaled, is toxic to aquatic life with long lasting effects and is a flammable liquid and vapour	X	X	X	3
Eucalyptol	A majority of data submitters agree this substance is Skin sensitising	This substance is a flammable liquid and vapour and may cause an allergic skin reaction	X	–	X	4

P—Physical; E—Environment; H—Health

References

Abbot, S.: HSP Basics. Disponível em https://www.stevenabbott.co.uk/practical-solubility/hsp-basics.php. Acesso em 13 July 2024

Byrne, F.P., Jin, S., Paggiola, G., Petchey, T.H.M., Clark, J.H., farmer, T.J., Hunt, A.J., McElroy, C.R., Sherwood, J.: Tools and techniques for solvent selection: green solvent selection guides. Sustain. Chem. Process. **4**, 7 (2016). https://doi.org/10.1186/s40508-016-0051-z

Capello, C., Fischer, U., Hungerbühler, K.: What is a green solvent? a comprehensive framework for the environmental assessment of solvents. Green Chem. **9** (2007). https://doi.org/10.1039/b617536h

Chawla, S., Varghese, B.S., A, C., Hussain, C.G., Keçili, R., Hussain, C.M.: Environmental impacts of post-consumer plastic wastes: treatment technologies towards eco-sustainability and circular economy. Chemosphere (2022). https://doi.org/10.1016/j.chemosphere.2022.135867

Evode, N., Qamar, S.A., Bilal, M., Barceló, D., Iqbal, H.M.N.: Plastic waste and its management strategies for environmental sustainability. Case Stud. Chem. Environ. Eng. **4**, 100142 (2021). https://doi.org/10.1016/j.cscee.2021.100142

Gao, Q., Zhu, P., Zhao, H., Farajtabar, A., Jouyban, A., Acree, W.E., Jr.: Solubility, Hansen solubility parameter, solvent effect and preferential solvation of benorilate in aqueous mixtures of isopropanol, N, N-dimethylformamide, ethanol and N-methyl-2-pyrrolidinone. J. Chem. Thermodyn. **161**, 106517 (2021). https://doi.org/10.1016/j.jct.2021.106517

Gu, Y., Jérôme, F.: Bio-based solvents: an emerging generation of fuids for the design of eco-efficient processes in catalysis and organic chemistry. Chem. Soc. Rev. **42**, 9550–9570 (2013). https://doi.org/10.1039/C3CS60241A

Ikegwu, U.M., Zavala, V.M, Van Lehn, R.C.: Screening green solvents for multilayer plastic films separation. Syst. Control Trans. **3**, 763–770 (2024). https://doi.org/10.69997/sct.162050

Jones, H., Saffar, F., Koutsos, V., Ray, D.: Polyolefins and polyethylene terephthalate package wastes: recycling and use in composites. Energies **14**(21), 7306 (2021). https://doi.org/10.3390/en14217306

Jubinville, D., Esmizadeh, E., Saikrishnan, S., Tzoganakis, C., Mekonnen, T.: A comprehensive review of global production and recycling methods of polyolefin (PO) based products and their post-recycling applications. Sustain. Mater. Technol. **25**, e00188 (2020). https://doi.org/10.1016/j.susmat.2020.e00188

Jung, H., et al.: Review of polymer technologies for improving the recycling and upcycling efficiency of plastic waste. Chemosphere **320**, 138089 (2023). https://doi.org/10.1016/j.chemosphere.2023.138089

Klotz, M., Oberschelp, C., Salah, C., Subal, L., Hellweg, S.: The role of chemical and solvent-based recycling within a sustainable circular economy for plastics. Sci. Total Environ. **906**, 167586 (2024). https://doi.org/10.1016/j.scitotenv.2023.167586

Maris, J., Bourdon, S., Brossard, J.-M., Cauret, L., Fontaine, L., Montembault, V.: Mechanical recycling: compatibilization of mixed thermoplastic wastes. Polym. Degrad. Stab. **147**, 245–266 (2018). https://doi.org/10.1016/j.polymdegradstab.2017.11.001

OECD: Global plastics outlook: economic drivers. In: Environmental Impacts and Policy Options. OECD Publishing. (2022). https://doi.org/10.1787/de747aef-en

Okunola, A.A., Kehinde, I.O., Oluwaseun, A., Olufiropo, E.A.: Public and environmental health effects of plastic wastes disposal: a review. J. Toxicol. Risk Assess. **5** (2019). https://doi.org/10.23937/2572-4061.1510021

Qui, Z., Zhou, Y., Höhne, Z.L., Bornscheu, U.T., Wu, S.: Production of biobased ethylbenzene by cascade biocatalysis with an engineered photodecarboxylase. Angew. Chem. Int. Ed. **2024**(63), e2023145 (2024). https://doi.org/10.1002/anie.202314566

Roosen, M., Mys, N., Kusenberg, M., Billen, P., Dumoulin, A., Dewulf, J., Van Geem, K.M., Ragaert, K., De Meester, S.: Detailed analysis of the composition of selected plastic packaging waste products and its implications for mechanical and thermochemical recycling. Environ. Sci. Technol. **54**(20), 13282–13293 (2020). https://doi.org/10.1021/acs.est.0c0337110.1021/acs.est.0c03371

Sánchez-Camargo, A. del P., Bueno, M., Parada-Alfonso, F., Cifuentes, A., Ibáñez, E.: Hansen solubility parameters for selection of green extraction solvents. TrAC—Trends Anal. Chem. **118**, 227–237 (2019). https://doi.org/10.1016/j.trac.2019.05.046

Singh, N., Hui, D., Singh, R., Ahuja, I.P.S., Feo, L., Fraternali, F.: Recycling of plastic solid waste: a state of art review and future applications. Compos. B Eng. **115**, 409–422 (2017). https://doi.org/10.1016/j.compositesb.2016.09.013

United Nations.: The Globally Harmonized System of Classification and Labelling of Chemicals. Tenth revised edition (2023). https://unece.org/sites/default/files/2023-07/GHS%20Rev10e.pdf

Wang, H.H., Jung, J.T., Kim, J.F., Kim, S., Drioli, E., Lee, Y.M.: A novel green solvent alternative for polymeric membrane preparation via nonsolvent-induced phase separation (NIPS). J. Memb. Sci. **574**, 44–54 (2019). https://doi.org/10.1016/j.memsci.2018.12.051

Winterton, N.: The green solvent: a critical perspective. Clean Technol. Environ. Policy. https://doi.org/10.1007/s10098-021-02188-8

Sustainable Mobility and Transportation

Smart and Cost-Effective System for Monitoring Electric Scooter Comfort Based on ISO 2631 Vibration Standards

Hélio Mendonça[1], Armando L. Ramalho[1,2](✉), Pedro M. B. Torres[1,3], and Volker Lohweg[4]

[1] Polytechnic University of Castelo Branco, Av. Pedro Álvares Cabral, nº 12, 6000-084 Castelo Branco, Portugal
aramalho@ipcb.pt

[2] CEMMPRE, ARISE - University of Coimbra, Rua Luís Reis Santos, 3030-788 Coimbra, Portugal

[3] SYSTEC - Research Center for Systems and Technologies, ARISE - Advanced Production and Intelligent Systems Associated Laboratory, 4200-465 Porto, Portugal

[4] Institute Industrial IT (inIT), Technische Hochschule Ostwestfalen-Lippe (TH OWL), Campusallee 6, 32657 Lemgo, Germany

Abstract. This study presents an in-depth experimental analysis of vibration exposure on electric scooters, focusing on its impact on rider comfort. The research uses a low-cost instrumentation setup based on MEMS accelerometers and microcontroller-based data acquisition systems to collect triaxial vibration data under real-world riding conditions. Measurements were conducted on various types of urban surfaces - including asphalt, cobblestones, and concrete pavement - to assess variations in whole-body vibration (WBV) levels. The analysis follows the ISO 2631-1 standard, which provides a framework for assessing human exposure to mechanical vibrations in terms of health, comfort and perception. Time-weighted root-mean-square (RMS) acceleration values and frequency-weighted spectral analysis were used to characterize the severity of vibrations and identify surface-specific vibration profiles. The results indicate significant variability in comfort levels depending on surface irregularities and scooter construction, highlighting the relevance of vibration attenuation mechanisms in the design of micromobility vehicles. This work demonstrates the feasibility and accuracy of using low-cost hardware for predictive comfort assessment and contributes to the development of safer and more ergonomic personal urban transportation systems.

Keywords: Vibration analysis · Whole-body vibration · E-Scooter riding comfort

© The Author(s), under exclusive license to Springer Nature Switzerland AG 2027
M. Calvete Gaspar et al. (Eds.): Regional Helix 2024, LNME, pp. 233–246, 2027.
https://doi.org/10.1007/978-3-032-36365-7_16

1 Introduction

Electric scooters (e-scooters) have emerged as an economical solution for short-distance urban transport, offering an energy-efficient and low-emission alternative to conventional vehicles [1]. Its growing popularity is driven by its portability, low operating cost, and ability to alleviate traffic congestion in densely populated cities [2]. Electric scooters are particularly suitable for short-distance mobility, bridging the gap between public transport stations and final destinations. In addition to their convenience, electric scooters contribute to sustainable urban development by reducing greenhouse gas emissions and urban noise levels, especially when integrated into shared mobility systems [3]. However, despite these benefits, users often report discomfort due to exposure to mechanical vibrations during operation, especially on uneven or poorly maintained urban surfaces [4]. Vibrations on electric scooters originate from the dynamic interaction between the scooter structure (including wheels, platform and frame) and the floor profile. These vibrations are transmitted to the cyclist's body, mainly through the hands and feet, which can affect not only comfort but also long-term health, depending on exposure levels and duration of riding [5, 6]. Unlike larger vehicles, electric scooters often lack advanced suspension systems, making them more susceptible to transmitting vibrations directly to the user [7]. Assessing these vibrations is essential, especially when considering ergonomic design and user acceptance of micromobility solutions. According to ISO 2631-1:1997 [8], human exposure to whole-body vibration (WBV) must be carefully evaluated in terms of effects on health, comfort and perception thresholds. Research shows that prolonged exposure to vibrations above certain limits can result in fatigue, discomfort and even musculoskeletal disorders [9]. Recent studies have investigated vibration exposure in bicycles [10], motorcycles [11] and electric wheelchairs [12], but the literature on electric scooters remains relatively limited. Furthermore, few studies have explored the use of low-cost sensing technologies for in situ vibration analysis in micromobility contexts. The integration of portable and affordable data acquisition systems using MEMS accelerometers and microcontrollers offers a promising path for real-time vibration monitoring and user comfort assessment [13].

This paper presents a vibration analysis methodology for electric scooters under various surface conditions using a low-cost sensing system. The aim is to quantify vibration levels based on ISO 2631–1 metrics and assess their impact on rider comfort. The findings aim to contribute to the design of more ergonomic and user-friendly electric scooters, while promoting sustainable and comfortable urban mobility.

The remainder of the paper is organized as follows: Sect. 2 presents the methodology adopted. Section 3 discusses the experimental results obtained from the analysis of vibrations on different pavement types. Finally, Sect. 4 provides the main conclusions and outlines proposals for future work.

2 Methodology

This section presents the followed methodology to evaluate vibration exposure and its potential impact on riding comfort during electric scooter riding, an experimental approach was developed following the ISO 2631-1:1997 standard for assessing human exposure to whole-body vibration. The solution was developed with low-cost hardware, but functional for collecting data in real driving conditions. The acquired data is subsequently processed and analysed on a computer.

2.1 Experimental Setup

A commercially electric scooter was selected for the study, considering parameters such as wheel size, frame stiffness, and absence of an advanced suspension system. The aim was to study common conditions encountered by typical users in urban environments. The Fig. 1 illustrates the experimental apparatus and the localization of each component.

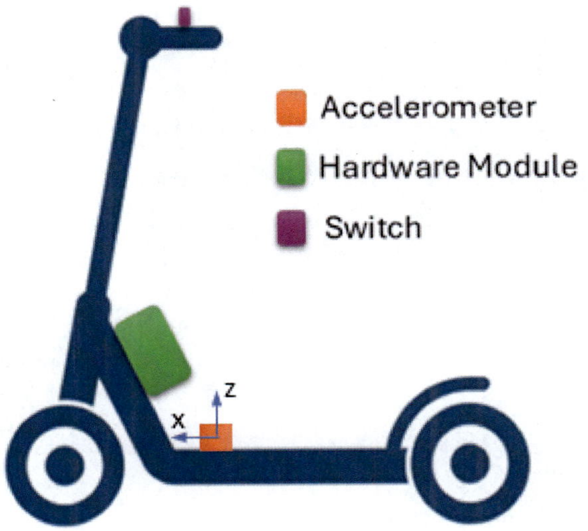

Fig. 1. Experimental setup of instrumented electric scooter.

The experimental setup developed and installed on the scooter consists of:

- **Microcontroller:** An Arduino Mega Wi-Fi R3 ATmega2560 was used as the central processing unit due to its integrated Wi-Fi capability, facilitating potential future applications in real-time data transmission.
- **Accelerometer:** An MPU-9250 inertial measurement unit (IMU), featuring a 3-axis accelerometer, gyroscope, and magnetometer. It was configured to measure acceleration data in three orthogonal directions (X, Y, Z), while driving.
- **Data Storage**: A MicroSD card module to enable local storage of time-stamped sensor data for post-processing. The Arduino was programmed to record vibration measures at fixed sampling intervals.

- **Power Supply**: A portable power bank was used to power the entire data acquisition system during the experimental tests.
- **User Interface**: A switch was installed on the handlebar to manually trigger the start and stop of data recording, allowing the user to synchronize data collection with the start and end of the test routes.

To accommodate and protect the hardware on the scooter, a custom-designed 3D printed enclosure made of PLA material was designed. The structure was fixed to the main frame of the scooter with anti-vibration padding to prevent internal movements and false readings.

The Fig. 2 illustrates the hardware components and corresponding connections setup.

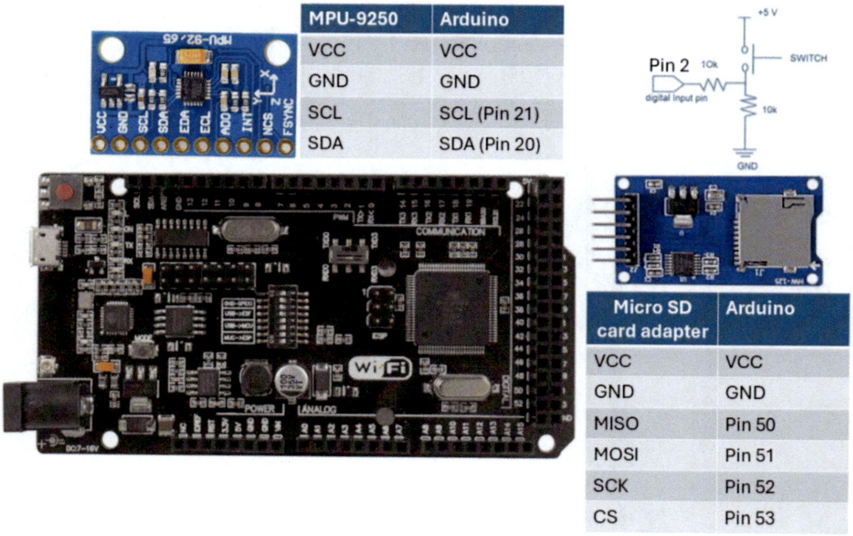

Fig. 2. Hardware connection diagram.

2.2 Measurement Chain Calibration

The calibration of the entire measurement chain, as referenced in Sect. 2.1, was a crucial step to ensure the accuracy and reliability of the vibration data collected during the study on electric scooter. This process was carried out using traceable equipment, thereby guaranteeing the integrity of the measurements across the full signal path - from sensor to data acquisition. Specifically, the setup included a DYTRAN 313-D 1955 piezoelectric accelerometer, known for its precision in dynamic measurements, which served as the sensor for acquiring vibration signals. The signal from the accelerometer was processed through a PCB Piezotronics 482A21 signal conditioner, responsible for proper amplification and filtering, as well as converting the raw charge signal from the sensor into a usable voltage output. This conditioned signal was passed to the computer trough a

National Instruments USB-6008 multi-function DAQ (Data Acquisition) module. The USB-6008 was chosen for its compatibility with LabVIEW and its ability to perform analog voltage measurements with sufficient resolution for the application. The measurement and visualization were performed in a LabVIEW application developed for this, as illustrated in Fig. 3. This software was specifically developed to facilitate real-time monitoring, signal logging, and preliminary data validation during the calibration and testing phases. Through this interface, it was possible to verify sensor response under controlled vibration conditions, ensuring the fidelity of the measurement chain prior to deployment on the scooter platform. The signal from the MPU-9250 accelerometer was collected from the MicroSD card to assure the calibration of the complete measurement chain.

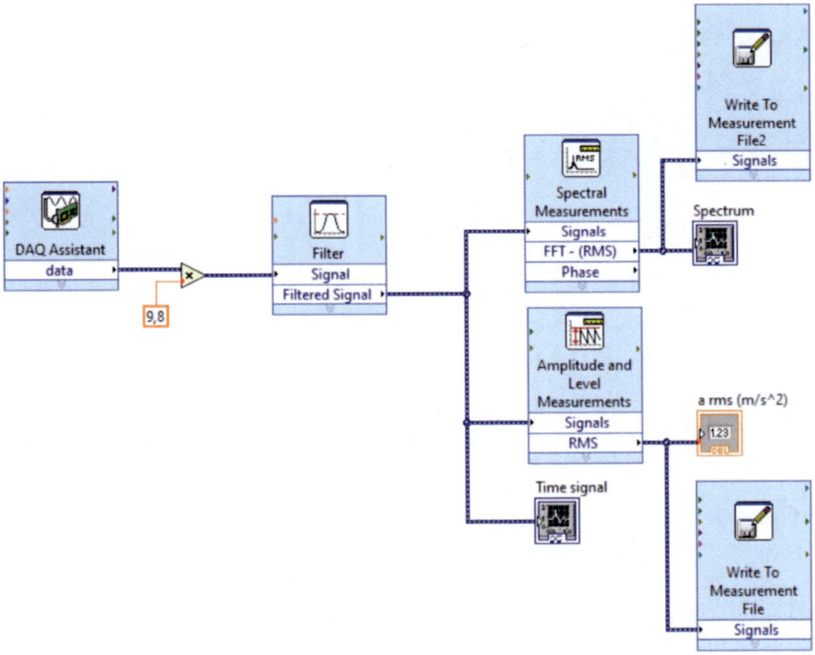

Fig. 3. Software used in data acquisition.

In the calibration process, the natural vibration (~25 Hz) of a cantilever beam, moved away from its equilibrium position, was used as an actuator. The process setup and details of the equipment used are shown in Fig. 4.

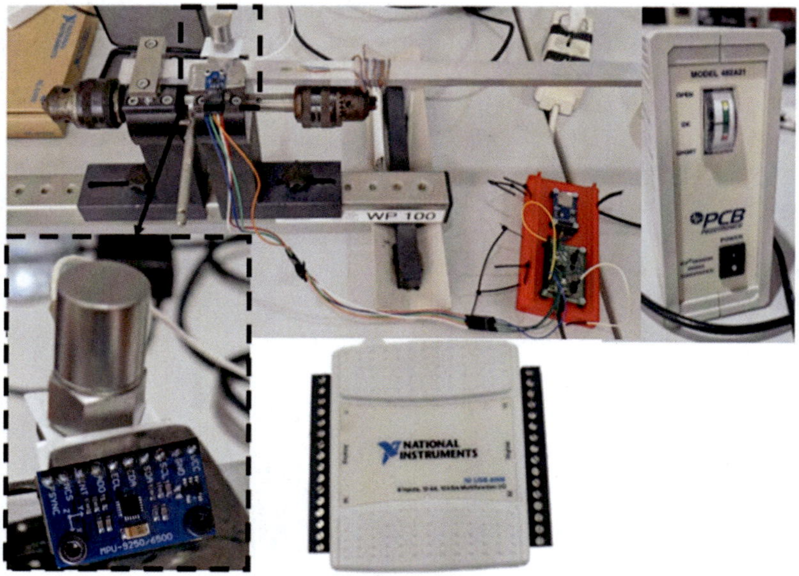

Fig. 4. Experimental setup for calibration.

2.3 ISO 2631–1 Vibration Evaluation Framework

ISO 2631-1 provides guidelines and methods for assessing the effects of vibrations and mechanical shocks on the human body, with a specific focus on whole-body vibration (WBV). It is widely used in professional and non-professional contexts, such as transportation systems, machinery and other equipment, to assess human exposure to vibrations based on their frequency and direction. As the physiological effects and the amplitude of perceived discomfort are significantly frequent, the ISO 2631-1 standard recommends the analysis of vibrations in the range of 0.5 to 80 Hz, which correspond to the frequencies that most affect driving comfort. To this end, the standard defines specific frequency weighting filters for each axis: *Wk* for vertical vibrations (x-axis) and *Wd* for horizontal axes (longitudinal and lateral, x and y). These filters consist of a low-pass filter, a high-pass filter, an acceleration-to-velocity transition function, and a scaling function. In this study, vibration analysis was performed considering the standing position of the driver, and accelerations were recorded along the three orthogonal axes (x, y, z) through the sensor integrated into the scooter structure. The evaluation followed the basic evaluation method based on the root mean square (RMS) value of the frequency-weighted acceleration over the exposure period. The total vibration level was calculated as the square root of the sum of the individual components along each axis. The resulting values were then compared with the limits for vibration intensity and their potential impact on comfort defined in ISO 2631-1 to assess driving comfort and potential health risks.

Table 1 presents the reference subjective comfort level, based on vibration intensity values, suggest in ISO 2631-1, and adopted in this study.

Table 1. Subjective perception of ride comfort based on acceleration intensity values.

Vibration (m/s^2)	Subjective comfort level
<0.315	Not uncomfortable
0.315–0.63	A little uncomfortable
0.5–1	Quite uncomfortable
0.8–1.6	Uncomfortable
1.25–2.5	Very uncomfortable
>2	Extremely uncomfortable

3 Experimental Results

This chapter presents the results obtained from vibration measurements carried out during driving tests with the electric scooter. Data was collected under different surface conditions to simulate real-world usage scenarios. The acquired acceleration values were analysed according to ISO 2631-1 standards, considering frequency-weighted RMS values along the three spatial axes. The analysis allows to assess the impact of whole-body vibrations on user comfort and to determine whether the measured vibration levels are within the recommended limits for human exposure or not.

The Fig. 5 shows the shows the assembly of the vibration measurement system on the scooter. In Fig. 5a the detail of the accelerometer assembly is represented and in Fig. 5b is presented the final setup of the instrumented scooter that was used for data collection.

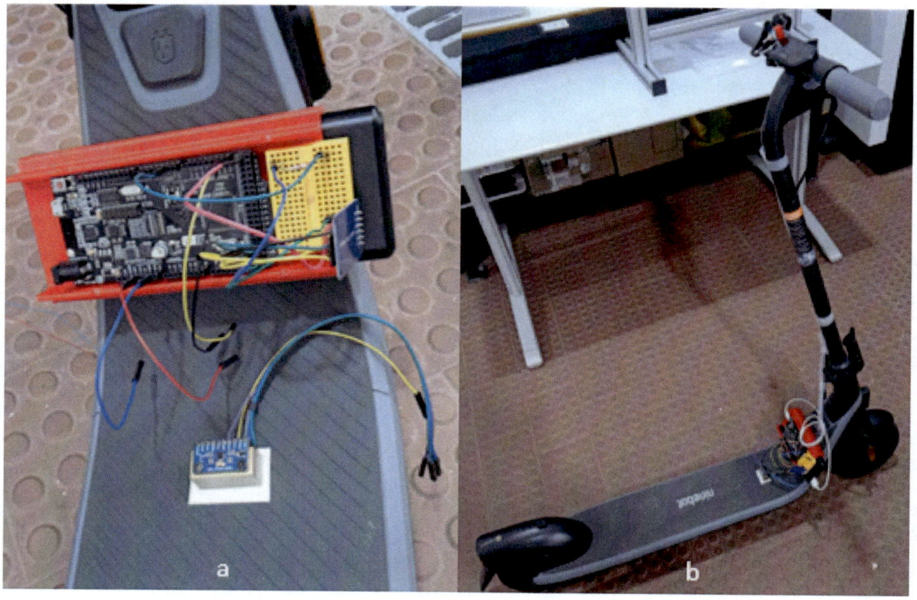

Fig. 5. Assembly of the vibration measurement system on the scooter.

To evaluate the influence of different road surfaces on whole-body vibration and driver comfort, three distinct test trajectories were defined, each corresponding to a specific type of road surface. All routes were covered three times by the same user, weighing 84 kg, resulting in a total of 9 datasets.

Path A is characterized by a pavement of concrete slabs, representative of common urban sidewalks or pedestrian areas. The total length of this route was approximately 202 m. This type of flooring is characterized by frequent joints between slabs and moderate irregularities in the surface, which can have a noticeable impact on the transmission of vibration to the user.

Figure 6 illustrates Route A.

Fig. 6. Route A - concrete slab pavement.

Route B corresponds to a longer route, with an approximate length of 425 m. The pavement type in this section was asphalt, commonly found on urban and suburban roads. This surface is generally more uniform than concrete slabs but may still have localized irregularities that affect driving comfort. The route has a slight incline of approximately 0.5%, which has little influence on the vehicle's dynamic response and, consequently, on the transmission of vibrations to the driver. A notable feature of this segment is the presence of an irregularity in the surface near the first exit of the roundabout. This localized imperfection introduces a distinct vibration profile, potentially amplifying acceleration values on specific axes. As with the other routes, this section was covered three times by the same user.

Figure 7 illustrates the second Route.

Fig. 7. Route B - asphalt route.

Trajectory C corresponds to a road characterized by an irregular pavement surface, made up of traditional Portuguese pavement. This type of surface, commonly found in historic urban areas, is made up of small, irregularly shaped stones placed manually, resulting in a discontinuous and irregular surface, with frequent unevenness and variations in height. The path taken is approximately 130 m long and has a steep incline of around 5%, which imposes additional effort on the user and can influence the transmission of vibrations. This type of pavement naturally generates high-intensity vibrations, especially for two-wheeled vehicles, such as scooters, which do not have suspension systems typical of larger vehicles. Constant surface irregularities induce continuous vibrational excitation in all three directions (longitudinal, lateral and vertical), directly impacting user comfort.

Figure 8 illustrates the third Route.

Fig. 8. Route C - traditional Portuguese pavement.

Figure 9 illustrates the time-domain plot obtained for the second trip on trajectory A.

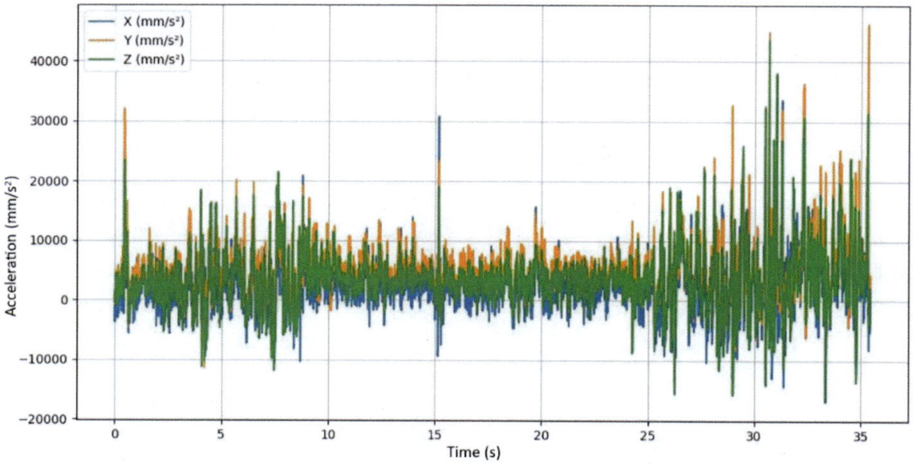

Fig. 9. Example of the time-domain plot obtained in the measurements.

With the measurements performed, the RMS weighted total acceleration was obtained, which provides information about the vibration exposure experienced during the three different paths corresponding to different pavement types. The processing of these values followed the guidelines defined by the ISO 2631–1 standard. In Table 2 is presented the obtained results.

Based on the results of Table 2, the following conclusions can be achieved.

Trajectory A – Concrete Pavers

Trajectory A presented the lowest average value of RMS weighted total acceleration, 7.67 m/s^2. The relatively uniform surface of the concrete pavements, despite some tight curves along the 202 m route, contributed to lower vibration levels. The driver maintained a consistent travel time (average of 46.33 s) and a maximum speed of 17 km/h. While this type of surface isn't perfectly smooth, overall vibration levels are the least uncomfortable in this study, with minimal abrupt changes in vertical movement.

Table 2. Experimental results.

	RMS weighted acceleration (m/s^2)		
	Trajectory A	Trajectory B	Trajectory C
1st Trip	7.35	8.44	9.31
2nd Trip	7.88	8.28	9.44
3rd Trip	7.79	8.33	8.73
Average	**7,67**	**8.35**	**9.16**

Trajectory B – Asphalt

Trajectory B had a slightly higher mean RMS weighted total acceleration of 8.35 m/s^2, which may seem counterintuitive given that asphalt is generally smoother than concrete pavements. However, an irregularity observed in the pavement near the exit of the first roundabout likely caused localized spikes in acceleration, which were captured in the RMS processing. This highlights the importance of not only the overall surface type but also localized imperfections in influencing perceived comfort. The higher average speed of 20 km/h may also have contributed to the increased dynamic response due to small surface imperfections.

Trajectory C – Portuguese Pavement

Trajectory C presented the highest vibration levels, with an average RMS weighted total acceleration of 9.16 m/s^2. This result is in line with expectations, since traditional cobblestone surfaces are highly irregular and discontinuous. The 130 m course included a significant 5% incline, where the downhill and uphill segments introduced further variation to the ride dynamics. The reduced speed in the final run (due to battery depletion) resulted in a lower RMS value (8.73 m/s^2) and was therefore excluded from the average calculation to maintain consistency. High levels of vibration on this surface indicate a low level of driving comfort and a potentially fatiguing experience for prolonged use.

4 Conclusions and Future Work

This study aimed to evaluate the whole-body vibration comfort associated with the use of electric scooters, using a model that does not have any type of suspension. The evaluation was conducted based on the ISO-2631–1 standard, using acceleration data collected during driving on three different types of pavements: concrete slabs, asphalt and Portuguese pavement. There were three tests on each floor. The results clearly showed that the scooter transmits considerable levels of vibration to the user, mainly due to its rigid wheels and the absence of a damping system. According to ISO-2631–1 guidelines, these vibration levels are high enough to affect comfort, especially when riding on uneven surfaces. Pavement type played a significant role in the results. In the parallelepiped (Trajectory C), the highest RMS weighted total acceleration were recorded, which demonstrates the impact of surface roughness. Even on smoother surfaces, such as asphalt (Trajectory B), vibrations were influenced by localized irregularities that caused discomfort. Furthermore, speed has been shown to be an important factor. Although the maximum speed on Trajectory A was 17 km/h and on Trajectory B 20 km/h, the RMS acceleration was notably higher on the latter, reinforcing the idea that increasing speed amplifies the effects of vibration.

As future work, it was intended to extend the study to include other models of electric scooters that incorporate front and rear suspension systems, to establish comparisons. Another proposal involves the integration of Machine Learning algorithms into the analysis and processing of vibration data. It can help to automatically classify pavement types based on vibration patterns, predict rider comfort levels, identify road anomalies in real-time, or even adjust scooter control parameters dynamically to increase comfort.

Acknowledgements. This research is sponsored by national funds through FCT – Fundação para a Ciência e Tecnologia, Portugal under the project UIDB/00285/2020 and LA/P/0112/2020.

References

1. Shaheen, S., Cohen, A.: Shared micromobility policy toolkit: Docked and dockless bike and scooter sharing. UC Berkeley (2019)
2. Gössling, S.: Integrating e-scooters in urban transportation: problems, policies, and the prospect of system change. Transp. Res. Part D: Transp. Environ. **79**, 102230 (2020)
3. Reck, D.J., Axhausen, K.W., Becker, H.: Mode choice, substitution effects, and the impact of free-floating e-scooters: an empirical panel study in Zurich. Transp. Res. Part D: Transp. Environ. **94**, 102783 (2021)
4. Askarzadeh, S., Cronin, K.: A study on rider exposure to vibration in electric scooters. Transp. Res. Interdis. Perspect. **14**, 100589 (2022)
5. Griffin, M.J.: Handbook of Human Vibration. Academic Press (1990)
6. Seidel, H.: Selected health risks caused by long-term, whole-body vibration. Am. J. Ind. Med. **23**(4), 589–604 (1993)
7. Petzke, F., et al.: Evaluation of ride comfort in electric scooters using vibration analysis. Procedia CIRP **91**, 371–375 (2020)
8. ISO 2631-1:1997. Mechanical vibration and shock – Evaluation of human exposure to whole-body vibration – Part 1: General requirements
9. Bovenzi, M.: Health effects of mechanical vibration. G. Ital. Med. Lav. Ergon. **27**(1), 58–64 (2005)
10. Jansen, K., et al.: Comfort assessment of city bicycles using vibration measurements. Appl. Ergon. **82**, 102936 (2020)
11. Marler, H., Frey, C.: Rider vibration exposure on motorcycles under typical urban riding conditions. SAE Int. J. Transp. Safety **6**(3), 157–166 (2018)
12. Lee, J., Kim, D.: Development of a vibration analysis system for comfort evaluation in powered wheelchairs. Assist. Technol. **31**(2), 101–109 (2019)
13. Ayaz, M., et al.: Low-cost vibration monitoring using MEMS sensors in intelligent transportation applications. IEEE Sens. J. **21**(9), 10314–10323 (2021)

Analysis of the Environmental Impacts Caused by Required Adjustments in Heavy-Duty Vehicles Maintenance Related to Biodiesel Consumption

Carlos Jacinto[1], Bruno Rodrigues[1], Diogo Silva[1,2], Ricardo Almeida[1,2], Fonseca Pereira[1,2], and Luis Serrano[1,2](✉)

[1] School of Technology and Management, Polytechnic of Leiria, 2411-901 Leiria, Portugal
luis.serrano@ipleiria.com

[2] ADAI - Association for Development of Industrial Aerodynamics, University of Coimbra, Coimbra, Portugal

Abstract. The problems of global energy dependency are currently evident, particularly from fossil energy sources. One possible solution to reduce crude oil dependency is the replacement of fossil fuels with renewable fuels in the road transport sector. Biofuels, as well as coming from renewable energy sources, offer a potential reduction in pollutant emissions when used. This work aims to study the cost-effectiveness of using biofuels in heavy-duty vehicles compared to traditional diesel (B7), focusing on the ecological footprint. This includes assessing the effects of necessary changes in maintenance intervals resulting from the use of biofuels and the resulting changes in CO_2 emissions associated with vehicle maintenance over a life cycle of 750000 km. A 'Concawe' simulation tool was used to determine CO_2 emissions in the production of the vehicle, in the production of the electricity consumed by the vehicle, as well as in the production and combustion of fuels. Regarding the environmental impact of maintenance, the plans of two leading vehicle brands on the market were considered. It can be concluded that the impact of maintenance on the vehicle's ecological footprint is negligible. It can be assumed that the major contributor to total CO_2 emissions is fuel consumption, due to the environmental impact associated with its use. It can therefore be deduced that among the alternative biofuels considered for B7, BtL, and e-Diesel presents the best results. The potential of electric propulsion for heavy vehicles in the near future is also evident, with the tendency towards adopting green energy sources for electricity production and increasing the efficiency of high-voltage batteries.

Keywords: Ecological footprint · GreenHouse gases · Biofuels · Renewable fuels · Heavy-duty vehicles · Vehicles maintenance

© The Author(s), under exclusive license to Springer Nature Switzerland AG 2027
M. Calvete Gaspar et al. (Eds.): Regional Helix 2024, LNME, pp. 247–261, 2027.
https://doi.org/10.1007/978-3-032-36365-7_17

1 Introduction

The use of biodiesel in heavy-duty vehicles has been promoted as a strategy to reduce greenhouse gas (GHG) emissions and dependence on fossil fuels. Derived primarily from vegetable oils, animal fats, or waste cooking oils, biodiesel offers the potential for lower lifecycle CO_2 emissions compared to conventional diesel, especially when sourced from sustainable feedstocks. However, the environmental benefits are not always straightforward. Factors such as land use change, feedstock cultivation practices, and the energy intensity of production can offset part of the anticipated GHG reductions.

Beyond environmental considerations, biodiesel can also affect the operational performance and maintenance requirements of heavy-duty engines. Higher blend ratios, particularly above B20 (20% biodiesel, 80% petroleum diesel), may lead to issues such as fuel filter clogging, injector deposits, fuel oxidation, and compatibility problems with certain materials in the fuel system. Cold-weather operability can also be a concern, as biodiesel generally has poorer low-temperature flow properties compared to petroleum diesel. These factors can increase maintenance frequency and operational costs if not properly managed.

Balancing potential GHG savings with the technical and maintenance implications is essential for determining the true viability of biodiesel as a sustainable fuel for the heavy-duty transport sector. A comprehensive assessment must therefore consider not only the fuel's carbon footprint but also its real-world impact on vehicle reliability and lifecycle costs. This was the main motivation of the present work.

1.1 Framework

The technological, economic and industrial development of world society is directly linked to the production of energy, especially fossil fuels. Since the end of the 18th century, following the industrial revolution, the world's demand for energy has increased exponentially. Currently, this demand shows no signs of slowing down (Fig. 1), with crude oil being the most exploited energy source. As well as being a non-renewable source of energy that could become extinct because of overuse, the emissions of polluting gases resulting from its use and production contribute to worsening global warming and atmospheric pollution.

Given the concerns about pollutant emissions from the use of crude oil and its potential depletion, as well as the increase in the feedstock price and the consequent increase in its byproducts (gasoline, diesel, etc.), various solutions to these problems have been pursued. The road transport sector accounts for around 20% of CO_2 emissions worldwide, which is the main gas responsible for the greenhouse effect, so there is a need to restructure this sector to reduce pollutant emissions and dependence on oil [2].

There are three ways of solving these problems: encouraging a reduction in the number of vehicles on the road, solutions applied to the vehicle itself by implementing new technologies to reduce emissions or adopting fewer polluting fuels.

The most viable solution with the most immediate potential will be the adoption of less polluting fuels, i.e. biofuels such as bioethanol, biodiesel and HVO (Hydrated Vegetable Oil). Because they are renewable fuels from non-fossil sources, they have lower pollutant emissions both in their use and in their production and transportation

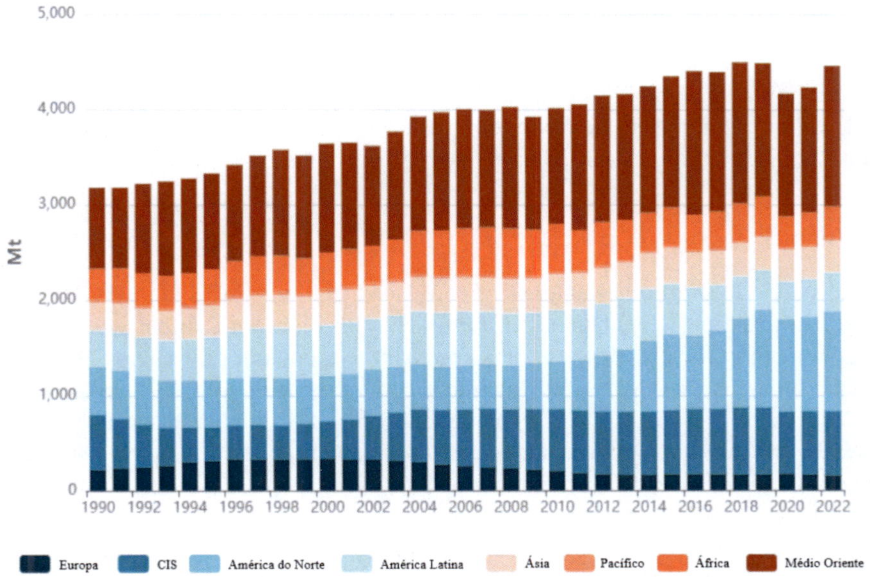

Fig. 1. World energy consumption (1990 e 2022, adapted from [1])

and are compatible with current vehicle technologies and supply network, making an immediate effect possible. However, this change to renewable fuels, especially in heavy-duty vehicles that typically use diesel engines, also implies changes in fuel consumption and maintenance plans, which is why it is necessary to assess these impacts in order to be able to accurately quantify the effects on reducing CO_2 emissions with this change to fuels of biological origin.

1.2 Objective

It is manifest the necessity to reduce the periodicity of changing the engine oil and oil filter when a vehicle runs on biofuel, compared to a vehicle consuming B7 diesel, as well as more frequent changes in fuel filter and preventive maintenance of the injection system. The aim of this work is to find out whether, given these conditions, the use of biofuels such as B15, B50, B100 and HVO has a sufficient reduction in environmental impact to justify switching from conventional diesel to a renewable diesel such as HVO or one with a higher Biodiesel blend. Knowing their properties, through simulation, a literature review and an analysis of the work carried out to date on this subject, it is intended to make a comparative assessment of the CO_2 emissions of each of the solutions. This will consider all the process, from their production to the final combustion phase and, with the results gathered, determine which is the most effective solution with a focus on reducing greenhouse gas emissions.

2 Bibliographic Research

The incorporation of biofuels in automobiles offers a attractive potential to accelerate the decarbonization of the transportation sector, but it needs thoughtful consideration given all of the characteristics and properties of each of these renewable energy sources.

2.1 Biofuels

Biofuels are a form of renewable energy produced from vegetable biomass, vegetable oils, urban waste and treated industrial waste. They have a better CO_2 balance compared to traditional fossil fuels. However, it is necessary to account for the energy required to produce and transport it, even if taking into account the capture of CO_2 by plants [4, 5].

They can be classified as primary or secondary biofuels. Primary biofuels use unprocessed biomass in its natural form, such as wood, animal fat, pellets, etc. Secondary biofuels, on the other hand, are produced by processing biomass using technology, such as biodiesel, bioethanol, biohydrogen, etc.

Secondary biofuels can be divided into four distinct generations, depending on the raw material of origin and the production process. These generations are as follows:

First generation biofuels: produced from agricultural food crops, such as cereals, starch crops, sugar and oil crops, etc. They are referred to as "FAME".

Advanced biofuels: from waste, non-food biomass and non-food crops (second generation), algae (third generation) and modified crops (fourth generation) [4, 5].

2.2 Biodiesel

Biodiesel is a renewable and biodegradable fuel that can be used in diesel internal combustion engines, either pure or as an additive in blends with diesel. It can be defined as a liquid mixture of long-chain alkyl esters, usually derived from vegetable oils or animal fat [4, 6].

Commercially it is designated by the acronym BXX, where XX represents the volume fraction of biodiesel in the fuel. For example, B15 means that the fuel contains 15% biodiesel and 75% diesel. In order to be marketed, the biodiesel found in petrol stations must meet the requirements found in the EN 14214 standard [7].

2.3 Hydrotreated Vegetable Oils (HVO)

HVO (Hydrotreated Vegetable Oils), also known as renewable diesel, are straight-chain paraffinic hydrocarbons, free of sulfur, oxygen and aromatic compounds. In terms of properties, it has the versatility to adjust its cold properties depending on the application, high cetane levels, higher calorific value and better storage stability. On the other hand, due to its paraffinic composition, it has lower lubricity and density than conventional diesel. Since its characteristics are similar to those of conventional diesel and meet the requirements of the standards (EN 590), its use does not require any adaptation [4, 8].

In the case of HVO, given its paraffinic composition, its properties must meet the requirements of EN 15940 [9].

2.4 BtL via FT

As an advanced secondary biofuel, BtL (Biomass-to-Liquid) can be produced from a wide range of non-food biomass, such as firewood, organic waste from urban and agricultural sources, animal fats, among others. Like HVO, BtL is a paraffinic fuel and therefore needs to comply with EN 15940.

Compared to traditional diesel, it has a high cetane index, high cold properties, higher calorific value and better stability, but due to its paraffinic composition it has a lower density. The main difference, which represents both an economic and environmental advantage over HVO, is the greater availability of organic matter, since this process starts with solid biomass, which contains CO and H_2 in its composition, and through a series of processes the desired hydrocarbon is obtained [10, 11].

2.5 e-Diesel

e-Diesel is a synthetic fuel with a reduced environmental impact, produced from water, CO_2 extracted from the air and electricity, through a process powered by renewable energy sources that produces blue crude oil (different from normal crude oil) which, after being refined, gives rise to e-Diesel. It is considered a CO_2 neutral fuel, since no new carbon is extracted and the energy sources for its production are from renewable sources [12, 13]. It is a paraffinic fuel and must comply with the EN 15940 standard in order to be marketed and used in internal combustion engines. It contains no sulfur or aromatic hydrocarbons and has a high cetane number compared to diesel [13, 14].

3 Methodology

To compare different displacement sizes of engines, two types of heavy-duty vehicles were selected: long haul and city bus. The "Concawe" simulation tool was employed to quantify the CO_2 (Carbon Dioxide) emissions associated with the vehicle manufacturing process, the generation of electricity consumed by the vehicle, and the production and combustion of fuels, across the VECTO (Vehicle Energy Consumption Tool) utilization cycles. Regarding the environmental impact of maintenance, the maintenance plans of two market leading brands, Scania and Volvo, were taken into consideration.

3.1 Simulation Tool

To evaluate the values for CO_2 emissions over the life cycle of a vehicle, the full life cycle CO_2 emissions simulator for heavy-duty vehicles developed by "Concawe" in collaboration with "*IFP Energies Nouvelles*" was employed. Two heavy-duty vehicle types were chosen for comparison: long haul (Class 5) and city bus, as mentioned before. The simulator itself, as shown in Fig. 2, consists of two main panels, identified by the letters "A" and "B". The panel marked with the letter "A" is the Results panel. In this panel there is a bar graph which allows the comparison of the LCA CO_2 emissions results of the different configurations.

The panel labelled with the letter "B" is the adjustment panel. This panel is subdivided into three sections marked "C", "D" and "E": the vehicle section, where the drive system

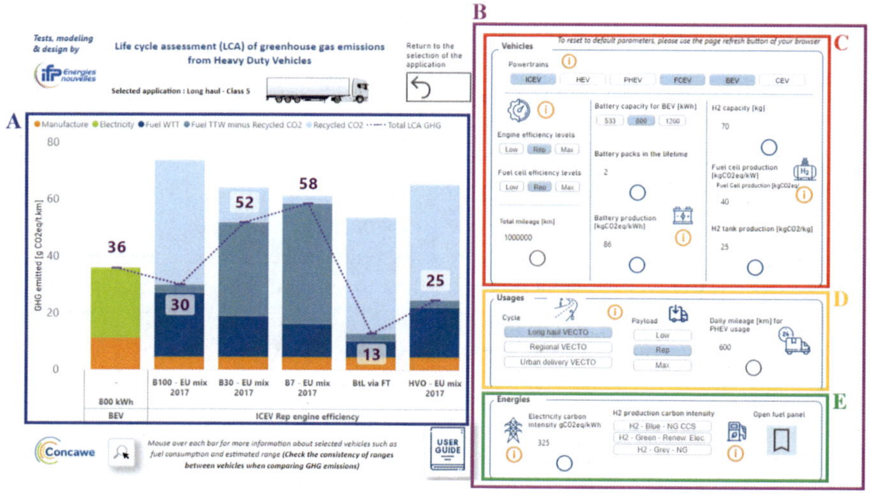

Fig. 2. Life cycle CO_2 emissions simulator, adapted from [3]

settings are adjusted, the use section, where the use cycles are configured; and the energy section, where the fuels to be considered and the CO_2 emissions during electricity generation are specified.

For each utilization cycle three simulations were made, for 1, 2 or 3 battery packs in each vehicle, considering different engine efficiency (46% or 50%) as well as different emissions in the production of batteries and electricity, as shown in the Tables 1 and 2, for the long haul and city bus respectively.

Table 2. City bus simulations

Simulation	ICEV Efficiency [%]	No. of battery packs (BEV)	Battery emissions [kgCO_2eq/kWh]	Electricity emissions [kgCO_2eq/kWh]	Utilization cycle
1	42,4	3	86	240	Urban VECTO
2	50	2	86	240	Urban VECTO
3	50	1	38	125	Urban VECTO

3.2 Maintenance

Vehicle maintenance encompasses the set of operations required to keep the equipment operational and functional, with the main aim of guaranteeing its reliability and longevity. Within the different branches of maintenance, preventive maintenance is the

Table 1. Long haul heavy duty vehicle simulations

Simulation	ICEV Efficiency [%]	No. of battery packs (BEV)	Battery emissions [$kgCO_2eq/kWh$]	Electricity emissions [$kgCO_2eq/kWh$]	Utilization cycle
1	46	3	86	240	Long Haul VECTO
2	50	2	86	240	Long Haul VECTO
3	50	1	38	125	Long Haul VECTO
4	46	3	86	240	Regional VECTO
5	50	2	86	240	Regional VECTO
6	50	1	38	125	Regional VECTO
7	46	3	86	240	Urban Delivery VECTO
8	50	2	86	240	Urban Delivery VECTO
9	50	1	38	125	Urban Delivery VECTO

methodology that is of greatest interest to this study, consisting of a plan of periodic inspection and overhaul routines for the vehicle to avoid possible breakdowns. Thus, disregarding the possible extraordinary faults that may arise in the systems during normal operation, preventive maintenance allows a detailed comparison of the differences between the plans of a vehicle operating on B7 diesel and that same vehicle operating on a biofuel. As HVO is similar to the other paraffinic biofuels, BtL and e-Diesel, the effects on maintenance are considered equivalent.

Considering the categories of commercial and passenger vehicles used in the "Concawe" software simulations, and to broaden the awareness of vehicle emissions, the maintenance plans used by Volvo [15] and Scania [16] will be taken into account for the purpose of comparing the different types of fuel and electric vehicles. It is important to note that the plans adopted are intended for vehicles with engines that are compatible with the use of biofuels, so possible changes and exceptional maintenance to the injection system are not directly accounted for in the vehicle's ecological footprint.

To evaluate the environmental impact associated with all maintenance activities conducted throughout the vehicle's life cycle, and to compare the outcomes of using biofuels versus conventional diesel, the analysis will take into consideration oil changes, fuel and oil filters, and, where applicable, particulate filters.

In terms of oil consumption, an estimate of the total amount of oil expected to be used over the vehicle's lifecycle was made, followed by a calculation of the associated CO_2 emissions to determine the environmental impact. For the filters, the constituent components of each filter were characterized and, knowing the material that makes up the component, a global estimate of the impact of each filter change was made, which was used to calculate the total CO_2 emitted as a result of all the expected filter changes. Finally, knowing the LCA (Life cycle assessment) of a particulate filter, the environmental impact of its changes was determined, ending with the sum of all the results obtained.

Scania

To determine the environmental impact of preventive maintenance for Scania's heavy-duty vehicles over a life cycle of 750000 km, the maintenance intervals in the bulletin [16] are taken into account. To get an idea of the differences between the maintenance of a vehicle running on B7, HVO and biodiesel, two vehicles will be considered, a heavy goods vehicle and a city bus both under the Euro 6 standard. For both vehicles, two Scania engines were chosen, with the code DC13 174 500 for the long haul heavy-duty vehicle and for the city bus with the code DC09 141 360.

Among the different types of vehicle use in Scania's maintenance plan, type 1 and type 4 operations were considered for the heavy-duty commercial and passenger vehicles, respectively. Type 1 operation considers fuel consumption of less than 42 l/100 km, loads typically less than 45 tons and average speeds above 60 km/h. Type 4 operation considers loads typically below 36 tons and average speeds under 40 km/h.

Volvo

For the Volvo heavy-duty vehicle brand. The maintenance intervals considered are taken from the pamphlet [15] for the "Normal Duty" type of operation, for both vehicles under study. This operation considers fuel consumption of less than 39 l/100 km. It also states that the particulate filter (DPF) is only cleaned, so it is not taken into account. Using the same methodology as the analysis carried out for Scania, two Volvo engines are considered: engine code D16K750 and D11K330 for the long-haul heavy-duty vehicle and city bus in the same order.

4 Results

This chapter presents the results obtained, considering both the life cycle analysis and the maintenance impacts.

4.1 Simulation Tool

The graphs of LCA (Life cycle assessment) CO_2 emissions for the Long Haul heavy-duty vehicle, presented in the Figs. 3, 4 and 5, are shown in relation to fuel type or high-voltage battery capacity for the VECTO Long Haul, Regional, and Urban Delivery use cycles, respectively. Figure 6 likewise depicts the CO_2 LCA emissions, specifically for the VECTO Urban utilization cycle pertaining to the city bus. The main point to take away from the analysis of the graphs is that in all simulations, BtL and e-Diesel biofuels consistently have the lowest emissions.

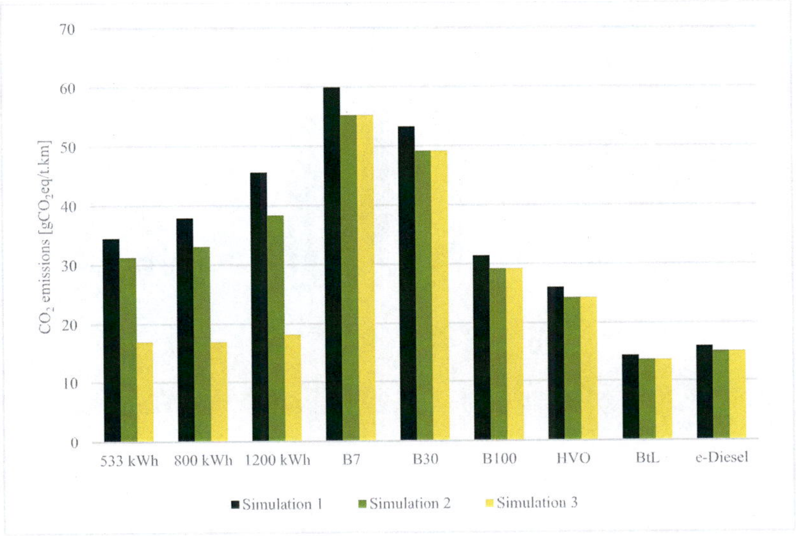

Fig. 3. LCA CO_2 emissions for the VECTO Long Haul cycle

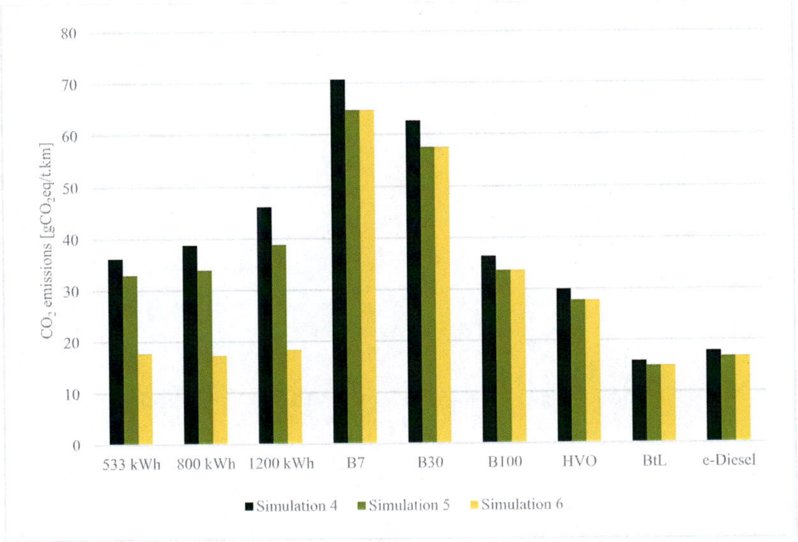

Fig. 4. LCA CO_2 emissions for the VECTO Regional cycle

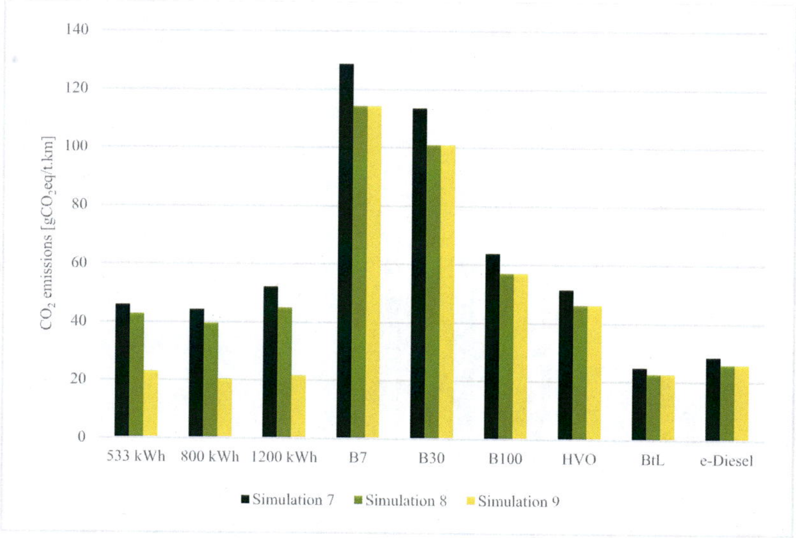

Fig. 5. LCA CO_2 emissions for the VECTO Urban Delivery cycle

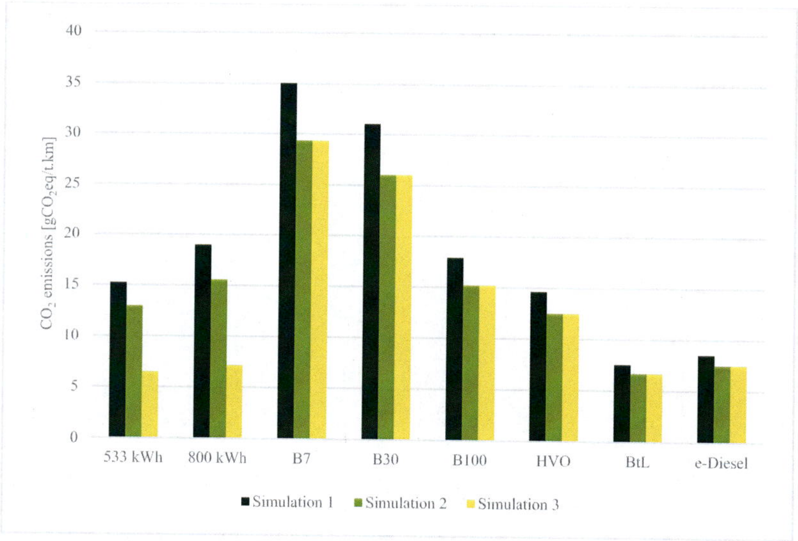

Fig. 6. LCA CO_2 emissions for the VECTO Urban cycle

4.2 Maintenance

The graphs in Figs. 7 and 9 show the total CO_2 emissions for the maintenance plan of the Scania heavy truck brand, based on a life cycle of 750000 km. The same is shown for Volvo's maintenance plan in Figs. 8 and 10. The analysis shows a twofold increase in emissions when using biodiesel compared to diesel or HVO.

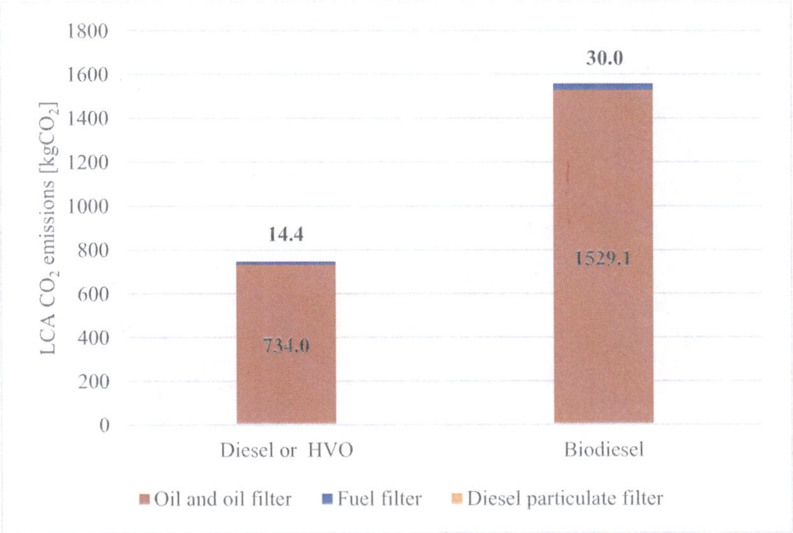

Fig. 7. Total CO_2 emissions - considering Scania's DC13 engine maintenance

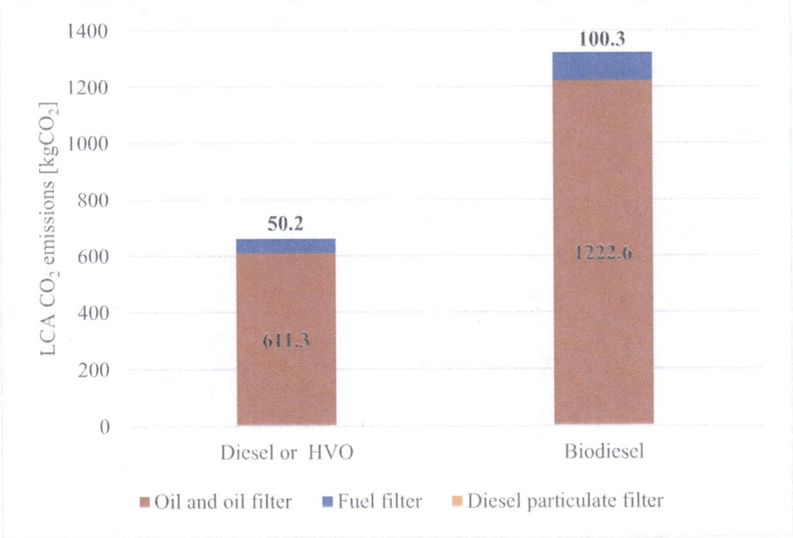

Fig. 8. Total CO_2 emissions - considering Volvo's D16K engine maintenance

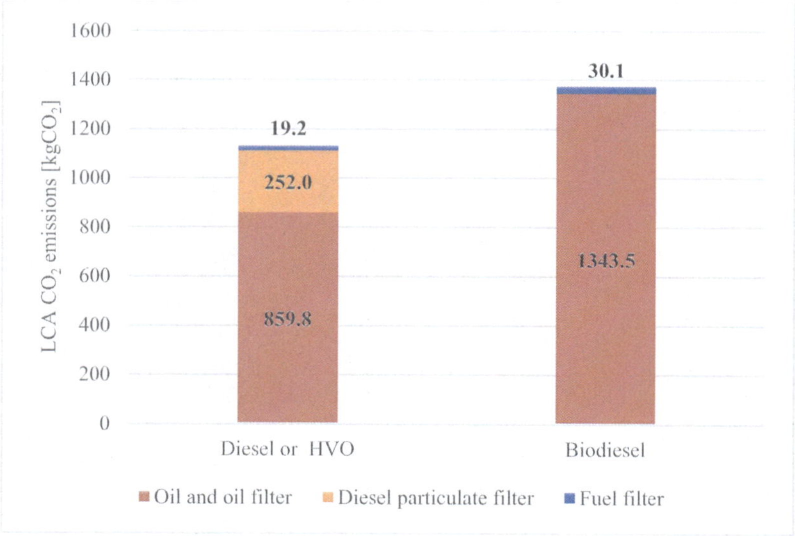

Fig. 9. Total CO_2 emissions - considering Scania's DC09 engine maintenance

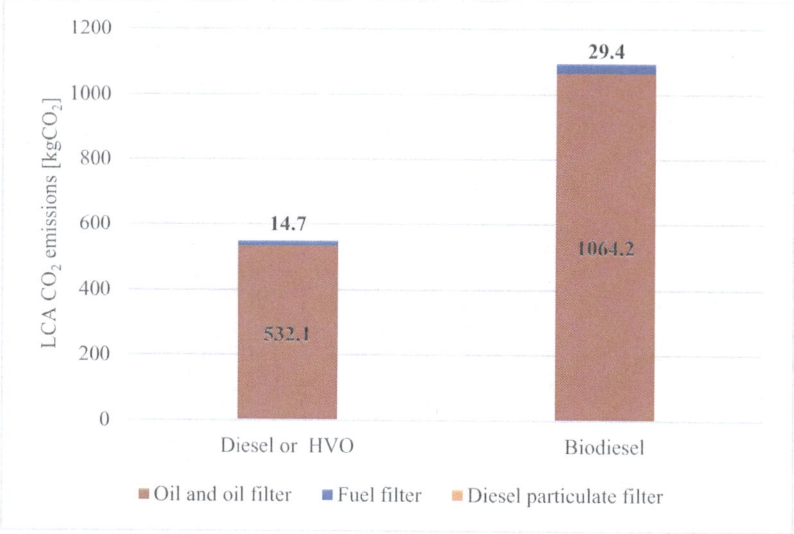

Fig. 10. Total CO_2 emissions - considering Volvo's D11K engine maintenance

4.3 Results Analysis

By consolidating the results obtained using the "Concawe" simulation tool, it was possible to gain insight into the CO_2 emissions generated during the vehicle production phase and emissions from the energy source required for propulsion, taking into account the specific type of use and the efficiency of the propulsion system in internal combustion vehicles and the batteries in electric vehicles, for a given distance travelled. In a primarily research-oriented approach, the impact of combustion vehicle maintenance on CO_2 emissions was validated over a 750000 km life cycle. In order to ascertain the impact of each factor on the vehicle's overall emissions, taking into account the type of fuel used throughout the vehicle's life cycle, the subsequent step will be to integrate the findings from the production and utilization phases with those from maintenance (Fig. 11), and draw the respective conclusions.

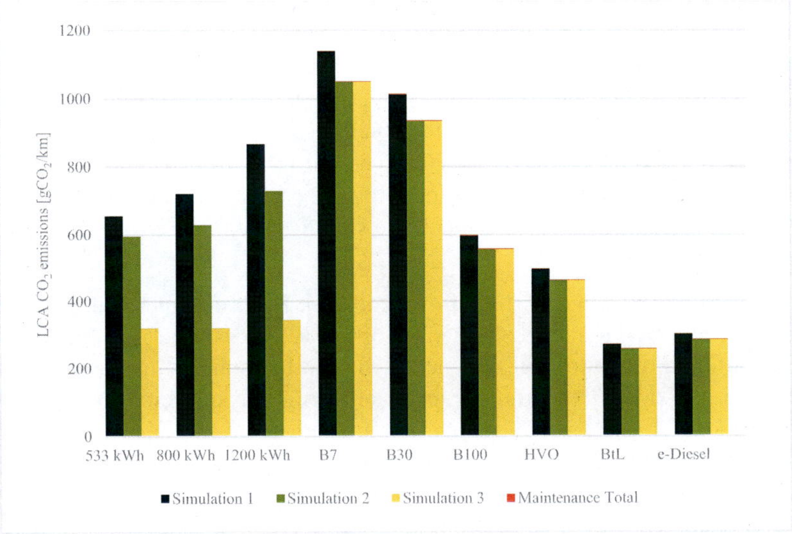

Fig. 11. Total LCA CO_2 emissions for the VECTO Long Haul cycle – considering Scania's DC13 engine

For the two vehicles studied, the best results in terms of CO_2 emissions emanate from the use of both BtL (Biomass to Liquid) and e-Diesel biofuels, with consistently lower emissions in all the simulations carried out. However, the results for electric vehicles were comparable to those for BtL and e-Diesel in the simulations where very low values were set for a 2030 forecast of CO_2 emissions in battery and electricity production (In yellow, Fig. 11). In the context of fuel consumption and vehicle autonomy, since they are interrelated, biofuels are undoubtedly the best option compared to electric vehicles, given their greater autonomy and availability of fuel stations.

Analyzing the results obtained regarding maintenance, and considering both Scania and Volvo disparities on the maintenance schedules and components involved, it was found that, in general, carbon dioxide emissions double when biodiesel is used compared to diesel or paraffinic fuels. However, comparing the emissions of maintenance with the emissions from the manufacture and usage of the vehicle, these values are practically insignificant, as can be confirmed by analyzing the results presented in Fig. 11, disregarding the type of use cycle considered since conclusions drawn are the same between all.

5 Conclusion

Given the high dependence of the transport sector on fossil fuels, the most viable solutions today are the implementation of biofuels or the electrification of vehicles. This study analyzes the feasibility of replacing diesel (B7) with biofuels in heavy-duty vehicles in terms of environmental impact, considering possible changes in the frequency of maintenance.

In the analysis of the results obtained, the carbon footprint generated by maintenance carried out throughout the life cycle is negligible when compared to global CO_2 emissions. It can be seen, however, that fuel consumption has a major influence on emissions, so the carbon neutrality of fuel production processes, as well as the efficiency of internal combustion engines, are the main objectives for reducing global carbon impact. Biofuels, on average, show slightly higher fuel consumption compared to B7, with a consequent reduction in vehicle range. However, this is a less significant question given that, in an overall assessment, the corresponding CO_2 emissions considering a life cycle assessment (LCA Life Cycle Assessment) are significantly reduced.

Summarizing it can be stated:

- The impacts of replacing fossil fuels by biodiesel, HVO or e-diesel allows to have a significant potential to reduce CO_2 emissions of vehicles to the atmosphere.
- The use of biodiesel introduces an increased need for vehicles maintenance requirements, but the overall economic impacts are insignificant facing the costs of fuel consumption.
- Biofuel, HVO and e-Diesel effectively represent an interesting solution for decarbonization of road transport sector, mainly for heavy-duty vehicles where eletric solution don't offer the required autonomy and freedom of movement and schedules required in transport of goods and services.

In conclusion, currently, the options that stand out most in terms of ecological footprint are BtL and e-Diesel, as they have substantially lower production emissions compared to the others. In near future some investment should be made to increase the production rates of these fuels, increasing availability and promoting a price reduction.

References

1. Crude Oil Production Statistics | Crude Oil | Enerdata. (n.d.). Retrieved March 18, 2024, from https://yearbook.enerdata.net/crude-oil/world-production-statistics.html#energy-consumption-data.html
2. Ritchie, H., Rosado, P., Roser, M.: Breakdown of carbon dioxide, methane and nitrous oxide emissions by sector (2024). Our World in Data. https://ourworldindata.org/emissions-by-sector
3. Home – HDV CO2 Comparator, https://hdvco2comparator.eu/, last accessed 2024/04/09
4. Análise do Mercado de Biocombustíveis 2018-2020, ERSE, Lisboa (2021)
5. Silva, L.: Processo de Produção de Biodiesel e Análise de Parâmetros de Qualidade, Instituto Politécnico de Bragança, Bragança (2014)
6. Diya'Uddeen, B.H., Abdul Aziz, A.R., Daud, W.M.A.W., Chakrabarti, M.H.: Performance evaluation of biodiesel from used domestic waste oils: a review. Process Safety Environ. Protection **90**(3), 164–179 (2012)

7. Vieira, M.: Projecto de uma Unidade Laboratorial de Produção de Biodiesel, Instituto Politécnico de Tomar, Tomar (2012)
8. Aatola, H., Larmi, M., Sarjovaara, T., Mikkonen, S.: Hydrotreated vegetable oil (HVO) as a renewable diesel fuel: trade-off between NOx, particulate emission, and fuel consumption of a heavy duty engine. SAE Int. J. Engines **1**(1), 1251–1262 (2008)
9. Robert Bosch GmbH: Automotive Handbook. 11th Edition. John Wiley & Sons Ltd (2022)
10. Chaves, H.: Problemática Ambiental do biodiesel: uma perspetiva química, Instituto Politécnico de Beja, Beja (2017)
11. Mesfun, S.A.: Biomass to liquids (BTL) via Fischer-TROPSCH a brief review. ETIP Bioenergy (2021)
12. Höpfner, S.: Audi steps up research into synthetic fuels. Technology and Innovation Communications (2017)
13. E-diesel – Wikipedia, https://en.wikipedia.org/wiki/E-diesel, last accessed 2024/04/24
14. Patricia, A., Karine, B., Chabrelie, M.-F., Sibylle, D.-D., Jean-Fritz, F., Olivier, G., Daphné, L., Mickaël, M., Alain, Q., Charlène, S., Boyang, X.: Future diesel-like renewable fuels-a literature review, (2022)
15. Service at a Glance Maintenance Intervals (2011)
16. Preface to periodic maintenance 00:17-30 issue 29. Scania CV AB

Implementation of CO_2 Emissions Assessment System in Heavy-Duty Vehicles in Portugal

Duarte Vieira[1], Paulo Carvalho[1,2], Ricardo Almeida[1,2], Diogo Silva[1,2], and Luis Serrano[1,2](✉)

[1] School of Technology and Management, Polytechnic of Leiria, 2411-901 Leiria, Portugal
{paulo.carvalho,luis.serrano}@ipleiria.com

[2] ADAI - Association for Development of Industrial Aerodynamics, University of Coimbra, Coimbra, Portugal

Abstract. The VECTO, or Vehicle Energy Consumption calculation Tool, is a simulation tool developed by the European Commission used to calculate the fuel consumption and CO_2 emissions of heavy-duty vehicles for type-approval purposes. Before any heavy-duty vehicle can be registered, a simulation must be carried out using this tool, which aims to ensure that there is a standardized simulation method for calculating fuel consumption and CO_2 emissions. A summary of the legislation related to the VECTO tool, and the CO_2 emission limits that vehicles need to comply with were presented and the VECTO tool and all the parameters needed to carry out a simulation were analyzed. Related to this analysis, the influence of certain parameters on fuel consumption and CO_2 emissions was investigated and it was concluded that aerodynamic resistance, the load carried by the vehicle and the rolling resistance are the parameters that most influence fuel consumption and consequently CO_2 emissions. Due to the low number of tractor units manufactured in Portugal and the higher number of trailers manufactured in Portugal, it was decided to consider the VECTO Trailer tool, which is specialized in calculating the influence of the trailer on the fuel consumption and CO_2 emissions of a heavy vehicle. The influence of certain parameters on fuel consumption and CO_2 emissions was also evaluated. It was also decided to simulate two real cases, using the VECTO Trailer tool for the homologation of a semi-trailer and VECTO tool to simulate the consumption of a known rigid truck on a known route and thus check the accuracy of the tool. The rigid truck was also optimized by fitting more efficient tires and by introducing aerodynamic devices, both with the aim of reducing fuel consumption and obtaining financial benefit.

Keywords: Ecological footprint · GreenHouse gases · Biofuels · Renewable fuels · Heavy-duty vehicles · Vehicles maintenance

1 Introduction

1.1 The Need for CO_2 Emissions Reduction in the Transport Sector

Climate change is one of today's biggest challenges. One of the main causes of climate change is the accumulation of greenhouse gases in the atmosphere, such as carbon dioxide, which increases the global temperature of the planet. To reduce the rise in global

© The Author(s), under exclusive license to Springer Nature Switzerland AG 2027
M. Calvete Gaspar et al. (Eds.): Regional Helix 2024, LNME, pp. 262–276, 2027.
https://doi.org/10.1007/978-3-032-36365-7_18

temperatures, the Paris Agreement came into force in 2016 with the main objective of limiting the rise in the average global temperature [1]. To limit this temperature increase, it is imperative to reduce carbon dioxide emissions in all sectors.

Over the years there has been an increase in vehicle traffic and, although the efforts to raise the efficiency of heavy vehicles, CO_2 emissions in the transport sector still trend upwards. The transport sector accounts for more than one-fifth of global greenhouse gas emissions in the energy sector. Of these 22.4% of emissions, the main contributors are passenger vehicles, which account for 44.4% of those results [2]. Another major contributor to greenhouse gas emissions is heavy vehicles, such as trucks and buses, accounting for 29.6%. It is estimated that the proportion of CO_2 emissions from heavy vehicles will continue to rise, achieving an increase of 9% between 2010 and 2030, as stated by the European Commission in Regulation 2017/2400 [3].

Looking at the evolution of Portuguese emissions for the transport sector represented in Fig. 1, it is possible to perceive that there was an increase in CO_2 emissions between 1990 and 2022 of around 47% [4].

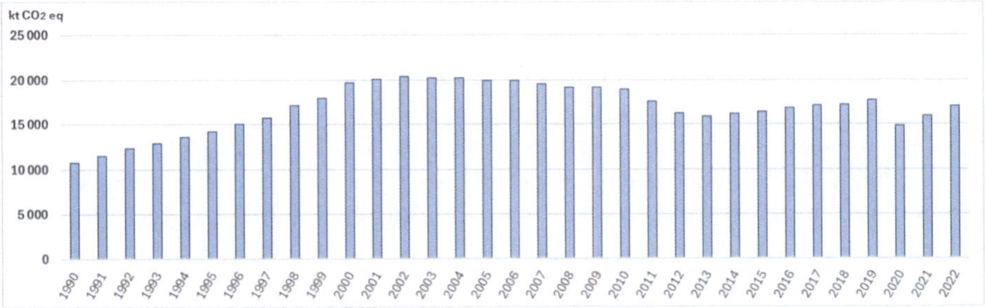

Fig. 1. CO_2 emissions from transport sector in Portugal (1990 e 2022, adapted from [4])

Until the early 2000s, there was a steady increase in CO_2 emissions, followed by a period of stabilization. From 2010 onwards, there was a slight decrease in emissions, but after 2013 this trend was reversed with a perceived increase. The significant reduction in emissions in 2020 was due to the major impact that COVID-19 had on the transport sector. The significant reduction in emissions in 2020 was due to the major impact that COVID-19 had on the transport sector [4].

As can be seen from the analysis of CO_2 emissions at global and national level, there is a trend towards increasing emissions in the transport sector. This increase in emissions in the transport sector counterbalances all the emissions reductions achieved in other sectors using renewable energy sources and energy efficiency measures.

To reduce CO_2 emissions and comply with the Paris Agreement, the European Commission has announced a series of initiatives aimed at achieving zero emissions by 2050. To fulfil these initiatives, it is essential to implement measures that encourage the transport industry, particularly the heavy-duty vehicle industry, to manufacture more efficient and even zero-emission vehicles. The European Commission has therefore established performance standards for CO_2 emissions from new heavy-duty vehicles. These performance standards have been established in all EU member states to maintain the smooth track of the transport sector and to avoid the fragmentation of the market [2].

Heavy-duty vehicles represented a sector where specific measures to reduce emissions were imperative. In order to comply with the European Green Deal agreement, Regulation 2019/1242 was introduced, establishing performance standards for CO_2 emissions. The regulation stipulates that new heavy-duty vehicles must reduce their specific CO_2 emissions by 15% from 2025, compared to the CO_2 reference emissions. From 2030, emissions must be reduced by 30%, also compared to the CO_2 reference emissions [3].

To ensure a smooth transition in reducing CO_2 emissions, the regulation initially only included new heavy-duty vehicles belonging to the N2 and N3 categories, which correspond to rigid lorries and tractors with 4×2 and 6×2 axle configurations, since these account for around 65% to 70% of CO_2 emissions from heavy-duty vehicles. These performance standards stipulate that for the years from 2025 onwards, CO_2 emissions from new heavy-duty vehicles (HGV) in these two categories must decrease by 15% compared to the CO_2 reference emissions [6].

For the years after 2030, CO_2 emissions must fall by 30% compared to the CO_2 reference values. Laboratory tests can be carried out to check the CO_2 emissions and fuel consumption values that must be met, but HGVs have a huge variety of possible configurations, from engine types, gear ratios, aerodynamic configurations, annual mileage, payload and trailer configurations. This wide variety of configurations makes the quantification of CO_2 emissions and fuel consumption very difficult, since laboratory testing is not a viable solution in practical and monetary terms, due to this variety of configurations. A much more practical method for calculating CO_2 emissions and fuel consumption in a standardized way is throughout vehicle simulation.

To effectively reduce CO_2 emissions, it is essential that manufacturers reduce emissions from their heavy-duty vehicles, but it is also important that vehicle buyers choose vehicles with lower emissions.

To do this, it is necessary that information on the CO_2 emissions performance of each vehicle be made public, allowing all buyers to make informed decisions. With this disclosure of CO_2 emissions, manufacturers are also able to compare their performance with that of other brands, which also increases progress in terms of reducing emissions. Thus, in July 2018, Regulation 2018/956 on the monitoring and reporting of CO_2 emissions and fuel consumption of new heavy-duty vehicles came into force. This regulation stipulates that manufacturers must report CO_2 emissions and fuel consumption figures to the European Commission. If manufacturers fail to report the data, or if it deviates from the actual values, a fine is imposed. The parameters to be reported and monitored are certain vehicle characteristics, such as engine power, cylinder capacity, fuel type, transmission type, CO_2 emissions in g/km, g/tkm, and fuel consumption in g/km, g/tkm, l/100 km, and other units of measurement [6].

Considering the heavy-duty vehicle sector high diversity, with the different possible configurations for each vehicle, the European Commission concluded that in order to obtain information on CO_2 emissions and fuel consumption for each heavy-duty vehicle at the lowest possible cost, simulation software should be used. According to the European Commission, this software should be made available to the public, be open source, and simulate the calculation of CO_2 emissions and fuel consumption based on input data that reflects the characteristics of the vehicle, such as the engine, transmission, axles, tires, aerodynamics, and auxiliaries.

Prior to registration, sale, or entry into service, the manufacturer must perform a simulation of the vehicle to determine the CO 2 emissions and fuel consumption of the heavy-duty vehicle, using the latest version of the simulation tool [3].

1.2 The VECTO Simulation Tool

VECTO, or Vehicle Energy Consumption calculation Tool, is the simulation tool developed by the European Commission used to calculate the fuel consumption and CO_2 emissions of heavy-duty vehicles [7]. The aim of this tool is to ensure that there is a standardized simulation method for calculating fuel consumption and CO_2 emissions during a given journey. Before any heavy-duty vehicle can be registered, it is necessary to carry out a simulation in this tool. VECTO offers two modes for simulation: Declaration mode, used for vehicle homologation, and Engineering mode, which can be used for development. In both modes, to carry out a simulation it is necessary to define in the tool, the various characteristics of the vehicle, such as its mass, the number of axles, the aerodynamic drag coefficient, the tire dimensions and so on. After these definitions, the routes to be travelled by the vehicle are automatically chosen by the tool correspondence of the vehicle characteristics. These routes represent different driving scenarios, with different average speeds and different number of stops. After defining all the vehicle's characteristics, the routes to be travelled and some simulation parameters, the tool automatically determines the fuel consumption and CO_2 emissions of the HGV.

The VECTO tool consists of three main phases. The first phase involves defining the input data through an interface, such as engine characteristics, transmission specifications, operating profiles, and auxiliary characteristics. This phase also includes a library with some predefined settings for heavy vehicles, already containing values for engine maps, component efficiencies, and some operating profiles. The second phase corresponds to the calculation of the simulation and the prediction of fuel consumption and CO_2 emissions for the heavy vehicle. Finally, the third phase corresponds to the generation and presentation of the results [8].

For a given operating profile and a defined payload, the VECTO tool uses input data in the form of single values, such as vehicle mass and rolling resistance, or in the form of maps, such as the engine map, fuel consumption map, and transmission loss map, to simulate the longitudinal dynamics of the heavy-duty vehicle. Input data in map form represents dynamic values that vary depending on the vehicle speed or engine rotation.

The VECTO tool follows a block model of the vehicle's longitudinal dynamics. This means that VECTO simulates the dynamic behavior of the vehicle along its longitudinal axis. With this model, the lateral forces acting on the vehicle are neglected. This simplification is acceptable, since the longitudinal forces are much greater in magnitude than the lateral forces. There are two different methods for operating a vehicle simulation: the forward-looking method and the backward-looking method.

In the forward-looking model, a target speed is defined which the vehicle attempts to reach. Thus, a forward-looking model can be characterized by an accelerator model for wheel. The torque produced by the engine is sent to the transmission and passes through the various gear ratios and is finally applied to the wheels. Since there is no speed imposed on the vehicle, there will be a small difference between the vehicle speed and the target speed. It is up to the simulation to reduce this difference by reducing or increasing the torque and engine speed.

In the backward-looking model, the target speed is defined, and the vehicle strictly follows that speed. Thus, a backward-looking model can be characterized by a wheel-to-engine model. The torque required to be applied to the wheel to meet the target speed is sent to the transmission, where the gear ratios are considered, and then the torque value is required from the engine. This defines the torque that the engine needs to produce in that specific condition. This model requires efficiency and consumption maps, which are usually obtained through real-world testing.

To calculate the fuel consumption and CO_2 emissions of a heavy vehicle, it is necessary to convert the operating cycle information into operating points on the engine map, considering the respective characteristics of the vehicle. The VECTO tool uses a backward-looking method to perform this operation.

The Driver Model converts the operating cycle information, such as target speed and road gradient, into a requested acceleration. This acceleration is transferred to the Vehicle module. The Driver Model converts the operating cycle information, such as target speed and road gradient, into a requested acceleration. This acceleration is transferred to the Vehicle module, which uses the vehicle mass, aerodynamic resistance, and dynamic tire radius information to convert the requested acceleration into a requested force. Through wheel inertia, this force is converted into a desired torque, Fig. 2 [9].

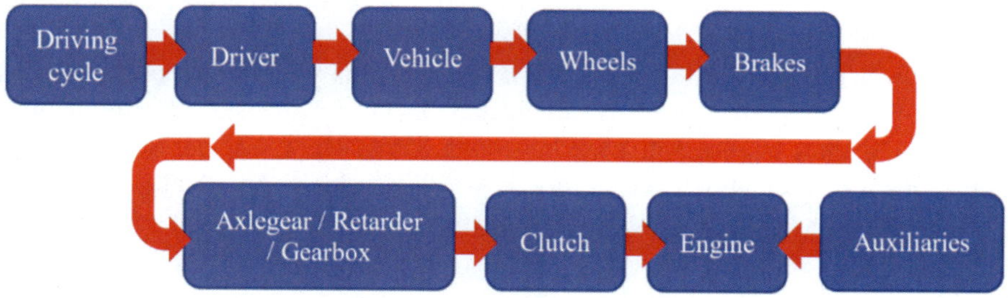

Fig. 2. Main structure of the VECTO simulation tool [9]

If this torque is in accordance with the vehicle's torque curve, a positive message is returned, and the simulation proceeds to the next interval. If the requested torque does not match the vehicle's engine torque curve, the engine module returns an error message to the Driver Model. With this, the Driver Model changes the requested acceleration until the torque requested in the engine module is within the values of the vehicle's torque curve. The Driver Model can change the requested acceleration based on the activation of the brakes. Figure 3 exposes a schematic representation of a simulation in the VECTO tool.

When the requested torque is in accordance with the vehicle's torque curve, a positive message is returned to the Driver Model, and fuel consumption can be calculated using the fuel consumption map. It should be noted that at this point, it is necessary to make all corrections to the consumption map, such as the WHTC correction factors [10].

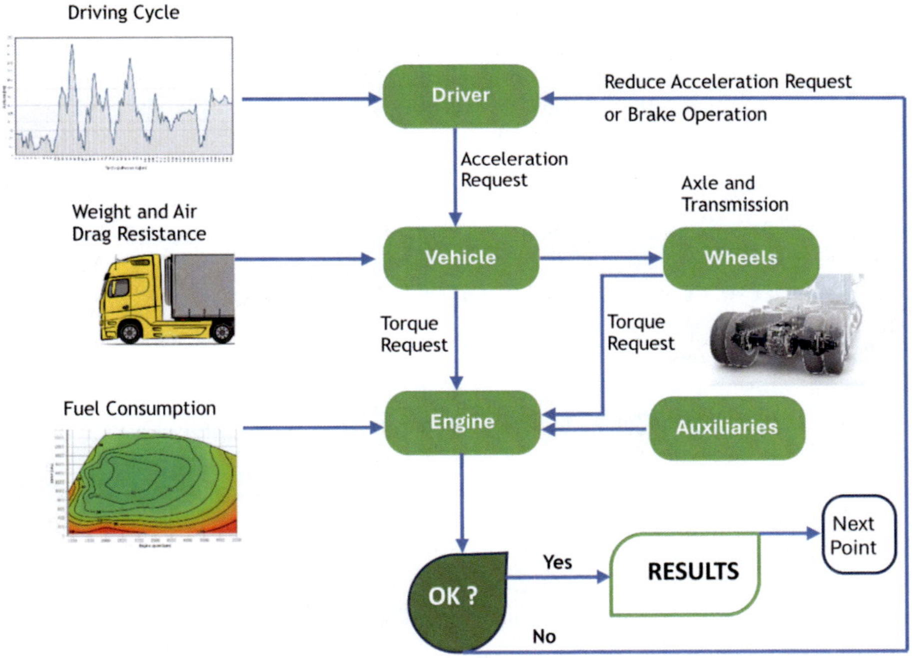

Fig. 3. Schematic representation of the VECTO simulation tool

2 Methodology

Although the VECTO considers various characteristics of the new heavy-duty vehicle, a fundamental parameter is not considered, which is the trailer carried by the vehicle. The simulation is carried out with a trailer, but it is a generic trailer, the same in all simulations. As there is a wide variety of trailers, all with different mass and dimensions, the results obtained by the VECTO would not be representative of the reality. Although this trailer produces no CO_2 emissions and consumes no fuel, it greatly affects the vehicle's fuel consumption. So, to encourage the development of lighter and more efficient trailers, the European Commission has issued a regulation that makes it compulsory to assess the performance of new semi-trailers and trailers in terms of CO_2 emissions. Any manufacturer must determine the performance of the semi-trailer or trailer before it can be sold. To determine the performance of trailers, a tool has been developed that considers the specific characteristics of the trailer. This tool is called VECTO Trailer. Therefore, before the trailer is put into circulation, the manufacturer must carry out a simulation in the VECTO Trailer and thus obtain the performance in terms of CO_2 emissions and fuel consumption. The main objective of the VECTO Trailer is to determine the influence that trailers placed into circulation have on fuel consumption and CO_2 emissions, so that it is possible to compare different trailers in terms of environmental performance.

To implement the VECTO tool in vehicles in Portugal, it was decided to use the tool's two simulation modes: 'Declaration mode' and 'Engineering mode'. To simulate a homologation case with 'Declaration mode', it was decided to simulate a semi-trailer homologation using the VECTO Trailer, since there is little or no production and homologation of tractor vehicles in Portugal.

In the case of the 'Engineering mode' it is possible to evaluate different development options of the vehicle. The VECTO tool was used and the fuel consumption of a known rigid truck travelling a known route was simulated to check the tool's accuracy. After this simulation, the rigid truck was adjusted by fitting more efficient tires and by fitting aerodynamic devices, both with the aim of reducing fuel consumption and obtaining a financial return.

The first step in the homologation of a semi-trailer using the VECTO Trailer was to enter all the characteristics of the semi-trailer into the tool.

The first parameters to be defined are the manufacturer's name, the manufacturer's address, the trailer model, the vehicle identification number, and the legislative category, whether it belongs to category O3 or O4. These five parameters are only required for documentation for approval processes and have no influence on the simulation.

In a further step, the trailer must be classified according to the number of axles, with trailers with 1, 2, or 3 axles being allowed. The types of trailers considered in VECTO Trailer are DA, which corresponds to a semi-trailer designed to be coupled to a tractor unit or a Dolly trailer, DB, which corresponds to a drawbar trailer with at least two axles, one of which must be steerable, and finally, a DC trailer, which is a center axle trailer. For this trailer, it is also necessary to indicate the coupling point, since this parameter affects the distance between the truck and the trailer, thus influencing the Cd x A value.

To classify a trailer, it is also necessary to identify the type of bodywork. The VECTO Trailer tool supports closed box bodies, refrigerated box bodies, conditioned bodies, bodies with side curtains, and open box bodies with folding side panels and tarpaulin covers. When classifying the trailer, it is possible to define whether it is designed to transport bulky goods, which indicates that the internal height is not less than 2.9 m. This parameter is not only important for the classification of the trailer but also influences the calculation of the Cd x A value [11].

When completing the trailer classification, it is essential to indicate the corrected mass in running order, which corresponds to the mass of the trailer without load. It is also necessary to indicate the maximum technically permissible laden mass of the trailer and axle. These two parameters limit the total mass in the simulation.

After defining these characteristics, it is essential to indicate the dimensions of the trailer. These dimensions play a very important role in calculating the value of CdxA. The dimensions to be indicated are the total length of the trailer, the maximum width, the total height, the exterior height of the box, the length from the front end of the trailer to the center of the first axle, and the load volume. These factors must always be specified, regardless of the number of axles on the trailer. For trailers with two or more axles, it is also necessary to specify the length between axles.

To encourage the implementation of components that improve the aerodynamic performance of trailers, an option has been included in the VECTO Trailer tool that allows to indicate whether aerodynamic devices are being used. If specific aerodynamic devices are used, these can be indicated using the "Certified aerodynamic device" option, and a file with the characteristics of that component must be attached. However, VECTO Trailer has four types of aerodynamic devices built in, namely short side skirt, long side skirt, short rear flap, and long rear flap, as shown in Fig. 4.

Fig. 4. Representation of aerodynamic devices accepted in the VECTO Trailer tool (adapted from [11])

The main purpose of these aerodynamic components is to reduce the drag coefficient, thereby lowering fuel consumption and CO_2 emissions. The side skirts limit airflow under the trailer and reduce turbulence caused by irregular geometries. Rear flaps reduce the wake formed behind the vehicle, which creates a low-pressure zone, increasing the drag force on the vehicle's movement [12]. According to the user manual for the VECTO Trailer tool, these aerodynamic components can reduce the CdxA value by up to 11%, as shown in Fig. 5 [13].

Allowed configurations					ΔCdxA reduction rates [%]			
Combination Number	Side cover SHORT	Side cover LONG	Rear flaps SHORT	Rear flaps LONG	Yaw 0.0	Yaw 3.0	Yaw 6.0	Yaw 9.0
0					0.0%	0.0%	0.0%	0.0%
1	1				1.4%	2.7%	3.0%	3.7%
2		1			4.0%	3.4%	3.5%	4.7%
3			1		2.8%	3.2%	3.8%	4.9%
4				1	3.9%	4.1%	5.1%	6.0%
5	1		1		3.8%	5.8%	8.3%	8.7%
6	1			1	4.7%	6.8%	9.2%	10.1%
7		1	1		6.5%	6.4%	7.8%	9.4%
8		1		1	7.6%	7.7%	9.2%	10.9%

Fig. 5. Reduction in CdxA value as a function of aerodynamic devices [13]

Once all the characteristics have been entered, a simulation is carried out with a generic tractor along the various predefined routes. As the VECTO Trailer tool specializes in trailers, it uses a generic truck, the same for all simulations, to calculate fuel consumption. Once the simulation is complete, two files are generated with the fuel consumption and CO_2 emissions figures, one of which needs to be presented to the customer at the time of the purchase, and the other needs to be sent to the European Commission.

As can be seen from the analysis in Fig. 5, the impact of aerodynamic devices varies according to the vehicle's steering angle, except for long side skirts, where the greater the steering angle, the greater the positive effect of these components. The analysis in Fig. 5 shows that the best configuration of aerodynamic devices is a combination of long side skirts and long rear flaps, reducing the CdxA value by more than 10%.

With this simulation of a homologation, it was concluded that using the VECTO Trailer tool for semi-trailer type-approval is a simple and straightforward process, since all that is needed is to define the parameters required by the tool, and all the results and files that are necessary be made available to the customer and the European Commission are automatically determined.

3 Results and Discussion

The results present the process of validation of the tool developed considering a real case scenario and further the use of this simulation model to achieve efficiency improvement and possible CO_2 reduction by using some modifications in the trailer.

3.1 Validation of the VECTO Toll

To apply the VECTO tool in optimizing a heavy-duty vehicle, it was decided to first check the tool's accuracy and validate the simulation carried out. This was done using a rigid lorry with known characteristics travelling along a known route. This route, represented in Fig. 6, has 147 km long and was traveled in 2 h and 17 min, with an average speed of 64 km/h. During this journey, the vehicle reached a maximum speed of 90 km/h.

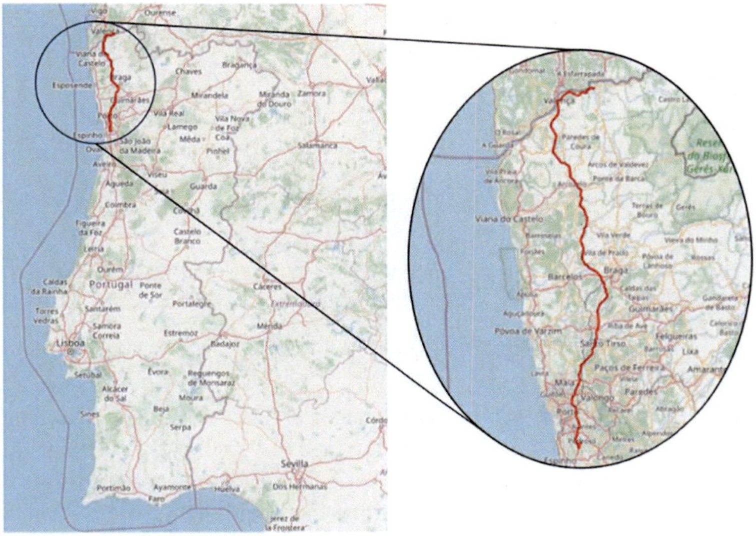

Fig. 6. Route considered for the vehicle to validate the simulation

To obtain the route data, a GPS system was used to record the coordinates of the route, the speed of the vehicle at each point, and the distance traveled.

When defining the route in the VECTO tool, a distance-based approach was chosen, considering the definition of a final distance, and a target speed is assigned for each point on the route. With this approach, the distance to be traveled by the vehicle in the simulation is exactly equal to the actual distance, although the time taken may vary.

Thus, the route is defined in VECTO by the distance traveled, the target speed for each distance point, and the road gradient, also at each point. The distance travel and the vehicle speed were determined from actual GPS data. To define the road gradient, it was necessary to determine the altitude for each coordinate point and calculate the road gradient using the distance traveled.

During the journey, as presented in Fig. 7, the maximum height reached was 330 m, around kilometer 39, while the minimum height was 8 m, around kilometer 18.

On this journey, the truck consumed 44 L, or an average of 29,85 l/100 km. In the VECTO tool, all the characteristics of the vehicle were defined, as well as the characteristics of the route, namely the distance, the target speed and the gradient of the road. In the simulation carried out using the VECTO tool, the vehicle took 2 h and 19 min to cover 147 km, with an average speed of 63,19 km/h. The fuel consumption obtained is presented in Table 1, was 29,12 l/100 km, representing a difference of just 0,73 l/100 km, or 2,4%.

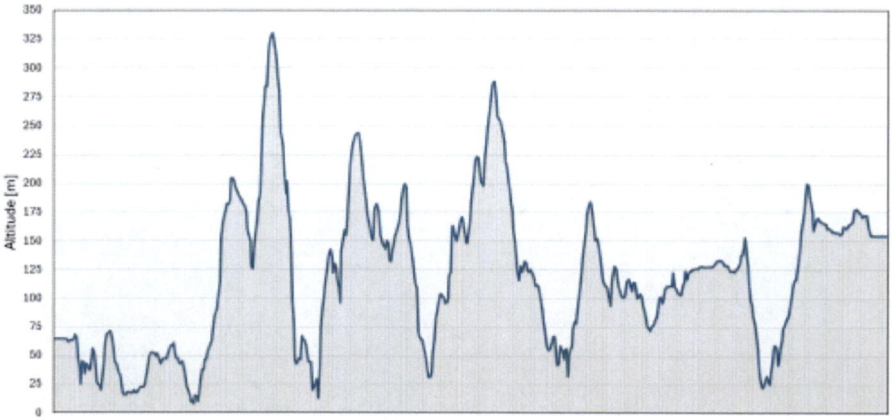

Fig. 7. Altitude profile of the route made by the vehicle

With these results, it was concluded that the VECTO tool can be used as a method to determine the fuel consumption of a heavy-duty vehicle, since the percentage difference obtained was quite small, considering that various vehicle and route parameters were assumed, since it was not possible to determine them with better rigor. As it was concluded that the VECTO tool can be used to accurately determine the fuel consumption of a heavy-duty vehicle, it was also concluded that it can be used to optimize the vehicle, as the changes made in the simulation will occur in a real situation.

Table 1. Fuel consumption of the vehicle in real situation compared with simulation results

Fuel consumption			
Real [l/100 km]	Simulation [l/100 km]	Difference [l/100 km]	Relative difference [%]
29,85	29,12	−0,73	−2,4

3.2 Evaluation of Possible Improvement Efficiency

Since the main objective was to apply the VECTO tool to vehicles in Portugal, it was decided to improve some parameters that companies in Portugal can perform, thus excluding any parameters related to production and development. In this way it was aimed at adapting parameters with the intention of reducing fuel consumption and consequently CO_2 emissions.

Two easily changed parameters that considerably reduce fuel consumption are the use of more efficient energy class tires and aero-dynamic devices. In the initial simulation, the vehicle was fitted with energy class C tires, and the aim was to determine the impact of fitting class A tires. It was found that using more efficient tires has a higher purchase cost, representing a difference of 1130€ per vehicle.

In the VECTO tool, to determine the effect of more efficient tires, the rolling resistance coefficient was altered, and it was found that class A tires save 1,76 l/100 km compared to class C tires. It was concluded that over the life of the tires, if the vehicle is fitted with energy efficiency class A tires, the total saving can reach 1445€, already considering the higher purchase cost. These results are represented in Fig. 8, where it can be perceived that fitting more efficient tires involves a higher purchase cost, although the total savings are considerably higher when compared to using less efficient tires. Based on these figures, it can be concluded that tires with higher energy efficiency ratings allow for considerable savings over their lifetime and clearly justify their purchase. In terms of CO_2 emissions, the reduction is 46 g/km with class A tires and so carbon dioxide emissions can be reduced by 4,6 tons of carbon dioxide per vehicle over the life of the tires.

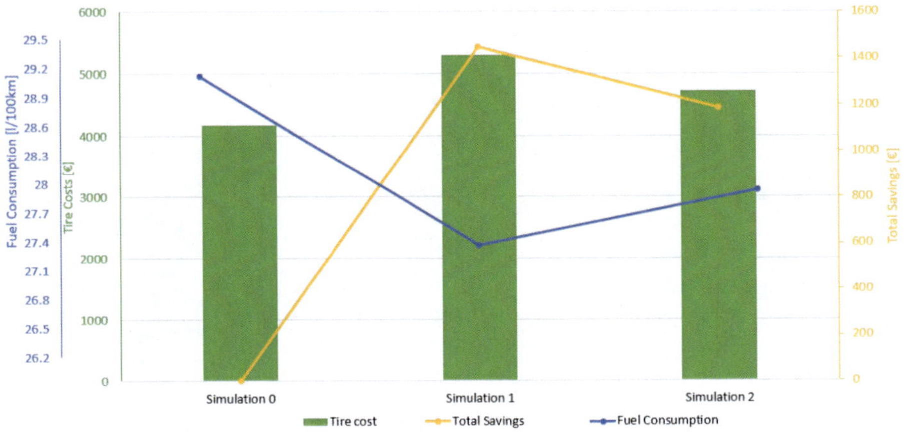

Fig. 8. Total savings and fuel consumption variations due to tires changes

Another modification that can be applied to vehicles in Portugal is the introduction of aerodynamic devices. For this study, the introduction of side skirts and rear flaps on rigid trucks was analyzed considering the expected reduction on aerodynamic drag offered by these devices. According to this tool, the introduction of side skirts reduces CdxA by 4%, rear flaps reduce CdxA by 3.9%, and these two devices together reduce CdxA by 7.6%. These reduction factors are used for semi-trailers, but in this study, it was decided to assume that for a rigid truck the factors remain the same, since to determine their actual contribution, it would be necessary to perform CFD simulations or experimental tests. However, the introduction of aerodynamic devices does not only have advantages, since they need to be purchased and add weight to the rigid truck structure. Table 2 shows the values for the CdxA of the vehicle with the various devices, their cost, and the additional mass that each device adds to the vehicle. According to the VECTO tool, and for approval purposes, the annual mileage of a vehicle in subgroup 9 is 108,000 km. Using these results and the average price of diesel at €1.463/l, the savings offered by these devices were calculated. In the case of aerodynamic devices, it was concluded with simulations made with the VECTO tool that the introduction of a side skirt reduces fuel consumption by 0,23 l/100 km and the introduction of a rear flap reduces fuel consumption by 0,26 l/100 km. Considering the cost of purchasing this equipment and the fuel savings, it was concluded that the most advantageous device is the rear flap, which allows gains to be made from 300000 km onwards.

Table 2. Aerodynamic devices analysis

	Baseline	Side skirt	Rear flaps	Side skirt + rear flaps
Cost [€]	–	2356,0	1175,6	3531,6
Fuel consump. [l/100 km]	29,1225	28,897	28,866	28,659
Fuel differ. [l/100 km]	–	0,226	0,256	0,463
Annual saves [€]	–	356,62	404,81	732,19
Distance [km]	–	713509	313633	520914

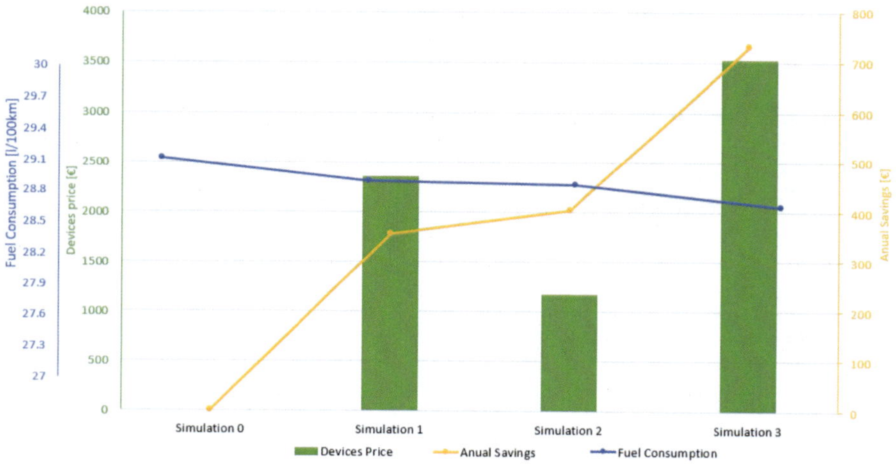

Fig. 9. Total savings and fuel consumption variations due to aerodynamic devices inclusion

Figure 9 reveal that the simulation with a rear flap and side skirt, represented by Simulation 3, has a higher acquisition cost, but the annual fuel savings are greater. However, the most advantageous simulation is Simulation 2, performed with only a rear flap, since the acquisition cost is much lower when compared to the other devices.

It can therefore be concluded that, for the introduction of aerodynamic components to pay off in reducing companies' costs, it is imperative that their mass is reduced and that their effect on reducing the drag coefficient is greater, thus requiring CFD simulations and experimental tests in the development of all components.

3.3 Extrapolation of the Results for Other European Circumstances

The analysis of the efficiency in the use of heavy vehicles in Portugal allows for the use of a specific context to understand the effects of using improvements that may be introduced in vehicles in order to reduce the fuel consumption and the emissions of CO_2 in the transportation of people and goods. This study does not consider factors like the amount of combustion consumed by ton/km, the age of the vehicles, as well as management and efficient driving techniques that are typically taken into consideration, but it is not as relevant in the context of this work, because this is a comparative analysis. In Portugal, the predominance of short and long-distance travel—often with topographical variations—directly affects efficiency indicators, leading to consumption performance that may differ from regions with more logistics density or more consistent travel.

When comparing these results with the European context, it is crucial to recognize that Europe is not homogeneous in terms of infrastructure, regulation, and operational standards. Countries such as Germany, France, and the Netherlands benefit from denser and more integrated road networks with advanced logistics centers, which tend to reduce downtime and optimize the use of cargo capacity. On the other hand, markets with geographies similar to Portugal"s, such as Spain or some southern European states, may have comparable dynamics, especially in regional transport corridors.

The extrapolation of Portuguese results to the whole of Europe should therefore be treated with caution. For example, the same improvements technologies may deliver significant efficiency gains in Portugal, but even better results in markets with more mature technological support and a larger scale of operations.

In summary, the efficiency results obtained in this study, considering a case applied in Portugal, are valuable as a basis for comparison and learning, but their extrapolation to Europe requires a contextualized analysis that should consider geographical, economic, and regulatory variables. Only then is it possible to transpose conclusions realistically and identify practices that promote efficiency gains adaptable to the diverse European realities.

4 Conclusions

Using the VECTO Trailer tool for trailer homologation is a simple and easy process, as it is only necessary to define the parameters required by the tool, and all the results and files required in the submission to the customer and the European Commission are automatically determined.

When using the VECTO tool to simulate the consumption of a rigid truck, comparing the real consumption of the known vehicle on a known route and thus validating the use of the tool and assessing its accuracy with the consumption obtained in the simulation, a difference of just 0.73 l/100 km, or 2.4%, was obtained. It was concluded that the VECTO tool can be used as a method to determine the fuel consumption of a heavy vehicle, since the percentage obtained was quite small, considering that various vehicle and route parameters were assumed, since it was not possible to determine them with the maximum rigor. Certain parameters assumed were in line with the tool's reference values, so it can also be said that the reference values and simulation parameters assumed by VECTO are very realistically defined in relation to the values of heavy vehicles.

To complement the study of the implementation of the VECTO tool on vehicles in Portugal, it was decided to optimize the rigid truck with more efficient tires and aerodynamic devices. With these two simulations, it was concluded that using tires in energy efficiency class A saves around €1,445 over their useful life, paying off after 44,000 km. The reduction in fuel consumption associated with the use of more efficient tires means that CO_2 reductions can reach 4.6 tons per year per vehicle, significantly reducing fleet emissions and reducing the possibility of fines for non-compliance with legislation. Regarding aerodynamic devices, it was found that they only provide an economic advantage after more than 300,000 km, representing a previsible investment amortization time of approximately 3 years.

However, the distinctive variables should be considered when applying this knowledge to a particular reality, such as Portugal or another European environment, since this would introduce major differences that affect the results.

It is therefore possible to conclude that, at this stage, the introduction of these components does not yet offer significant advantages and that it is necessary to reduce the weight of these components and increase their effect on reducing the drag coefficient.

References

1. Agência portuguesa do ambiente: Acordo de Paris 2015–2020 (2020). https://www.portugal.gov.pt. Accessed in November 2023
2. Ruiz, A.: 45 Carbon Dioxide, Greenhouse Gas & Climate Change Statistics (2023). https://theroundup.org/co2-greenhouse-gas-emission-statistics/
3. Comissão Europeia: Regulamento 2017/2400 da Comissão (2017). https://eur-lex.europa.eu/legal-content/PT/TXT/PDF/?uri=CELEX:32017R2400 Accessed in November 2023
4. Agência portuguesa do ambiente: Inventário Nacional de Emissões. Emissões nacionais de gases com efeito de estufa (GEE) por setor - Inventário Nacional 2023 (2024). https://apambiente.pt/clima/inventario-nacional-de-emissoes-por-fontes-e-remocao-por-sumidouros-de-poluentes-atmosfericos. Accessed in September 2024
5. European Commission: Regulation 2019/1242 European Parliament (2019). https://eur-lex.europa.eu Accessed in November 2023
6. Comissão europeia: Regulamento 2018/956 do Parlamento Europeu (2018). https://eur-lex.europa.eu/legal-content/PT/TXT/HTML/?uri=CELEX:32018R0956 Accessed in November 2023
7. European Commission: Vehicle Energy Consumption calculation Tool VECTO (2023). European Commission: https://climate.ec.europa.eu/eu-action/transport/road-transport-reducing-co2-emissions-vehicles/vehicle-energy-consumption-calculation-tool-vecto_en
8. Franco, V.: Heavy-Duty Vehicle Fuel-Efficiency Simulation: A Comparison of US and EU Tools (2015). Accessed in January 2024
9. Holmberg, J.: Heavy duty vehicle simulation tool for the calculation of fuel consumption and development of a new VECTO gearshift algorithm (2018). Obtido em janeiro de 2024, de https://www.diva-portal.org/smash/get/diva2:1218050/FULLTEXT01.pdf
10. Rodríguez, F.: Fuel Consumption Simulation of HDVs in the EU: Comparisons and Limitations (2018). Obtido em janeiro de 2024, de https://theicct.org/wp-content/uploads/2021/06/VECTO_Simulation_Comparison_final_20180308.pdf
11. Comissão Europeia: Regulamento 2022/1362 do Parlamento Europeu (2022). https://eur-lex.europa.eu/legal-content/PT/TXT/PDF/?uri=CELEX:32022R1362
12. Rodríguez, F.: Trailer CO2 Certification in the European Union (2018). https://theicct.org/wp-content/uploads/2021/06/Trailer_CO2_certification_20180921.pdf
13. Comissão Europeia: VECTO Trailer User Manual (2023)

Development of an Optimization Framework for Suspension Parameters in Automotive Vehicles Using Genetic Algorithms

Iúri Silva[1], Fernando Batista[1,2], and Olivier R. Gouveia[1,2,3(✉)]

[1] School of Technology and Management (ESTG), University of Leiria and Oeste, Leiria, Portugal
olivier.gouveia@ipleiria.pt

[2] Centre for Rapid and Sustainable Product Development (CDRSP), University of Leiria and Oeste, Leiria, Portugal

[3] IDMEC—Institute of Mechanical Engineering, Instituto Superior Técnico, University of Lisbon, Lisbon, Portugal

Abstract. This study presents an optimization framework for automotive suspension systems using Genetic Algorithms (GA) to improve ride comfort and handling performance. Traditional suspension tuning relies on trial-and-error methods, which are time-consuming and suboptimal. The proposed method ad-dresses these limitations by optimizing key suspension parameters–stiffness and damping–using GA, which mimics natural selection processes. The framework is applied to a quarter-car model representing the vertical dynamics of a vehicle. GA optimizes the parameters based on performance metrics such as minimizing body displacement and acceleration. Simulations using sinusoidal and step road profiles demonstrated significant improvements, with reductions of 8.1% and 22.8% in displacement and 5.6% and 66.2% in acceleration for sinusoidal and step disturbances, respectively. These results highlight GA's potential to enhance suspension tuning, outperforming conventional methods. Future work will expand this approach to multi-objective optimization and more complex vehicle models.

Keywords: Vehicle suspension · Optimization · Genetic algorithm

1 Introduction

The vehicle suspension system has a direct influence on performance, stability, and comfort for both the vehicle and passengers. Usually consisting of springs and dampers, its characteristics vary depending on the type of vehicle and road, requiring specific optimizations to ensure good performance and an optimal driving experience. The complexity of optimizing these suspensions arises from the interaction between multiple variables, with stiffness and damping being key parameters.

© The Author(s), under exclusive license to Springer Nature Switzerland AG 2027
M. Calvete Gaspar et al. (Eds.): Regional Helix 2024, LNME, pp. 277–287, 2027.
https://doi.org/10.1007/978-3-032-36365-7_19

Traditionally, the optimization of these parameters has relied on expert knowledge and trial-and-error methods, which are time-consuming, demand significant resources, and cannot ensure a global optimal solution. However, new modeling and optimization technologies, such as Genetic Algorithms (GA), simplify this process by identifying optimal parameters for different driving scenarios. Genetic Algorithms are powerful search heuristics inspired by the principles of natural selection and genetics [1]. Albadr et al. [2] highlighted the efficiency of genetic algorithms in optimizing engineering systems, including vehicle suspensions.

2 Numerical Model

Although there are many mathematical models to represent vertical movement of automotive suspension, the quarter car model is the simplest model, Jazar and Marzbani [3]. This model simplifies the whole problem by analyzing each wheel isolated. A representation of this model is presented in Fig. 1.

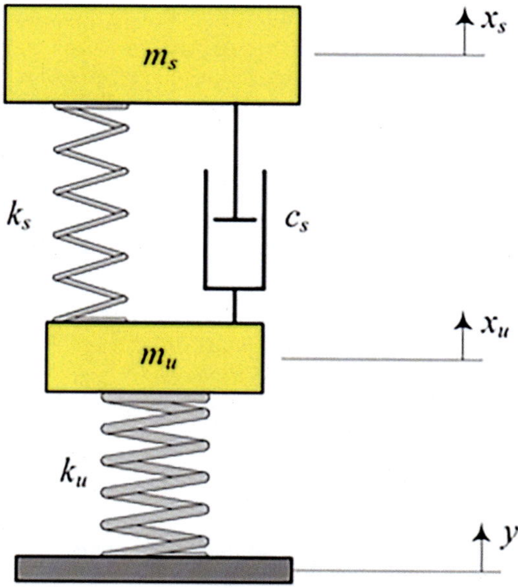

Fig. 1. Quarter-car model [3].

It is represented by two masses, a sprung mass m_s and an unsprung mass m_u. Their vertical displacements of each mass is represented by x_s and x_u for the sprung and an unsprung mass, respectively, while y represents the road profile input. The tire stiffness is denoted by k_u which connects the unsprung mass to the ground. The sprung mass represents one quarter of the total vehicle mass, while the unsprung mass represents one of the vehicle wheels. A spring with stiffness, k_s, and a damper with damping coefficient, c_s, supports the sprung mass.

The equations of motion that describe the behavior of this model are given by [3]:

$$m_s\ddot{x}_s = -k_s(x_s - x_u) - c_s(\dot{x}_s - \dot{x}_u), \tag{1}$$

$$m_u\ddot{x}_u = k_s(x_s - x_u) + c_s(\dot{x}_s - \dot{x}_u) - k_u(x_u - y). \tag{2}$$

With this numerical model it is possible to analyze the vertical movement of the suspension and optimize its parameters. Due to the simplicity of this model, it does not require high computational resources, which makes a good option for optimiza-tion.

3 Optimization

Optimization can be defined as the science that determines the best solution for a particular mathematical problem that usually represents a real physical system. The objective is to identify the ideal combination of variables that minimize, or in some cases maximize, a determined objective function $f(x)$ [4]. According to Fletcher [1], the definition of the optimum criteria, development of algorithms, the study of the structure and computational experiments are part of problem optimization. The canonical form of optimization problems can be represented as:

$$\begin{aligned} \min_{\boldsymbol{x}} \quad & f(\boldsymbol{x}) \\ \text{s.t.} \quad & h_i(\boldsymbol{x}) = 0, \quad i = 1, \dots, m \\ & g_j(\boldsymbol{x}) \leq 0, \quad j = 1, \dots, l \\ & \boldsymbol{x} \in \Omega. \end{aligned}$$

In this formulation $\boldsymbol{x}$ is an n-dimensional vector with unknown values, $\boldsymbol{x} = [x_1, x_2, ..., x_n]$ and $f(\boldsymbol{x})$ is the objective or cost function; $h(\boldsymbol{x})$ and $g(\boldsymbol{x})$ are equality and inequality constraints that depend on the values of $\boldsymbol{x}$. The numerical set Ω represents all possible $f(\boldsymbol{x})$ values inside the imposed search bounds.

3.1 Meta-Heuristic Methods

Meta-heuristic based methods explore large-scale solutions and escape local minima more efficiently than gradient based methods. These methods add an aleatory component in the search of the best solution that minimizes $f(\boldsymbol{x})$. Methods such as evolutionary algorithms are widely used to optimize complex, real world problems. These methods, such as the Genetic Algorithm, simulate Darwinian natural selection [5]. The algorithm begins with an initial population of solutions and, in each iteration, generates a new population based on genetic operators such as crossover and mutation, until it finds a global optimal solution.

3.2 Genetic Algorithms

According to Alhijawi and Awajan [6] a simple genetic algorithm starts with an initial population of individuals, each one representing a possible solution to the problem. These individuals have chromosomes composed of genes, any of the individuals are a possible solution of the problem. For example, in binary encoding, a chromosome could look like, $x = [10011]$, where each bit represents a gene. The initial population is randomly generated, allowing the algorithm to explore different regions of the search space based on the imposed restrictions. The algorithm iterates through generations, selecting individuals based on their performance in a fitness function that evaluates how good each solution is. The individuals with best ranking on the fitness function, in each generation, are more likely to be selected to exchange information (genes), with other individuals through crossover and mutation process which will generate more individuals like the best ones of the last generation, obtaining better solutions in each generation. This fitness function is crucial, as it determines which individuals are more likely to reproduce and pass on their genetic information to the next generation. Figure 2 illustrates a flow chart representing the operation of a simple genetic algorithm.

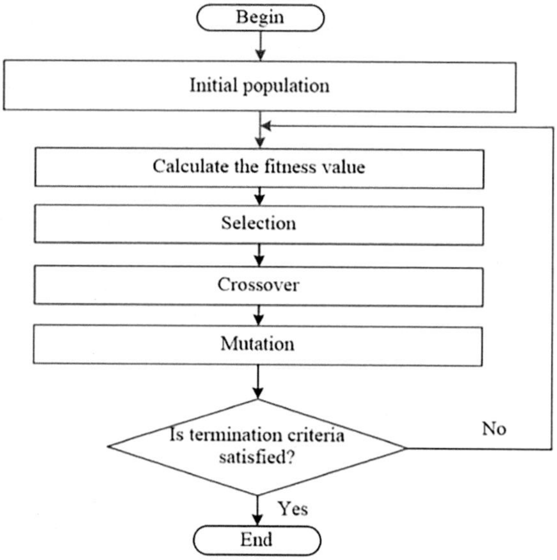

Fig. 2. Flowchart of the Genetic Algorithm [2].

For sake of completeness, we briefly summarize each genetic operator employed in a general Genetic Algorithm.

Selection: This operator prepares the individuals for reproduction, choosing them based on their fitness scores. Higher fitness individuals are more likely to be selected, but even weaker individuals can be chosen occasionally to maintain diversity. Various selection methods can be used, such as probabilistic selection, where fitter individuals have a greater chance, or stochastic selection, which ensures a wide search by selecting individuals from different areas of the search space.

Crossover: Once individuals are selected, crossover is used to create new individuals by combining genetic material from two parents. In a single-point crossover, a random point on the chromosome is chosen, and the genetic material is swapped between the two parents to generate two new individuals. A double-point crossover works similarly but swaps genes between two random points, increasing diversity.

Mutation: During mutation, random changes are introduced to some genes in the new individuals to maintain diversity in the population. It consists in swapping one or more genes randomly in a chromosome, which helps the algorithm avoid getting stuck in local optima by exploring new areas of the search space. The mutation rate is typically kept low to avoid turning the algorithm into a random search.

Replacement Techniques: After generating a new set of individuals, the algorithm must decide which individuals to retain for the next generation. One approach is to replace the entire population with the offspring, focusing on the assumption that new individuals will improve overall fitness. Alternatively, some of the best individuals from the previous generation can be retained, or the worst individuals in the new generation can be discarded in favor of those generated by crossover and mutation.

4 Case Study

The vehicle model was optimized according to different objective functions, one being to minimize acceleration RMS (Root Mean Square) value of the sprung mass and the other being minimize displacement RMS value of the sprung mass. Additionally, two types of excitations were assessed as road profiles: sinusoidal and step profiles.

4.1 Road Inputs

To analyze the vertical movement of the vehicle suspension, the first excitation analyzed was proposed by [7] and consists in a step road with a pulse of 0.05 m as depicted in Fig. 3. The vehicle test speed used was 50 km/h. The second excitation used was a sinusoidal function that depends on the vehicle velocity v_{car}, with an amplitude of $A = 0.01\,\mathrm{m}$ and wavelength $\lambda = 5\mathrm{m}$. The sinusoidal road profile can then be modelled as:

$$y(t) = A \sin\left(\frac{\mathrm{v}_{car}}{3.6\,\lambda} 2\pi t\right) \quad [m] \tag{3}$$

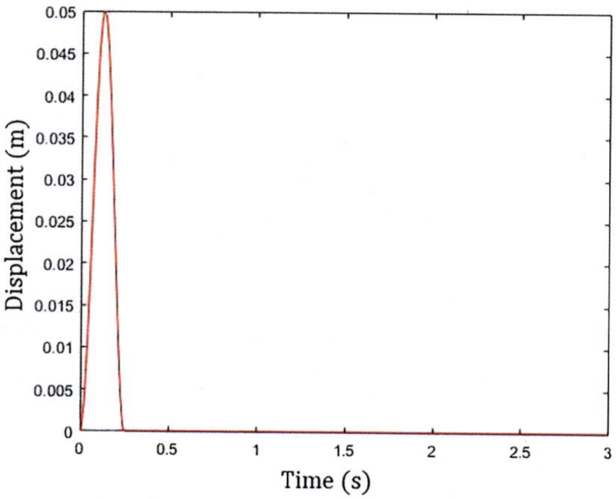

Fig. 3. Step-impulse road profile

4.2 Vehicle Parameters

The properties of the quarter car model are adapted from a Mercedes-Benz AMG SLC43 [7] and listed in Table 1.

Table 1. Quarter car model properties

Design variable	Symbol	Value
Sprung mass	m_s	395 kg
Unsprung mass	m_u	38 kg
Spring stiffness	k_s	29300 N/m
Damping coefficient	c_s	3000 Ns/m
Tire stiffness	k_t	290000 N/m

4.3 Optimization Problem

The definition of the optimization problem with the objective of the optimization, minimize vertical acceleration and displacement, respectively, and the restrictions are presented as:

$$\text{Objective Function 1:} \quad \min_{k_s,\, c_s} \text{RMS}(\ddot{x})$$
$$\text{s.t.} \quad 10000 \leq k_s \leq 200000 \ \text{N/m},$$
$$500 \leq c_s \leq 10000 \ \text{Ns/m}$$

$$
\begin{aligned}
\text{Objective Function 2:}\quad & \min_{k_s,\,c_s}\ \mathrm{RMS}(x) \\
& \text{s.t.}\quad 10000 \le k_s \le 200000\ \ \mathrm{N/m}, \\
& \qquad\ \ 500 \le c_s \le 10000\ \ \mathrm{Ns/m}
\end{aligned}
$$

A script was created in MATLAB, which utilized the software's own native libraries, *the Global optimization Toolbox*, to solve the optimization problem using a genetic algorithm. The optimization was implemented in MATLAB using the `ga()` function. This function evaluates and ranks the individuals of each generations and provides the best solution to the problem. The problem's fitness function is called via a Simulink model representative of the quarter car model and evaluates the selected variable of interest, either accelerations or displacements. The optimization variables considered were the suspension stiffness k_s and damping coefficient c_s with constrain ranges between 10000 N/m and 200000 N/m, and 500 Ns/m and 10000 Ns/m, respectively. Both variables were treated as continuous real numbers within their respective bounds, allowing the genetic algorithm to explore the full real-valued search space rather than restricting the search to integer or discretized values. This approach enables the GA to sample solutions with high precision, including any fractional values within the intervals. The GA parameters, such as population size, crossover fraction, mutation rate, and selection method, were configured to efficiently search the continuous domain, facilitating convergence toward an optimal solution that balances ride comfort and suspension performance. The genetic algorithm parameters used in this study are presented in Table 2.

Table 2. Genetic Algorithm parameters

Property	Value
Population size	80
Crossover	70%
Mutation	10%
Max number of generations	100
Tolerance	0.0001 [m or m/s^2]

The choice of these parameters directly influences the effectiveness of the algorithm. Mitchell [8] outlines several studies using genetic algorithms, indicating that the optimal parameters include a population size between 50 and 100 individuals, a crossover rate between 60 and 90%, and a recommended mutation rate below 10%. A high crossover rate means that this percentage of the population from the previous generation exchanges information, generating new individuals similar to the existing ones. In regions where the algorithm has found good solutions, more points are formed. Those that perform better in the fitness function survive and are selected to form new points in the next generation, contributing to the algorithm's convergence. Maintaining a low mutation fraction allows the algorithm to escape local minima by randomly altering a portion of the information, which promotes exploration of new areas in the search space and preserves genetic diversity. This prevents the algorithm from becoming too random, which could otherwise difficult convergence.

5 Results and Discussion

In this chapter are presented and discussed the results obtained in the optimization of the quarter car model for both excitations: step-impulse road and sinusoidal excitation.

Step-impulse road. Table 3 showcases the results obtained for step-road optimization. Significant improvements were obtained for both excitation types.

Table 3. Quarter-car model optimization through step-road excitation

	OEM	RMS($\ddot{x}$)	RMS(x)
RMS($\ddot{x}$) [m/s^2]	1.1032	0.3728	–
RMS(x) [m]	0.00987	–	0.00762
k_s [N/m]	29300	10000	10194
c_s [Ns/m]	3000	516	2461
Improvement [%]	–	66.21	22.79

As we can observe in both optimizations, considering acceleration and displacement responses, the spring stiffness parameter is relatively close to the minimum limit set in the problem definition. This suggests that the drastic reduction of this parameter contributes significantly to minimizing both acceleration and displacement. The proximity of this parameter to the minimum constraint value indicates that a better solution could potentially be achieved if this value were further reduced. However, extremely low values might not be physically feasible for automotive suspensions, as components with such characteristics might not exist. In Fig. 4(a) is presented the difference between the acceleration value throughout the simulation in the original (OEM) and optimized model for step road excitation. The difference between the highest peak of both models (original and optimized) is easily detected. The optimized model reaches a maximum vertical acceleration just below 2 m/s^2 against the almost 6 m/s^2 in the original model. Although the system with the original parameters reaches equilibrium sooner, around 1.5 s compared to the 4.5 s for the optimized system, the latter shows a better RMS response in reducing vertical acceleration. The optimized model demonstrates a 66.21% improvement over the original. Figure 4(b) presents the difference between the displacement value throughout the simulation in the original (OEM) and optimized model for the step-impulse road excitation.

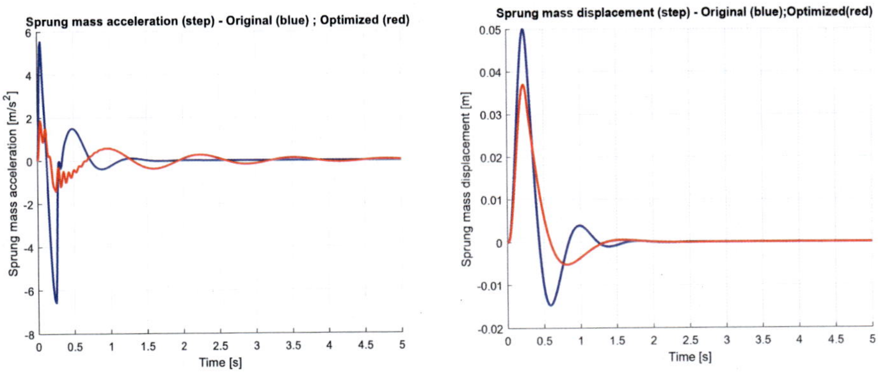

(a) Acceleration response (b) Displacement response

Fig. 4. Comparison between OEM and optimized models for acceleration and displacement response under step-impulse road excitation

Once again, the difference between the maximum values obtained in each system, original and optimized, is clear. In the original configuration, the maximum displacement of the sprung mass was around 0.05 m, while in the optimized model, the maximum displacement is slightly above 0.035 m, representing a reduction of approximately 30% in the maximum displacement value. It is also observed that, although both systems return to equilibrium at almost the same time, around 2 s after the excitation, the movement of the sprung mass in the optimized system is much smoother allowing the RMS response of the sprung mass displacement to be lower than in the original configuration.

Sinusoidal road. Table 4 presents the results obtained for sinusoidal excitation optimization.

Table 4. Quarter-car model optimization through step-road excitation

	OEM	Case 1	Case 2
RMS($\ddot{x}$) [m/s^2]	0.07294	0.06885	–
RMS(x) [m]	0.007808	–	0.007172
k_s [N/m]	29300	10000	199994
c_s [Ns/m]	3000	4836	8936
Improvement [%]	–	5.61	8.14

As we can observe in the acceleration optimization the spring stiffness value coincides once again with lower bound limit for this parameter, while the spring damping coefficient value rises slightly to achieve the best solution. This suggests that lowering the spring stiffness has more impact in minimizing acceleration for the sinusoidal excitation. In the displacement optimization the spring stiffness value is relatively close with the upper bound limit of this parameter as well as the damping coefficient. For minimizing the displacement of the sprung mass, both parameters have a high impact on solution. In Fig. 5(a) is presented the difference between the acceleration value throughout the simulation in the original (OEM) and optimized model for the sinusoidal excitation.

This time, although the optimized model initially shows a higher peak acceleration of the sprung mass, around 0.3 s compared to approximately 0.25 s in the original configuration, the optimized model quickly adjusts to the sinusoidal excitation conditions, reducing the downward overshoot and providing a 5.61% reduction in the RMS response of the sprung mass acceleration to this excitation. Figure 5(b) presents the difference between the displacement value throughout the simulation in the OEM and optimized model for the sinusoidal excitation. In this case the difference between the peak displacements of the sprung mass is not so clear. However, the maximum displacement in the original configuration reaches 0.012 m, while in the optimized model, the maximum displacement is around 0.01 m. The optimized system provided an 8.14% improvement in the RMS response of the sprung mass displacement compared to the original configuration, contributing to a smoother ride.

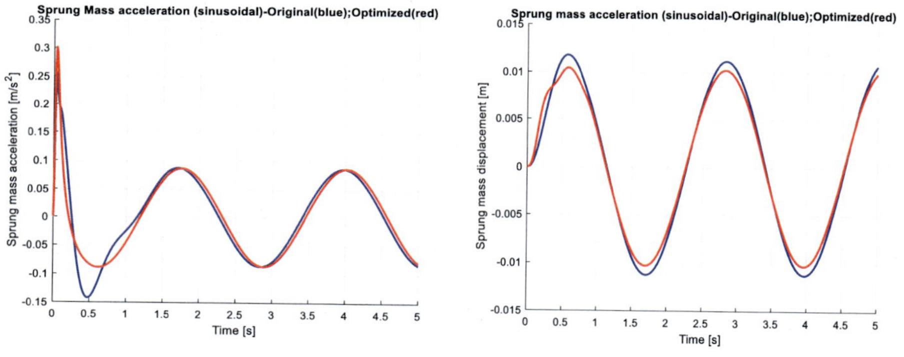

(a) Acceleration response (b) Displacement response

Fig. 5. Comparison between OEM and optimized models for acceleration and displacement response under step-impulse road excitation

6 Conclusions

A quarter car model representing a vehicle suspension system was successfully optimized with genetic algorithms. Significant improvements in the performance were presented in response to the proposed excitations. Something that could have been done is conducting laboratory tests on the suspension systems under study to corroborate the results obtained from the optimization. It was found that the effectiveness of the suspension parameter optimization process directly depends on the variable being optimized. For the step road excitation, the spring stiffness had more direct impact to mini-mize both RMS responses on acceleration and displacement, where the recommended values coincide with the lower boundaries of the optimization problem. The damping coefficient was slightly lower for the displacement optimization. For the acceleration optimization the damping coefficient was lowered until the lower boundary value pre-established on algorithm parameters. For the sinusoidal function excitation, although the spring stiffness had a bigger difference in the optimized models in relation to the original characteristics, the in-crease of damping coefficient allowed the optimization of the system. So, in this case both parameters had impact on the optimization, for both excitations, with more emphasis in spring stiffness variation. On acceleration optimization the spring stiffness value recommended is close to the minimum limit of this parameter established on the algorithm properties. But to optimize the model minimizing the displacement of the sprung mass, it is indicated a bigger spring stiffness value relatively close to the maximum limit of the parameter. It is important to keep in mind that, although the parameter values in the optimized systems can coincide with the constraint values and better solutions could exist, some mechanical elements for suspension systems with that characteristics might not exist in real life. Future work will focus on the development of more complex models and consider multi-objective optimization of the system. Analysis of other types of excitations, such as optimizing suspension parameters in the context of motorsport can be possible as well.

References

1. Fletcher, R.: Practical Methods of Optimization, 2nd edn. Wiley (1987)
2. Albadr, M.A., Tiun, S., Ayob, M., Al-Dhief, F.: Genetic algorithm based on natural selection theory for optimization problems. Symmetry **12**(11), 1–31 (2020)
3. Jazar, R.N., Marzbani, H.: Vehicle Vibrations, 1st edn. Springer International Publishing, Cham (2024)
4. Gouvêa, E.: Métodos Convergentes de Otimização Global Baseados no Vetor Q-Gradiente. Ph.D. Thesis, Instituto Nacional de Pesquisas Espaciais (INPE), São José dos Campos (2016)
5. Darwin, C.: The Origin of Species. P. F. Collier & Son (1909)
6. Alhijawi, B., Awajan, A.: Genetic algorithms: theory, genetic operators, solutions, and applications. Evol. Intel. (2023)
7. Ebrahimi-Nejad, S., Kheybari, M., Borujerd, S.V.N.: Multi-objective optimization of a sports car suspension system using simplified quarter-car models. Mech. Indus. **21**(4) (2020)
8. Mitchell, M.: An Introduction to Genetic Algorithms. MIT Press, Cambridge, MA (1996)

New Trends on Landing Gear Systems for Aircrafts: Case Study

Rodrigo Amorim[1], Luís Coelho[1,4,5], and M. L. Alves[1,2,3](✉)

[1] School of Technology and Management, Polytechnic University of Leiria, 2411-901 Leiria, Portugal
leopoldina.alves@ipleiria.pt
[2] INESC Coimbra - Institute for Systems Engineering and Computers at Coimbra, Polytechnic Institute of Leiria, Leiria, Portugal
[3] School of Technology and Management, Polytechnic of Leiria, 2411-901 Leiria, Portugal
[4] Department of Physics, Univ. Coimbra, CFisUC, 3004-516 Coimbra, Portugal
[5] School of Technology and Management, Polytechnic of Leiria, 2411-901 Leiria, Portugal

Abstract. The focus of this paper is the development of a retractable landing gear system for aircraft, with the goal of improving overall efficiency, safety and performance. The project emphasizes in dimensioning a realistic solution for an initial draft of an aircraft, with several parameters defined considering current market requirements and the regulations and guidelines imposed by European Union Aviation Safety Agency. The research encompasses various critical aspects such as the landing gear arrangement, retraction mechanisms, shock absorption solutions and braking systems. This project serves as the initial definition of several parameters to then be further applied to continue the development of a complete landing gear system, suitable for a lightweight aircraft. The final arrangement ended up being a tricycle one. Concerning the retraction type, the retractable landing gear being a possible feature of the aircraft, with additional bays added to the sides of the aircraft, providing a storage solution that would have no impact on the current design. For the shock absorption, the option was the oleo pneumatic with a telescopic structure due to several benefits compared to the alternatives. For the braking system, there are still many other steps to take when deciding the final type and solution for braking the aircraft, however two of those presented, the anti-skid system and material of the brakes will be considered when picking the appropriate brake from manufactures datasheets, and if necessary, air brakes will be further implemented if the brakes selected end up not being powerful enough for the missions operated.

Keywords: Retractable landing gear · Aircraft design · Shock absorption system · Braking system

1 Introduction

As one of the key factors differentiating a successful flight from an unsuccessful one, landing gear, whether retractable or not, must be extensively researched and tested, to ensure that no single mishap or failure happens. Throughout the last years, several accidents have occurred in commercial aviation due to issues related to landing gears, with

© The Author(s), under exclusive license to Springer Nature Switzerland AG 2027
M. Calvete Gaspar et al. (Eds.): Regional Helix 2024, LNME, pp. 288–299, 2027.
https://doi.org/10.1007/978-3-032-36365-7_20

varied problems, landing gears not locking down due to a broken spring and immediate landing gear cylinder failures upon touchdown, both caused by fatigue. Another accident occurred due to ruptured tires on take-off, which upon landing spun with the ground producing similar natural frequencies of the main landing gear, eventually leading to a full collapse of the right main landing gear. These accidents serve as a reminder of the consequences of a final error, of a fatigued component, of a missed check or evaluation, reminding us of the importance of the landing gear and its purpose.

For these reasons, a thorough project is required for the development of a landing gear for an aircraft, from a design standpoint to an engineering analysis, to an extensive testing procedure and constant maintenance. Due to the systematic need for longer flight times, heavier payloads and better aerodynamics, a retractable landing gear is seen as the ultimate solution for aircraft. The overall efficiency of the aircraft is improved with the use of retractable landing gear systems, enabling the reduction of drag throughout the flight, resulting in less fuel being used and better performance [1]. The landing gear positioning is the first decision that must be made regarding the development. A huge variety of configurations have been used throughout history, but the four most common arrangements are illustrated in Fig. 1. These four arrangements cover around 99% of all aircraft ever built.

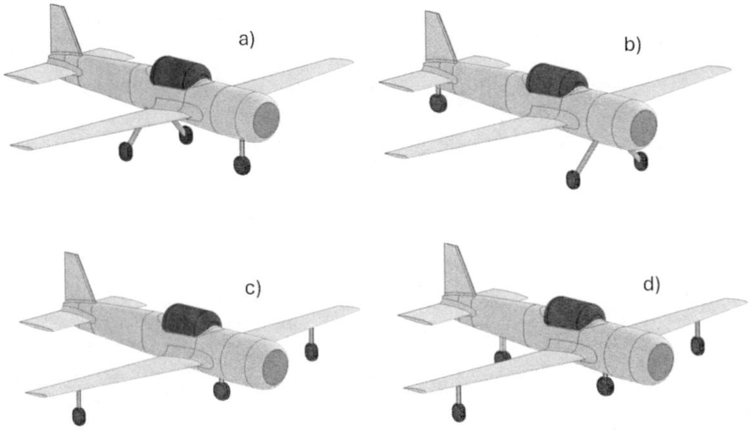

Fig. 1. Landing gear arrangement: (a) Tricycle, (b) Taildragger, (c) Monowheel with outriggers and (d) Tandem with outriggers.

The type of retraction plays a major role in the efficiency and aerodynamics of the aircraft as opting to use a retractable landing gear has several advantages to the aircraft, with the main benefit being the added efficiency, if the system is correctly dimensioned and the structure accounted for, as a retractable solution is always heavier and more complex than a stationary one. The goal is to determine the best retraction system and orientation, considering several parameters, as well as some restrictions in the fuselage. Some existing solutions of different retraction systems are presented below, as well as the main advantages and disadvantages of each.

• Electro-Mechanical Systems:

One of the most common retraction solutions is based on the use of an electric motor or a hydraulic actuator to both retract and deploy the landing gear. This solution is highlighted by its reliability, simplicity, and weight efficiency [3]. However, depending on the aircraft dimensions, this solution may require higher power consumption than desired, especially when compared to the other solutions presented, which may end up having an overall negative impact on the aircraft energy efficiency.

• Hydraulic Systems:

Already established as one of the main solutions for landing gear retractions, their precise control and quick response make it a perfect solution for a variety of aircraft configurations [4]. Mainly applied to heavier aircraft due to their power requirements and load capacity. Another offset of the hydraulic system is the complexity behind the components, requiring a huge number of additional parts like pumps, valves, and fluid lines, which not only increases the weight but also provides a challenge for maintenance purposes.

• Pneumatic Systems:

Despite being less common, pneumatic alternatives have been studied mainly due to their reduced complexity and weight compared to the previous solutions. However, due to their reduced precision and low power density, it ends up being a solution only appropriate to small aircraft with low landing gear loads [5].

There are several solutions and orientations for the retraction system, with some possibilities presented in Fig. 2.

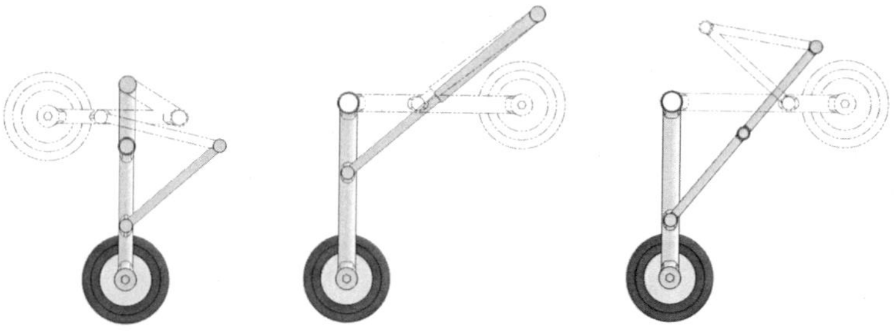

Fig. 2. Side view retraction solutions.

The main purpose of a shock absorber is to minimize the impact of a quick deceleration over a given distance on the aircraft [6]. On an ideal scenario, the deceleration happens through the force acting on the absorber, changing the applied force from zero to a maximum load just as the rate-of-descent approaches zero. Ideally, this deceleration would take place over a long distance, to provide a soft landing, void of any shock. However, due to the size constraints, a solution must be engineered to solve this problem in only a few inches. The answer relies on a spring action that is neither too flexible nor too stiff to enable the reduction of the vertical speed to 0 comfortably.

The relation between deflection and load, defined by the variable Ks, varies according to the type of absorption systems and their efficiency. If a value of Ks = 1 is achieved, then the entire landing load will have reacted without any deflection, which would be the perfect scenario. The Oleo-Pneumatic system is one of the most efficient solutions of slowing down the aircraft vertically. Most of the load is absorbed in a very short distance [5].

The main types of landing gear shock absorption solutions are: Rigid axle, Solid spring and Oleo-Pneumatic Strut.

Behind every good flight is a good landing, however the landing is not finished until the aircraft is at full stop. Despite this, in the beginning, aircraft brakes were not a necessity, however with the increase of weight and speed of the aircraft, the need for them became a necessity and as such, braking solutions were developed to meet the newfound requirements [7].

Three additional aspects will also be addressed, all of which are relevant towards the final decision regarding the available braking solutions. The different options are as follows: Anti-skid systems, Brakes materials and Air brakes.

The European Union Aviation Safety Agency (EASA) stands as the central authority in establishing and enforcing the regulation behind the design, production, and operation of all the aircraft within the European Union. Due to this, their guidelines and standardizations must be carefully studied to ensure that airworthiness of an aircraft is achieved [8]. The required EASA regulations for the aircraft were followed as well as the requisites needed to validate a landing gear system, from the redundancies needed to the simulations that must be passed before a maiden flight. The aircraft portrayed for this project is classified as a "Light Sport Aircraft" (CS-LSA), with the precise category of a Small Rotorcraft (CS-27) due to the type of engine used. For the CS-LSA category, the following criteria is defined [9]: - A maximum take-off mass of no more than 600 kg; - A maximum stalling speed of no more than 83 km/h when the aircraft is at its maximum take-off mass and most critical CG; - A single, non-turbine engine or electric propulsion unit fitted with a propeller; - A non-pressurized cabin is used.

The main purpose of the present research is to develop a functioning retractable landing gear suitable for an aircraft while ensuring that the solution is compact, lightweight, and sturdy. One of the main goals is to ensure the minimization of the occupied space when the landing gear is stowed while still meeting the requirements defined regarding structural integrity. Additionally, some more aspects need to be considered regarding the landing gear, for instance the need for a braking and damping system.

When designing the landing gear several aspects must be considered, and this is further compounded when the gear is retractable. Both the kinematics of the landing gear mechanism and space-claims inside the airframe must be accounted for, so the following issues must be defined before progressing: (i) Landing gear arrangement; (ii) Type of retraction; (iii) Shock absorption; (iv) Braking system.

2 Methodology

By explaining how a project is expected to be carried out, the development becomes logical and systematic, ensuring that the aims and objectives previously defined are properly met. Here, the methods, techniques, and tools used for the selection of each component will be detailed.

2.1 Landing Gear Arrangement

As the landing gear represents one of key components of the aircraft, several features must be taken into consideration upon the initial design. Some of the most vital aspects to be accounted for are both the positioning and ground clearance of the landing gear. The former must be paid crucial attention to as the main and auxiliary landing gears must be positioned respectively to the center of gravity thus reducing the risk of certain problems like ground looping, overturning, crosswind canting and to enable the aircraft to maneuver as needed during take-off and landing procedures. The latter is also of huge importance as a constant clearance is needed between the ground and all the aircraft components, even during worst case scenarios like flat tires or shock absorber abnormalities. An example of the steps considered can be seen in Fig. 3 where h_{cg}, corresponding to the height of the centre of gravity in relation to the ground.

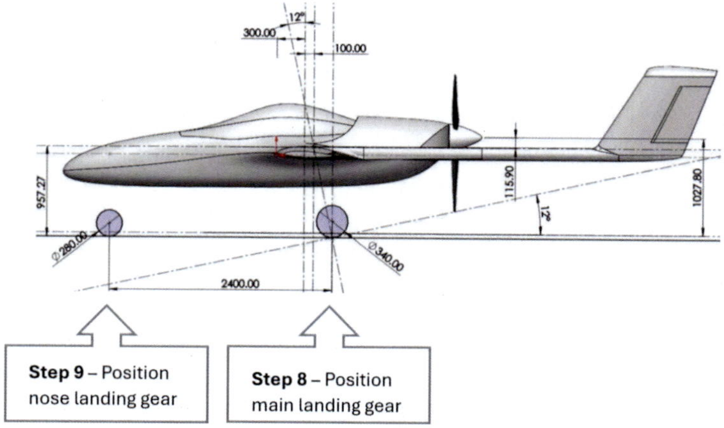

Fig. 3. Steps 8 and 9 of the landing gear geometry definition.

For this project, the tricycle arrangement was selected after evaluating the most common arrangements due to the significant advantages it presents compared to the other options.

2.2 Type of Retraction

The type of retraction plays a major role in the efficiency and aerodynamics of the aircraft as opting to use a retractable landing gear has several advantages to the aircraft, with the main benefit being the added efficiency, if the system is correctly dimensioned and the structure accounted for, as a retractable solution is always heavier and more complex than a stationary one. The goal is to determine the best retraction system and orientation, considering several parameters, as well as some restrictions in the fuselage. Some existing solutions of different retraction systems are: Electro-Mechanical Systems; Hydraulic Systems; Pneumatic Systems. Despite being less common, pneumatic alternatives have been studied mainly due to their reduced complexity and weight compared to the other solutions. However, due to their reduced precision and low power density, it ends up being a solution only appropriate to small aircraft with low landing gear loads [10].

Defining the best place to store the wheels is not difficult, the difficulty lies in designing a simple, light, and robust mechanism to reposition the wheels from point A to point B. An evaluation of the several available positions and the complexity for each of the movements must be made. Despite a position appearing initially favorable, the required complexity of the mechanism to retract the landing gear may invalidate it in favor of another area.

The retraction movement has also a great impact on the travel of the center of gravity of the aircraft, so travel must be considered as the landing gear usually represents from 2.5 to 5% of the maximum takeoff weight. Due to this, an evaluation must be done with the landing gear deployed and retracted to analyze the impact on the aircraft. Another major issue is identifying zones where the landing gear cannot move through. One clear example of this issue arises when a frame exists in the path of retraction, making it almost impossible to have enough clearance for the landing gear to pass through without compromising the structure. Another example is a requirement for a certain zone of the aircraft to be occupied with a payload or component that leaves no space for the landing gear to retract to that position. The most common fuselage structure components that must be considered upon deciding the retraction path are: frame, longeron, skin, stringer and bulkhead [11].

The aircraft geometry is also of huge importance, as some solutions are deemed suitable or not depending on it. For instance, a high or mid wing almost certainly invalidates the option to retract the landing gear to the wing, as the required length and weight would be a major disadvantage even if the wing retraction seemed ideal, causing most aircraft with high wings to retract the landing gear to the fuselage. A visual aid is present in Fig. 4 to differentiate between the three most common wing positions. As it can be seen, the landing gear length greatly differentiates from one to another.

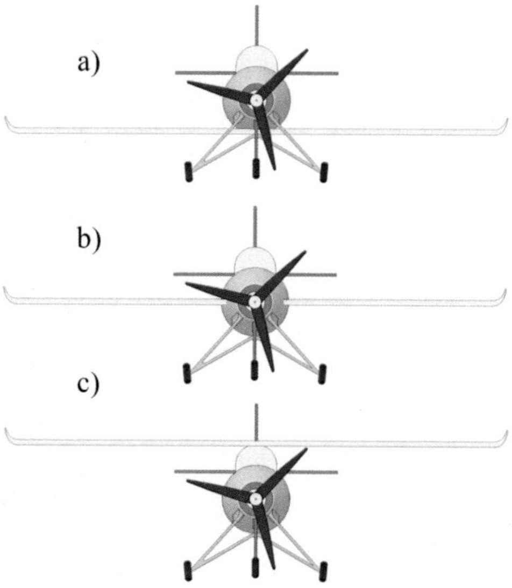

Fig. 4. Common wing positions: (a) Low wing, (b) Mid wing, (c) High wing.

2.3 Shock Absorption

To be able to pick an existing solution for the shock absorption, an initial dimensioning must be made to find out the required diameter and stroke length. To proceed with this, the touchdown kinetic and potential energy must be determined, through Eq. 1 which is only appliable to the telescopic strut configuration [13]:

$$E_t = 0.5(W_L)\left(\frac{v_z^2}{g}\right) + (W_L - L)s_s \tag{1}$$

In which, E_t – Touchdown energy in J; W_L – Weight of the aircraft at landing in kg; v_z – Vertical touch rate in m/s; g – gravity in m/s^2; L – Lift at landing in N; s_s – Shock absorber stroke length in m.

An assumption is made that all the energy at touchdown ends up being absorbed by the main landing gear, so that respective energy absorbed can be calculated with the following equation:

$$E_{absorbed} = W_L N_g(\eta_s s_s) \tag{2}$$

In which: $E_{absorbed}$ – Energy absorbed by the main landing gear in J; N_g – Landing gear load factor; η_s – Shock absorber efficiency.

To continue with the dimensioning, W_L can be defined as follows:

$$W_L = n_s P_m \tag{3}$$

where: n_s – Number of main gears struts; P_m – Maximum static load per main strut in kg.

Which results in the following equation:

$$E_{absorbed} = n_s P_m N_g (\eta_s s_s) \tag{4}$$

By having these two equations, we can then relate both by equating the energy absorbed to the touchdown energy to obtain the shock absorber stroke length.

$$n_s P_m N_g (\eta_s s_s) = 0.5(W_L)\left(\frac{v_z^2}{g}\right) + (W_L - L)s_s \tag{5}$$

A simplification can be made, if the lift generated during landing equals the weight of the aircraft, making the potential energy term inconsequential. With this simplification applied, the shock absorber stroke length can be calculated by:

$$s_s = \frac{\left[\frac{0.5(W_L)\left(\frac{v_z^2}{g}\right)}{n_s P_m N_g}\right]}{\eta_s} \tag{6}$$

Using the previously presented Eq. 6, the equation can be further simplified as follows:

$$s_s = \frac{\left[\left(\frac{v_z^2}{2gN_g}\right)\right]}{\eta_s} \tag{7}$$

With this final equation, the value required for the stroke length will be obtained in meters and further converted to inches, as it is the most common unit in datasheets.

To finalize the dimensioning of the shock absorber, the diameter of the inner struct can be estimated by the following equation, with the variable P_m in pounds for this equation:

$$d_s = 0.041 + 0.0025{P_m}^{0.5} \tag{8}$$

where: d_s – diameter in ft.

As was done with the stroke length, this value will also be further converted to inches to facilitate searching for an adequate solution in datasheets.

With the diameter and stroke length obtained, a solution can be implemented for the appropriate aircraft if all the required parameters are obtained previously.

2.4 Braking System

To decide which braking solution to implement on the aircraft, some values must be acquired previously, like the previous equations. The methodology involves calculating the required energy absorption and braking torque based on the aircraft's mass, touch-down speed and geometry of the brakes. Properly dimensioning a brake is critical to ensure a safe and effective stop during landing.

The first step is to determine the landing kinetic energy (LK_e) to be absorbed by the brakes when landing. To get this value, the Maximum Takeoff Weight (MTOW) of 600 kg and the landing speed of 29.85 m/s must be applied to the following formula:

$$LK_e = \frac{1}{2}mv^2 \tag{9}$$

where: LK_e – Landing kinetic energy in J; m – mass in kg; v – landing speed in m/s.

The value obtained will be the required energy to be absorbed by the braking system. As the braking system is composed of two brakes, the LK_e will be divided by 2 to account for each one. The selected brakes must be able to handle at least the value achieved. The second value to be calculated is the braking force (F_b) necessary to achieve a desired deceleration rate. To obtain this force, Newton's second law will be used:

$$F_b = m * a \tag{10}$$

where: F_b – Braking force required in N; m – mass in kg; a – deceleration rate in m/s^2.

With this value obtained, the next step is to calculate the required braking torque (T_b).

$$T_b = F_b * r_{eff} \tag{11}$$

where: T_b – Braking torque in Nm; r_{eff} – braking effective radius in m.

The braking effective radius is obtained through contact with a brake supplier.

These calculations will provide the two main values required for selecting a braking solution, the absorbed energy and braking torque, which the final option must surpass.

3 Results

The results presented followed the four major factors to be decided: (1) Landing gear arrangement; (2) Type of retraction; (3) Shock absorption; (4) Braking system.

3.1 Landing Gear Arrangement

The landing gear arrangement was defined at a very preliminary stage of the development of the aircraft, without a well-defined structure which would impact the overall progress as it was critical to have a side and top view, as well as the CG envelope. The final arrangement was defined to be a tricycle one, as the plane has the engine rear mounted with a pusher configuration, making this solution a much better option when compared to a taildragger. The final height of the aircraft, from the propeller line to the ground ended up around 1073 mm, mainly due to the dimension of the propeller, identified as the first component of the aircraft to touch the ground in case of a higher angle. The distance between the front and main landing gear is 2400 mm. Both values can easily undergo changes, mainly from tire modifications, payload configurations as well as mission requirements, but the current proposal offers a solution suitable for the parameters defined. The final landing gear arrangement position is presented in Fig. 5.

Fig. 5. Final landing gear arrangement position.

3.2 Type of Retraction

Concerning the retraction type, a different approach was idealized, with the retractable landing gear being a possible feature of the aircraft, with additional bays added to the sides of the aircraft, providing a storage solution that would have no impact on the current design. Several other problems, such as the proximity to the engine and the choice of a mid-wing design made other retraction movements difficult to imagine, thus leaving only one possible alternative, with the wheels retracting to the front of the fuselage, against the wing, but with the added benefit of being easier to deploy in case of emergency.

The mechanism for retracting the whole system will be an electric actuator, with a great focus on backup systems and a state of always deployed in case of emergency. The trajectory of retracting ended up being a simple rotation to the front, as any other solution involving folding struts proved to be way too complex and heavy for the aircraft requirements.

3.3 Shock Absorption

For the shock absorption, the focus was to pick and dimension a suitable solution, to then select an appropriate product capable of the loads calculated. To do this, an initial analysis was done regarding the available options and the advantages and disadvantages of each.

After this step, the option that seemed more suited for the requirements was the oleo pneumatic with a telescopic structure due to several benefits compared to the alternatives. With the option selected, the next steps involved defining the required variables for the dimensioning and solving the equations to find out the necessary travel and diameter of the cylinder.

The final solution, with a stroke length of 8.8 inches or 223.52 mm and a diameter of 0.1 feet or 30.48 mm, was then drafted in a 3D model to then be integrated along the continuation of the landing gear development. This component, as well as the others, must undergo several simulations as well as validation tests to ensure it meets the requirements of the aircraft.

3.4 Braking System

Regarding the braking system, a similar approach to the one used for the shock absorption was used, with an initial analysis of the required steps to take and which aspects to include in the system.

However, the main goal was to dimension the brakes, which resulted in a series of equations with parameter such as the mass, landing speed and deceleration rate used to calculate the energy absorption required as well as the torque, thus enabling the next step of procuring products that fit these values. One of the products suitable for the current dimensions is the brakes manufactured by Beringer Aero [14].

After finding some alternatives, the next step is to compare them in terms of the additional features, such as the material of the brakes and whether the system includes an anti-skid solution, as both features can be significantly advantageous.

4 Conclusion

The current paper presents the initial definition of several parameters to then be further applied to continue the development of a complete landing gear system, suitable for a lightweight aircraft. The final arrangement used was a tricycle one. Concerning the retraction type, the retractable landing gear being a possible feature of the aircraft, with additional bays added to the sides of the aircraft, providing a storage solution that would have no impact on the current design. For the shock absorption, the option was the oleo pneumatic with a telescopic structure due to several benefits compared to the alternatives. For the braking system, there are still many other steps to take when deciding the final type and solution for braking the aircraft, however two of those presented, the anti-skid system and material of the brakes will be considered when picking the appropriate brake from manufactures datasheets, and if necessary, air brakes will be further implemented if the brakes selected end up not being powerful enough for the missions operated.

Acknowledgements. This work was funded by national funds through FCT/MCTES (PIDDAC), UIDB/00308/2020 (INESCC), AERO.NEXT Portugal-Portugal's Aeronautical Agenda and Tekever UAS and was also supported by FCT - Fundação para a Ciência e Tecnologia, I.P. through the project UIDB/04564/2020 (https://doi.org/10.54499/UIDB/04564/2020) and UIDP/04564/2020 (https://doi.org/10.54499/UIDP/04564/2020).

References

1. Raymer, D.P.: Aircraft Design: A Conceptual Approach (2012)
2. Gudmundsson, S.: General aviation aircraft design: applied methods and procedures. General Aviation Aircraft Design: Applied Methods and Procedures, pp. 1–1034, https://doi.org/10.1016/C2011-0-06824-2 (2013)
3. Crane, A., Johnson, B.: Advancements in electro-mechanical landing gear systems. J. Aircraft Technol. **35**(2), 112–128 (2018)
4. Smith, P., et al.: Hydraulic retraction system: design and performance analysis. Aerospace Eng. J. **42**(4), 245–263 (2019)
5. Anderson, M., Brown, K.: Pneumatic landing gear systems: challenges and opportunities. Int. J. Aviati. Technol. **28**(3) (2020)
6. "Shock absorber | Suspension, Dampening & Vibration | Britannica." Last accessed: Jan. 04, 2024. [Online]. Available: https://www.britannica.com/technology/shockabsorber (2024)

7. "Aircraft Brakes: The Ultimate Guide for Airplane Brakes." Last accessed: Jan. 04, 2024. [Online]. Available: https://www.pilotmall.com/blogs/news/aircraft-brakes-the-ultimate-guide-for-airplane-brakes (2024)
8. "European Union Aviation Safety Agency | European Union." Last accessed: Jan. 03, 2024. [Online]. Available: https://european-union.europa.eu/institutions-lawbudget/institutions-and-bodies/search-all-eu-institutions-and-bodies/europeanunion-aviation-safety-agency-easa_en (2024)
9. "CS-LSA Amendment 1 | EASA." Last accessed: Jan. 03, 2024. [Online]. Available: https://www.easa.europa.eu/en/document-library/certification-specifications/cs-lsaamendment-1 (2024)
10. Anderson, M., Brown, K.: Pneumatic landing gear systems: challenges and opportunities. Int. J. Aviat. Technol. **28**(3) (2020)
11. FAA, Amateur-Built Aircraft & Ultralight Flight Testing Handbook (2015)
12. "2.2.2: Wing - Engineering LibreTexts." Last accessed: Jun. 26, 2024. [Online]. Available: https://eng.libretexts.org/Bookshelves/Aerospace_Engineering/Fundamentals_of_Aerospace_Engineering_%28Arnedo%29/02%3A_Generalities/2.02%3A_Parts_of_the_aircraft/2.2.02%3A_Wing (2024)
13. Roskam, J.: Airplane Design: Part IV. DAR corporation (1989)
14. Beringer Aero, "Wheel & Brake Kits," Last accessed: Dec. 20, 2023 [Online]. https://www.beringer-aero.com/kits-rouesfreins (2023)

Author Index

© The Editor(s) (if applicable) and The Author(s), under exclusive license
to Springer Nature Switzerland AG 2027
M. Calvete Gaspar et al. (Eds.): Regional Helix 2024, LNME, pp. 301–302, 2027.
https://doi.org/10.1007/978-3-032-36365-7

FSC
www.fsc.org
MIX
Papier | Fördert
gute Waldnutzung
FSC® C083411

Kolibri 360 GmbH
Ferdinand-Jühlke-Straße 7
99095 Erfurt, Deutschland
produktsicherheit@kolibri360.de